Thomas Scott

The Marine Fishes and Invertebrates of Loch Fyne

Thomas Scott

The Marine Fishes and Invertebrates of Loch Fyne

ISBN/EAN: 9783337420253

Printed in Europe, USA, Canada, Australia, Japan

Cover: Foto ©berggeist007 / pixelio.de

More available books at **www.hansebooks.com**

II.—THE MARINE FISHES AND INVERTEBRATES OF LOCH FYNE. By Thomas Scott, F.L.S. Mem. Soc. Zool. de France. (Plates I.-III.)

CONTENTS.

INTRODUCTORY.

In the following catalogue an endeavour has been made to include, as far as possible, the various species of marine fishes and invertebrates that are known to have been obtained in Loch Fyne. But though the records brought together here—the most of which are believed to be authentic,—comprise a large number of fishes, and include representatives of almost every group of the marine invertebrates, the catalogue can only be considered as preliminary to further research. The investigation of the lower forms of life existing in Loch Fyne has already been in some respects very successful, and has yielded interesting results; still, there are several groups of the Invertebrata, such as the Tunicata, the Polyzoa, the smaller Crustacea, the Annelida, the Cœlenterata, and the Porifera, that have only been very partially studied, and it may reasonably be expected that numbers of species belonging to these various groups will yet be discovered in Loch Fyne.

In this catalogue the total number of species enrolled is 837, which includes several not before recorded from the Clyde, and also one Copepod new to Britain, and one new to science.

An appendix at the end of the catalogue of the Loch Fyne fauna contains records and descriptions of several new and rare species that have been observed during the past year in the seaward part of the Clyde area, and which, therefore, could not find a place in the catalogue.

These records include three apparently undescribed species of Copepoda two Amphipods, and a parasitic Isopod, new to Britain; and a few others that are either new records for the Clyde, or that refer to species that are comparatively rare.

As *Calanus finmarchicus* is a very important constituent of the food of fishes, attention is also directed in the appendix to three different kinds of parasites that have been found more or less infesting the *Calani* in the Firth of Clyde.

The following is a tabular view of the principal groups represented in the catalogue, their principal sub-divisions, and the number of species belonging to each :—

Names of the Principal Groups represented in the Catalogue.	Names of the Principal Sub-divisions.	Number of Species belonging to each of the Principal Sub-divisions	Number of Species belonging to each of the Principal Groups.	Remarks.
Fishes	Teleostei	56	62	
	Ganoidei	1		
	Elasmobranchii	5		
Tunicata	Larvacea	1	9	
	Ascidiacea	8		
Mollusca	Cephalopoda	5	221	* One is doubtful.
	Pteropoda	1		
	Opisthobranchiata	20		
	Nudibranchiata	18		
	Pulmonata	2		
	Prosobranchiata	84*		
	Polyplacophora	5		
	Scaphopoda	2		
	Pelecypoda	84		
Branchiopoda			2	
Crustacea	Brachyura	27	319	† Two are doubtful.
	Macrura	18		
	Schizopoda	18		‡ One is doubtful.
	Cumacea	15		
	Isopoda	18		§ Includes a new genus and two new species.
	Amphipoda	76†		
	Phyllocarida	1		
	Cladocera	3‡		
	Ostracoda	67		
	Copepoda	99§		
	Cirripedia	7		
Polyzoa	Cheilostomata	15	22	‖ One is doubtful.
	Cyclostomata	5		
	Ctenostomata	2		
Vermes	Chætopoda	47‖	52	
	Gephyrea	2		
	Chætognatha	1		
	Nemertea	1		
	Turbellaria	1		
Echinoderma	Crinoidea	1	32	¶ One is doubtful. ** Two are doubtful.
	Asteroidea	9¶		
	Ophiuroidea	11**		
	Echinoidea	5		
	Holothurioidea	6		
Actinozoa	Alcyonaria	3	12	
	Actiniaria	9		
Hydrozoa	Hydroida	14	14	
Spongozoa			7	
Foraminifera	Miliolidæ	14	55	
	Astrorhizidæ	1		
	Lituolidæ	5		
	Textulariidæ	3		
	Lagenidæ	18		
	Globigerinidæ	1		
	Rotuliidæ	8		
	Nummulinidæ	5		
Total number of Species			837	

I desire to explain, though it may hardly be necessary to do so, that the information contained in this catalogue is partly at least a compilation from the recorded results of past investigations, and has only to a limited extent been obtained as the result of independent study. The following are the principal sources from whence the information contained in the catalogue has been derived :—

(1) An extensive and valuable series of MS. records from the steam yacht 'Medusa,' which Dr John Murray has kindly placed at my disposal.*

(2) *The Decapod and Schizopod Crustacea of the Clyde*, by Dr (now Professor) J. R. Henderson.

(3) *A Contribution towards a Catalogue of the Amphipoda and Isopoda of the Clyde* (in two parts), by the late Dr Robertson of Millport.

(4) 'A List of the Marine Fauna collected at the Tarbert (Loch Fyne) Laboratory during 1885,' by George Brook and Thomas Scott (published as an Appendix in the *Fourth Annual Report of the Fishery Board for Scotland*).

(5) *The Mollusca of the Firth of Clyde*, by Alfred Brown.

(6) 'Notes on the Copepods of Loch Fyne,' by W. L. Calderwood (published as an Appendix in the *Fourth Annual Report of the Fishery Board for Scotland*).

(7) The scientific researches in Loch Fyne during the past year of the Fishery steamer 'Garland.' These researches have yielded a considerable number of new records.

The following abbreviations are used in the catalogue :—

(M.)   Refers to the MS. Records of the steam yacht 'Medusa,' as the source of information.

(H.)   Dr Henderson's *Higher Crustacea of the Clyde*.

(R.)   Dr Robertson's *Amphipoda and Isopoda of the Clyde*.

(B. & S.) The List of Tarbert Fauna for 1885.

(G.)   Records obtained by means of the Investigations of the 'Garland.'

Other sources of information are, wherever necessary, duly acknowledged.

I have to acknowledge my indebtedness to Captain Campbell of the 'Garland' for his unwearied interest in the work, and for his efforts to render it successful. I have also, as in former years, been greatly indebted to the Rev. T. R. R. Stebbing, F.R.S., for the kindly manner in which he has from time to time endeavoured to assist me with the identification of Crustacean species. Professor G. S. Brady, F.R.S., has also kindly assisted me on various occasions. My son, Mr Andrew Scott, has prepared the series of interesting drawings which illustrate this paper.

FISHES OF LOCH FYNE.

Note.—In this catalogue the names and arrangement in Dr Day's work on the British Fishes are followed.

TELEOSTEI.

Cottus scorpius, Bloch.—Common in Loch Fyne and generally distributed ; the large and richly coloured variety, *Grœnlandicus*, is also more

---

* A very large number of the specimens referred to in these lists were collected by Captain Alexander Turbyne and Mr Fred. G. Pearcey in Dr Murray's steam yacht 'Medusa,' and the species were determined by Dr A. Günther, F.R.S. (Fishes); Professor F. Jeffrey Bell, F.R.S. (Echinodermata) ; Professor A. C. Haddon, F.R.S. (Actiniaria) ; Professor Arthur Dendy, F.L.S. (Sponges) ; E. A. Smith, Esq., F.L.S. (Mollusca and Tunicata) ; Miss F. Buchanan (Worms) ; R. I. Pocock, Esq. (Crustacea) ; and R. Kirkpatrick, Esq. (Hydrozoa and Polyzoa).

or less frequent (B. & S.).—Upper Loch Fyne at Minard, and on east side in 10 to 20 fathoms (M.).

Cottus bubalis, Euph.—Moderately common in Loch Fyne (B. & S.).

Trigla Gurnardus, Linné.—Upper Loch Fyne (M.).   In trawl-net near the head of the loch, and in the vicinity of Furnace and Minard (G.).

Trigla lineata, Gmelin.—East Loch Tarbert, Loch Fyne, not common (B. & S.).

Agonus cataphractus, Linné.—East Loch Tarbert, occasionally (B. & S.). Off Castle Lachlan, rare (G.).   Ardrishaig (Dr Scouler).

Scomber scomber, Linné.—Abundant during the summer, particularly along the west shore of Loch Fyne (B. & S.).

Zeus faber, Linné.—Usually enters East Loch Tarbert in September to feed on herring offal (B. & S.).

Gobius Ruthensparri, Euph.—Common amongst *Zostera* in East Loch Tarbert (B. & S.).   Upper Loch Fyne (M.).   Off Ardno, Upper Loch Fyne (G.).   Loch Fyne (Dr Scouler).

Gobius minutus, Gmelin.—Frequent in East Loch Tarbert (B. & S.). Upper Loch Fyne at Minard, west and east sides, and centre in 10 to 70 fathoms (M.).   Loch Gair, off Ardno, Kinglass Spit, and near the head of the loch (G.).

Gobius niger, Linné.   Upper Loch Fyne (M.).   Taken by the fishery steamer 'Garland' in various parts of, and at various depths in, Upper Loch Fyne.

Gobius, sp.—West side of Upper Loch Fyne in 10 to 15 fathoms (M.).

Calionymus lyra, Linné.—East Loch Tarbert, frequent but small (B. & S.).   Upper Loch Fyne, east side, in 15 fathoms (M.).   Loch Gair, Tarbert Bank, and other places (G.).

Cyclopterus lumpus, Linné.—Adult specimens are not common, but the young are moderately frequent (B. & S.).   Upper Loch Fyne, in the centre near the head (M.).   Otter Spit, one 45 mm. in length (G.).

Liparis vulgaris, Linné.—Taken between tide marks in East Loch Tarbert (B. & S.).   Upper Loch Fyne, on the shore (M.).

Liparis Montagui, Donovan.—Taken in East Loch Tarbert between tide marks (B. & S.) (?).   Upper Loch Fyne, west side and shore (M.).

Carelophus Ascanii, Walh.—One specimen taken amongst boulders at low water in East Loch Tarbert (B. & S.).

Lumpenus lampretæformis (Walbaum).—Four specimens taken with shrimp-trawl between Dunderave and Carndow (G., April 1896).

Centronotus gunnellus, Linné.—Plentiful between tide-marks East Loch Tarbert (B. & S.).   Upper Loch Fyne on the shore (M.).   In Inveraray Bay (G.).

Atherina presbyter (Jenyns).—Frequent amongst *Zostera* in East Loch Tarbert in the spring of 1885, but not met with later (B. & S.).

Gasterosteus spinachia, Linné.—Frequent in the in-shore water amongst sea-weed and *zostera;* East Loch Tarbert and other places (B. & S.). Upper Loch Fyne, not uncommon among the contents of the shrimp-trawl (G.).

Labrus maculatus, Blainville.—Frequent in Loch Fyne in the autumn (B. & S.).

Labrus mixtus, Frees and Eks.—Occasionally in Loch Fyne, at the mouth of East Loch Tarbert (B. & S.).

Ctenolabrus rupestris, Linné.—Common, especially near Skate Island, Loch Fyne (B. & S.).

Centrolabrus exoletus, Linné.—Taken occasionally in Loch Fyne (B. & S.).

Gadus morhua, Linné.—Not very abundant in Loch Fyne (B. & S.). Upper Loch Fyne, east side, in 10 to 20 fathoms (M.). In Loch Gair, off Ardno, Kinglass Spit, and off Carndow (G.).

Gadus æglefinus, Linné.—Not abundant in Loch Fyne (B. & S.). Upper Loch Fyne, in the centre, in 65 to 70 fathoms (M.). Between Penmore and Inveraray (G.).

Gadus luscus, Linné.—Occasionally in Tarbert Harbour (B. & S.). Tarbert Bank, Lower Loch Fyne ; and between Dunderave and Carndow, Upper Loch Fyne (G.).

Gadus minutus, Linné.—Frequent in Tarbert Harbour (B. & S.). Upper Loch Fyne, in the centre, near the head (M.). Off Lowburn, Upper Loch Fyne (G.). Tarbert, Loch Fyne (Dr Scouler).

Gadus merlangus, Linné.—Not abundant in Loch Fyne (B. & S.). Upper Loch Fyne, in the centre (M.). Two nearly ripe males were taken in the trawl-net in May 1896 (G.).

Gadus virens, Linné.—Abundant, but mostly immature ; these immature forms—termed 'podlies'—frequent the shallow in-shore water, especially in the neighbourhood of wharfs or piers.

Gadus pollachius, Linné.—This species is also of more or less frequent occurrence, but is usually of small size ; large specimens are, however, occasionally brought to East Tarbert.

Molva vulgaris, Fleming.—Frequently brought into Tarbert by the fishermen (B. & S.). A ripe male, 38 inches in length, was taken by trawl-net near the head of Loch Fyne in May 1896 (G.).

Merluccius vulgaris, Cuvier.—Upper Loch Fyne, in the centre, in 34 to 36 fathoms (M.). In the vicinity of Furnace, and between Dunderave and Carndow, taken with the beam-trawl (G.).

Motella mustela, Linné.—Taken between tide-marks in East Loch Tarbert, not common (B. & S.).

Raniceps trifurcatus, Yarrell.—Loch Fyne (Dr Scouler, in *Trans. of Nat. Hist. Soc. of Glasgow*, vol. i., p. 8).

Ammodytes lanceolatus, Lesauvage.—Occasionally in the neighbourhood of East Loch Tarbert (B. & S.).

Hippoglossus vulgaris (Fleming).—Occasionally taken in Loch Fyne (B. & S.).

Hippoglossoides limandoides (Bloch).—Upper Loch Fyne, east side and centre, in 10 to 70 fathoms (M.).   Tarbert Bank, Lower Loch Fyne; off Minard; off Castle Lachlan; Inveraray Bay; and between Dunderave and Lowburn, Upper Loch Fyne (G.).   A common and widely-distributed species.

Rhombus maximus (Linné).—Occasionally taken in Loch Fyne (B. & S.).

Zeugopterus unimaculatus (Risso).—A few specimens of this interesting species were taken near Barmore (B. & S.).

Zeugopterus punctatus (Bloch).—Upper Loch Fyne, west side, in 10 to 25 fathoms (M.).

Pleuronectes platessa (Linné).—More or less common in Loch Fyne where the conditions are suitable.   Upper Loch Fyne, on west and east sides and centre, in 10 to 36 fathoms (M.).   A number of large plaice, measuring from 14 to 18 inches in length, and most of which were either spawning or spent, were captured in the trawl-net near the head of Loch Fyne in May 1896 (G.).

Pleuronectes microcephalus, Donovan.—Frequent in suitable localities (B. & S.).   On May 6, 1896, a scarcely ripe female, 13 inches in length, was taken near the head of the loch; two nearly ripe specimens—one a male 13 inches in length, the other a female 17 inches in length—were captured at Kinglass Spit; and another, 11 inches, was taken off Ardno. On the 7th, several other specimens were trawled on Tarbert Bank, Lower Loch Fyne (G.).

Pleuronectes cynoglossus (Linné).—Upper Loch Fyne, at west side and centre, in 15 to 36 fathoms (M.).   Moderately common in deep water, from off Inveraray to the head of the loch (G.).

Pleuronectes limanda, Linné (= (?) *Hippoglossoides platessoides*). — Barmore Bay, Lower Loch Fyne (B. & S.).   Upper Loch Fyne, west and east sides and centre, in 10 to 70 fathoms (M.).   Lowburn to Dunderave, off Inveraray, Kinglass Spit, off Castle Lachlan, off Minard, Ardrishaig Bay, Tarbert Bank.   Common all over the Loch (G.).

Pleuronectes flesus, Linné.—Common in Tarbert Harbour (B. & S.). Occasionally in Upper Loch Fyne.   One specimen, a scarcely ripe male, was taken near the head of the loch in May 1896, and another off Largymore (G.).

Solea vulgaris, Quensel.—Small specimens have been captured in Barmore Bay (B. & S.).

Salmo salar, Linné.—Taken occasionally in herring-nets between Tarbert and Barmore (B. & S.).

Salmo trutta, Linné.—Regularly taken in small quantities just outside East Loch Tarbert (B. & S.).

Clupea harengus, Linné.—The herring usually enters Loch Fyne in May, and the fishing usually continues till November (B. & S.).

Anguilla vulgaris, Turton.—The common eel affords a small but regular fishery in Tarbert Harbour (B. & S.). One specimen was taken in the shrimp trawl-net in Inveraray Bay in May 1896 (G.).

Conger vulgaris, Cuvier.—Frequent in Loch Fyne, but small (B. & S.).

Siphonostoma typhle (Linné).—Captured in East Loch Tarbert amongst *zostera* (B. & S.).

Syngnathus acus, Linné.—Generally distributed throughout the loch, especially inshore, where the water is shallow.

Nerophis æquoreus (Linné).—East Loch Tarbert, amongst *Zostera* (B. & S.).

Nerophis lumbriciformis (Willughby).—In East Loch Tarbert, amongst *zostera*, with the previous species (B. & S.).

GANOIDEI.

Acipenser sturio, Linné.—Fine specimens are often noticed during the herring fishery, but are seldom captured (B. & S.).

ELASMOBRANCHII.

Pristiurus melanostomus (Bonaparte).—Occasionally brought in amongst the *Acanthii* in the winter fishing (B. & S.). Upper Loch Fyne, east side, in 10 to 20 fathoms (M.).

Acanthius vulgaris, Risso.—Frequent during the herring fishing. They are occasionally the cause of serious injury to the herring-nets. Upper Loch Fyne, east side, in 10 to 20 fathoms (M.).

Raia clavata, Linné.—Frequent during the winter fishing (B. & S.). Occasionally in Upper Loch Fyne (G.).

Raia maculata, Montagu.—Upper Loch Fyne, east side, in 10 to 20 fathoms (M.).

Raia circularis, Couch.—Between Loch Gair and Largymore. Taken with the beam-trawl (G.).

TUNICATA OF LOCH FYNE.

Comparatively little appears to have yet been done towards the investigation of the Loch Fyne Tunicates. This group is a somewhat difficult one to study; and in the discrimination of many of the species, and especially of the *Botrylledæ*, it is all but absolutely necessary that living specimens should be available. Nine species are here recorded, and, with one exception, these records are all obtained from the MS. notes of the

H

steam yacht 'Medusa.' The number of species will doubtless be considerably increased when the Tunicata come to be more thoroughly worked up.

LARVACEA.

Oikopleura (?) *flabellum*, J. Müller.—Lower Loch Fyne in surface tow-net; Upper Loch Fyne, near Minard, in bottom tow-net (G.).

ASCIDIACEA.

Botrylloides, sp.—Upper Loch Fyne, west side, in about 15 fathoms.

Ciona intestinales, Linné.—Upper Loch Fyne, in the centre, in 60 to 65 fathoms (M.).

Ascidiella virginea, O. F. Müller.—Upper Loch Fyne, at Minard Narrows; on both sides and centre, in 10 to 70 fathoms; and also on the shore (M.).

Ascidiella scabra, O. F. Müller.—Upper Loch Fyne, at Minard Narrows, and on the east side and centre, in 12 to 65 fathoms (M.).

Ascidia mentula, O. F. Müller.—Upper Loch Fyne, at Minard; and on both sides and centre, in about 10 fathoms (M.). Tarbert Bank, Lower Loch Fyne; and in East Loch Tarbert, frequent and large (Mihi).

Corella parallelogramma, O. F. Müller.—In the centre of Upper Loch Fyne, in 60 to 65 fathoms (M.).

Styelopsis grossularia (Van Beneden).—At Minard Narrows, Upper Loch Fyne, on the east side, and in the centre, in 10 to 36 fathoms; and also on the shore (M.).

Polycarpa rustica, Linné.—At Minard Narrows, Upper Loch Fyne, west and east sides and centre, in 10 to 70 fathoms (M.).

MOLLUSCA OF LOCH FYNE.

Note.—The arrangement and nomenclature of Part IV. of the Rev. A. M. Norman's Catalogues (Museum Normanianum) are followed for the Mollusca. The *Revision of British Mollusca*, by the same author, as far as, and inclusive of, Order III. (Nudibranchiata),* has also been consulted; also the *Mollusca of the Firth of Clyde*, by Alfred Brown.

CEPHALOPODA.

Octopus vulgaris, Lamk.—In herring-net in the summer, Laggan Bay (B. & S.).

Scæurgus cirrhosa (Lamk.). Taken in herring-net in Laggan Bay, in December; young specimen (B. & S.); east side of Upper Loch Fyne, 30 fathoms (M.).

---

* *Revision of British Mollusca, Ann. and Mag. Nat. Hist.*, sixth series, vol. v. pp. 452-481, and vol. vi. pp. 60-91 (1890).

Sepiola Rondeletii, Leach.—Laggan Bay ; eggs frequent, attached to Ascidians and other objects (B. & S.) ; west and east sides of Upper Loch Fyne, 10 to 30 fathoms (M.).

Rossia macrosoma (D. Ch.).—Laggan Bay (B. & S.).   Mid-channel, near the head of Upper Loch Fyne (M.).

Rossia Oweni, Ball.—Off Minard, and at east side of Upper Loch Fyne, 10 to 30 fathoms (M.).   Dr Jeffreys says that Steenstrup considers *R. Oweni* of Ball to be the male of *R. macrosoma.**

PTEROPODA.

Limacina retroversa (Flem.).   Occasionally taken with the tow-net in Loch Fyne.

OPISTHOBRANCHIATA.

Actæon tornatilis (Linné).—Laggan Bay, Loch Fyne (B. & S.). Ardrishaig, dead specimens (Alf. Brown).   Upper Loch Fyne, at Minard Narrows, in 12 fathoms (M.).   Variety, *bullæformis*, Jeffreys, in Loch Fyne, in 40 to 50 fathoms, muddy bottom (M'Nab ; see *Jeffreys' Brit. Conch.*, vol. iv., p. 435).

Tornatina obtusa (Montagu).—Off Silvercraigs, Loch Fyne (Alf. Brown).   Loch Fyne (Robertson).

Tornatina mamillata (Philippi).—Off Silvercraigs, Loch Fyne ; scarce in the living state ; dead shells not uncommon (Alf. Brown).

Tornatina truncatula (Bruguiere).   East Loch Tarbert (B. & S.).   Off Silvercraigs (Alf. Brown).

Tornatina umbilicata (Montagu).   In Upper Loch Fyne, at Minard Narrows, and on the west side and centre in 15 to 70 fathoms.   Var. *strigella*, Lovén, has been taken on the east side in 20 fathoms (M.).

Tornatina nitidula, Lovén.—Loch Fyne (Mr Barlee and Dr J. G. Jeffreys).   East side of Upper Loch Fyne, in 15 fathoms (M.).

Volvula acuminata (Bruguiere).—Loch Fyne (Barlee, Alder, J. G. Jeffreys, Robertson).

Cylichna cylindracea (Pennant).—Tarbert Bank (Mihi).   In Upper Loch Fyne at Minard Narrows, and at both sides and centre in 10 to 70 fathoms (M.).   Loch Fyne (Jeffreys ; var., *linearis*).

Diaphana hyalina (Turton).—East Loch Tarbert (B. & S.).   Moll Dhu and Silvercraigs, Loch Fyne (Alf. Brown).

Scaphander lignarius (Linné).   Furlong Bay, and off Moll Dhu Point in 15 fathoms (B. & S.).   Off Silvercraigs and Otter Spit, Loch Fyne (Alf. Brown).   In Upper Loch Fyne, on both sides, in 10 to 15 fathoms (M.).

* *British Conchology*, vol. v. p. 134.   See also the *Revision of British Mollusca*, already referred to

Bulla utriculus, Brocchi.—Tarbert Bank, Loch Fyne (Mihi). Dr Jeffreys dredged a single specimen of the variety *oblonga* in Loch Fyne, (*Brit. Conch.*, vol. iv. 441).

Haminea hydatis (Linné).—Has hitherto been found only in the deep water of Upper Loch Fyne (M.).

Acera bullata, Müller.—East Loch Tarbert—living (Mihi).

Philine scabra (Müller).—Tarbert Bank, Loch Fyne—living (Mihi). In Upper Loch Fyne, both sides, in 10 to 20 fathoms (M.).

Philine catena (Montagu).—In the centre of Upper Loch Fyne, in 35 fathoms (M.).

Philine punctata (Clark).—Found in deep water (70 fathoms), Upper Loch Fyne (M.).

Philine pruinosa (Clark).—Loch Fyne (Barlee).

Aplysia punctata, Cuvier.—White Shore, E. Loch Tarbert, rare (B. & S.).

Pleurobranchus plumula (Montagu).—Between tide-marks and dredged, East Loch Tarbert (B. & S.).

Runcina coronata (Quatrefages) (= *R. Hancocki,* Forbes).—Shallow water amongst weed in East Loch Tarbert, rare (Mihi).

NUDIBRANCHIATA.

Doris tuberculata, Cuvier.—Frequent in East Loch Tarbert (B. & S.). Shore, Upper Loch Fyne (M.).

Doris repanda, Alder and Hancock.—East Loch Tarbert, rare (B. & S.). This species is readily distinguished by its white colour, and by the row of spots along each side.

Doris Johnstoni, Alder and Hancock.—East Loch Tarbert—frequent (B. & S.). Shore, Upper Loch Fyne (M.).

Acanthodoris pilosa (Müller).—Occasionally in East Loch Tarbert (B. & S.).

Goniodoris nodosa (Montagu).   East Loch Tarbert, between tide-marks (B. & S.).

Triopa clavigera (Müller).—Between tide-marks, East Loch Tarbert ; not common (B. & S.).

**Polycera quadrilineata* (Müller).—East Loch Tarbert (B. & S.). Both the ordinary form and a dark-coloured variety were occasionally obtained.

Ægirus punctilucens (D'Orbigny).—Obtained occasionally on the under-side of stones in East Loch Tarbert (B. & S.).

---

* *Polycera ocellata,* Alder and Hancock, has also been obtained in East Loch Tarbert.

Tritonia Hombergi, Cuvier.—Obtained in the vicinity of Minard, Upper Loch Fyne, in 11 to 25 fathoms (M.).

Dendronotus frondosus (Ascanius) [= *D. arborescens* (Müller)].—Fre· quent in Loch Fyne in 10 to 20 fathoms (B. & S.).

Eolis papillosa (Linné).—Frequent between tide-marks in East Loch Tarbert (B. & S.).

Galvina picta, Alder and Hancock (= *Eolis picta*), Alder and Hancock. —Obtained in East Loch Tarbert ; rare.

Galvina Farrani, Alder and Hancock (= *Eolis farrani*), Alder and Hancock).—Rare in East Loch Tarbert (B. & S.). One specimen, of a rich orange colour.

Favorinus albus, Alder and Hancock (= *Eolis alba*), Alder and Hancock.—Frequent amongst *Zostera* in East Loch Tarbert (B. & S.).

Facelina Drummondii (W. Thompson) (= *Eolis Drummondii,* W. Thompson).—Moderately common amongst *zostera* in East Loch Tarbert (B. & S.). On the shore between tide-marks, Upper Loch Fyne (M.).

Hermæa bifida (Montagu).—East Loch Tarbert ; rare (B. & S.).

Alderia modesta (Lovén).—Amongst *Zostera* near the head of East Loch Tarbert (Mihi).

Elysia viridis (Montagu).—Common amongst *Zostera* in East Loch Tarbert. Colour variable from pale green to brown (B. & S.).

PULMONATA.

Alexia bidentata (Montagu).—Common under stones between tidemarks, East Loch Tarbert, especially round towards Garvald Point (B. & S.).

Oncidium celticum, Cuvier.—Shore, Upper Loch Fyne (M.). Only two British localities are mentioned by Dr Jeffreys for this species—viz., Lantivet Bay, Cornwall ; and Whitsand Bay, near Plymouth.

PROSOBRANCHIATA.

Clathurella linearis (Montagu).—Off Battle Island, Loch Fyne, and in East Loch Tarbert ; dead shells occasionally containing small hermit crabs (B. & S.).

Clathurella purpurea (Montagu).—West side of Upper Loch Fyne, 10 to 20 fathoms (M.).

Clathurella reticulata (Renier).—West side of Upper Loch Fyne, 12 to 15 fathoms (M.).

Clathurella teres (Forbes).—In the vicinity of Minard, Upper Loch Fyne, 12 to 15 fathoms (M.). East Loch Tarbert (Mihi).

Mangelia attenuata (Montagu).—In the vicinity of Minard, in 15 to 20 fathoms (M.).

Mangelia lævigata (Philippi).—In the same locality as the last (M.).

Bela turricolla (Montagu).—East Loch Tarbert (living); off Battle Island, dead (B. & S.); Upper Loch Fyne, in the vicinity of Minard, on both east and west sides and in the centre, in 11 to 70 fathoms (M.).

Bela Trevellyana (Turton).—Upper Loch Fyne, in 11 to 20 fathoms (M.).

Neptunea antiqua (Linné).—Taken in the vicinity of Minard; on the east and west sides and centre, and near the head of the loch, in 10 to 70 fathoms, and between tide-marks (M.). East Loch Tarbert, living (Mihi).

Sipho gracilis (Da Costa).—Loch Fyne (B. & S.). Vicinity of Minard; on the east and west sides and centre; and near the head of the loch in 10 to 70 fathoms (M.).

Buccinum undatum, Linné.—East Loch Tarbert (B. & S.). At various places from the vicinity of Minard to near the head of the loch, and from between tide-marks down to 70 fathoms (M.). In deep water, Upper Loch Fyne (G.).

Nassa incrassata (Ström).—East Loch Tarbert (B. & S.). Minard, in 12 to 20 fathoms, and between tide-marks (M.).

Nassa reticulata (Linné).—East Loch Tarbert (B. & S.). Minard, in 12 to 20 fathoms, and between tide-marks (M.).

Nassa pygmœa (Lamarck).—In the vicinity of Minard, in 11 to 20 fathoms (M.).

Trophon truncatus (Ström).—Tarbert Bank, Loch Fyne; not very rare in suitable localities.

Trophon muricatus (Montagu).—In Upper Loch Fyne, in 12 to 25 fathoms (M.). Tarbert Bank, Loch Fyne (Mihi). Alfred Brown does not include this species in his work on the *Mollusca of the Clyde*, probably because *Trophon clathratus* may sometimes have been mistaken for it. I have both species in my collection, and when they are placed side by side the difference between them is quite apparent. There can be no doubt that the specimen from Tarbert Bank is *Trophon muricatus*.

Purpura lapillus (Linné).—Commonly distributed, especially in shallow water. Vicinity of Minard, in 12 to 20 fathoms (M.).

Trivia europœa (Montagu).—On rocks at extreme low water in Loch Fyne, also in East Loch Tarbert, living (B. & S.). West side, Upper Loch Fyne, in 10 to 25 fathoms, and also on the shore (M.).

Aporrhaïs pes-pelicani (Linné).—Furlong Bay; off Battle Island, &c. (B. & S.). Minard; west and east side and centre; and near the head of the loch, in 10 to 70 fathoms (M.).

Triforis perversa, Linné.—Tarbert Bank, Loch Fyne, dead shell (Mihi).

Cerithiopsis tubercularis (Montagu).—East Loch Tarbert (B. & S.). Living shells not very common.

Trichotropis borealis, Broderip and Sowerby.—Furlong Bay, Loch Fyne (B. & S.). Minard, west side and centre, in 11 to 35 fathoms (M.).

Turritella terebra (Linné).—East Loch Tarbert, dead shells (B. & S.). Minard, west and east sides, in 10 to 30 fathoms (M.).

Cæcum glabrum (Montagu).—Loch Fyne and East Loch Tarbert; not common (B. & S.).

Littorina littorea (Linné).—Common, especially between tide-marks (B. & S.). On west and east sides of Upper Loch Fyne, 10 to 25 fathoms, and on the shore (M.).

Littorina rudis (Maton).—Common between tide-marks (B. & S.).

Littorina obtusata (Linné).—Common between tide-marks. The variety *ornata,* Jeffreys, is also occasionally obtained; specimens of the variety were collected at 'White Shore,' East Loch Tarbert (B. & S.). West and east sides and centre, and near the head of Upper Loch Fyne, in 10 to 20 fathoms; and also on the shore (M.).

Lacuna pallidula (Da Costa).—Taken on *Laminaria* in Loch Fyne, in shallow bays in the vicinity of East Loch Tarbert (B. & S.). The variety *albescens,* Jeffreys, was also occasionally obtained (B. & S.).

Lacuna divaricata (Fabricius).—Off Battle Island, Loch Fyne, and East Loch Tarbert. The variety *canalis* (Montagu) was obtained in East Loch Tarbert (B. & S.). West and east sides and centre of Upper Loch Fyne, and near the head of the loch, in 10 to 20 fathoms (M.).

Skenea planorbis (Fabricius).—Common; Loch Fyne and East Loch Tarbert (B. & S.).

Homologyra atomus (Philippi).—East Loch Tarbert; not very common (B. & S.).

Zippora membranacea (Adams).—Common amongst *Zostera* in East Loch Tarbert (B. & S.). The specimens were usually thin, and without, or with nearly obsolete, ribs; they are probably the variety *elata* of Philippi.

Rissoa parva (Da Costa).—Loch Fyne and East Loch Tarbert, amongst weed in shallow water.

Rissoa violacea, Desmarets.—East Loch Tarbert, between tide-marks, and dredged (B. & S.). Off Silvercraigs, Loch Fyne (Alf. Brown). Head of Upper Loch Fyne (G.).

Alvania cancellata (Da Costa).—Loch Fyne (Jeffreys and Barlee).

Alvania reticulata (Montagu).—Off Battle Island, Loch Fyne (B. & S.). In the vicinity of Minard, in 12 to 15 fathoms (M.). Off Silvercraigs, Loch Fyne (Alf. Brown).

Alvania abyssicola, Forbes.—Loch Fyne, amongst mud, in 50 to 100 fathoms (M'Andrew and Forbes, A. M. Norman, and others).

Alvania carinata (Da Costa) [= *Rissoa striatula* (Montagu)].—' I obtained a single dead specimen and some fragments off Silvercraigs, Loch Fyne, in about 12 fathoms nullipore and sand ' (Alf. Brown, in *Mollusca of the Firth of Clyde*, p. 62).

Flemingia zetlandica (Montagu).—Tarbert Bank, Loch Fyne (Mihi). In 20 fathoms, off Silvercraigs, Loch Fyne (Alf. Brown).

Cingula trifasciata (Adams) [= *Rissoa cinguillus* (Montagu)].—Common in Loch Fyne and East Loch Tarbert (B. & S.).

Onoba striata (Adams).—Common between tide-marks and in shallow water, especially where there is a weedy bottom, Loch Fyne and East Loch Tarbert (B. & S.).

Hydrobia stagnalis (Bast.) [= *H. ulvæ* (Penn.)].—At the head of Upper Loch Fyne (G.).

Capulus hungaricus (Linné).—Furlong Bay in 15 fathoms (B. & S.). Vicinity of Minard, west and east sides, in 10 to 25 fathoms (M.).

Velutina lævigata (Pennant).—East Loch Tarbert (B. & S.). In Upper Loch Fyne, at Minard, on both the west and east sides, in the centre, and near the head of the loch, in 10 to 70 fathoms (M.). Off Moll Dhu and Silvercraigs (Alf. Brown).

Velutella flexilis (Montagu) [= *Velutina plicatilis* (Müller)].—' One specimen in Loch Fyne, on stony ground, in 25 fathoms' (Forbes and M'Andrew ; see *Mollusca of the Firth of Clyde*, by Alf. Brown, p. 83).

Lamellaria perspicua (Linné).—East Loch Tarbert, on stones between tide-marks, frequent (B. & S.).   Off Silvercraigs (Alf. Brown).

Lunatia (Natica) sordida (Philippi).—Taken in the vicinity of Minard in 15 to 20 fathoms (M.).

Lunatia pulchella (Risso) [= *Natica Alderi*, Forbes].—East Loch Tarbert and off Battle Island, Loch Fyne (B. & S.).—In the vicinity of Minard, west and east sides and centre, and also near the head of Upper Loch Fyne, in 5 to 30 fathoms (M.).

Adeorbis subcarinatus (Montagu).—A few dead specimens were obtained in the coralline zone off Silvercraigs, Loch Fyne, by Alfred Brown (*Mollusca of the Firth of Clyde*, p. 82).

Aclis supranitida (S. Wood).—Tarbert Bank, Loch Fyne ; rare (Mihi).

Eulima polita (Linné).—West and east sides and centre of Upper Loch Fyne, in 10 to 65 fathoms (M.).   Off Moll Dhu and Silvercraigs (Alf. Brown).

Eulima intermedia, Cantraine ; var. *rubro-tincta* (Jeff.).—Loch Fyne (Jeffreys, see *Brit. Conch.*, vol. iv. p. 204).

Eulima incurva (Ren.) [= *Eulima distorta* (Deshayes)]. East Loch Tarbert (B. & S.). West and east sides and centre of Upper Loch Fyne, in 10 to 35 fathoms (M.).

Eulima bilineata, Alder.—West side of Upper Loch Fyne, in 10 to 15 fathoms (M.). Off Silvercraigs, one living, and a few dead specimens (Alf. Brown).

Turbonilla rufa (Philippi).—Taken alive off the pier at Ardrishaig in 5 fathoms, where it is rather common (Alf. Brown).

Parthenia interstincta (Montagu).—Tarbert Bank, Loch Fyne (Mihi).

Parthenia rufescens (Forbes).—Off Silvercraigs, Loch Fyne, dead shells only (Alf. Brown as *Odostomia scalaris*; variety, *rufescens*). Upper Loch Fyne, in 12 to 25 fathoms (M.). Dr Jeffreys in *British Conchology*, vol. iv. p. 161, and Alfred Brown in the *Mollusca of the Firth of Clyde*, p. 77, consider *Parthenia* (*Odostomia*) *rufescens* to be only a variety of *Parthenia* (*Odostomia*) *scalaris* (Philippi).

Parthenia spiralis (Montagu).—Loch Fyne and East Loch Tarbert ; not uncommon (B. & S.).

Odostomia conspicua, Alder.—Loch Fyne (M'Nab ; see Jeffreys' *Brit. Conch.*, vol. iv. p. 133).

Odostomia unidentata (Montagu).—Loch Fyne and East Loch Tarbert (B. & S.).

Odostomia acuta, Jeffreys.—Loch Fyne (Barlee).

Odostomia pallida (Montagu).—Off Moll Dhu and Silvercraigs, Loch Fyne (Alf. Brown). Usually found on the shells of living *Pecten opercularis*, and sometimes on *Pecten maximus*.

Odostomia umbilicaris (Malm).—Loch Fyne (Barlee, see Jeffreys' *op. cit.*, vol. iv. p. 130).

Auriculina (*Odostomia*) *obliqua* (Alder).—Tarbert Loch Fyne.

Auriculina insculpta (Montagu).—Off Silvercraigs, Loch Fyne, living shells (Alf. Brown).

Eulimella (*Odostomia*) *Scillæ* (Leacchi).—Off Silvercraigs, Loch Fyne, in sand and nullipore (Alf. Brown).

Eulimella acicula (Philippi).—Tarbert, Loch Fyne (Robertson ; see Brown's *Mollusca of the Firth of Clyde*, p. 78).

[*Molleria costulata* (Möller).—A single specimen, dead but perfect, off Silvercraigs, Loch Fyne, in 12 fathoms nullipore (Alf. Brown). Probably a post-tertiary fossil, as suggested by Mr Brown.]

Cyclostrema nitens (Philippi).—Off Silvercraigs, Loch Fyne, living (Alf. Brown).

Zizyphinus zizyphinus (Linné).—Laggan Bay and other localities in Loch Fyne (B. & S.). Off Moll Dhu, Loch Fyne (Alf. Brown).

Zizyphinus millegranus (Philippi).—Off Battle Island (B. & S.). Fairly abundant at Minard Narrows, Upper Loch Fyne, and along the sides of the loch, in 10 to 25 fathoms (M.). Fine specimens obtained off Silvercraigs, where the species is plentiful (Alf. Brown).

Zizyphinus Montagui (W. Wood).—Off Otter, Loch Fyne ; a few dead specimens (Alf. Brown).

Gibbula magus (Linné).—White Shore, East Loch Tarbert and Loch Fyne (B. & S.). Has been found in Upper Loch Fyne in 10 fathoms . (M.).

Gibbula umbilicata (Montagu).—Common near low-water mark East Loch Tarbert (B. & S.). Taken in Upper Loch Fyne, near the head, in shallow water (M.).

Gibbula cineraria (Linné).—Common in East Loch Tarbert (B. & S.). Vicinity of Minard, west and east sides and centre, and near the head of Upper Loch Fyne, in 5 to 70 fathoms (M.).

Gibula tumida (Montagu).—East Loch Tarbert (B. & S.). Vicinity of Minard, west and east sides and centre of Upper Loch Fyne, in 10 to 20 fathoms (M.). Off Silvercraigs (Alf. Brown).

Margarita helicina (Fabricius).—East Loch Tarbert (B. & S.). This species was not uncommon at extreme low-water on the fronds of *Laminaria* and on boulders, both in East Loch Tarbert and in the neighbouring bays in Loch Fyne.

Emarginula crassa, J. Sowerby.—Off Battle Island, dead shells (B. & S.). Upper Loch Fyne at Minard Narrows, and on both sides, in depths of 10 to 20 fathoms (M.).

Emarginula fissura (Linné).—Off Battle Island, and in other localities ; dead shells (B. & S.). Upper Loch Fyne at Minard Narrows, and on both sides and in the centre of the loch, as well as on the shore (M.).

Puncturella noachina·(Linné).—Laggan Bay (B. & S.). In Upper Loch Fyne at Minard Narrows, and on both sides, in 10 to 30 fathoms (M.).

Acmæa testudinalis (Müller). — White Shore, East Loch Tarbert (B. & S.). In Upper Loch Fyne, on the shore (M.).

Pilidium fulvum (Müller).—Tarbert Bank, Loch Fyne (Mihi). Off Moll Dhu and Silvercraigs (Alf. Brown).

Helcion pellucidum (Linné).—White Shore, East Loch Tarbert ; var. *lœvis* (B. & S.).

Patella vulgata, Linné.—Common throughout the district on stones between tide-marks.

POLYPLACOPHORA.

Chiton fascicularis, Linné.—East Loch Tarbert ; rare (B. & S.).

Chiton cinereus, Linné.—In Upper Loch Fyne at Minard Narrows, in 15 to 20 fathoms (M.).

Chiton marginatus, Pennant.—White Shore, East Loch Tarbert (B. & S.). In Upper Loch Fyne at Minard Narrows, on the west and east sides and centre, in 10 to 36 fathoms, and on the shore (M.).

Chiton ruber, Linné.—Under stones between tide-marks (B. & S.). In Upper Loch Fyne at Minard Narrows, on both sides, in 10 to 20 fathoms, and on the shore (M.). Off Moll Dhu (Alf. Brown).

Chiton marmoreus, Fabricius.—East Loch Tarbert (B. & S.). Has been found abundantly in Achagoil Bay, in Upper Loch Fyne (M.).

SCAPHOPODA.

Dentalium entalis, Linné.—Common off Barmore and in other localities (B. & S.). In Upper Loch Fyne at Minard Narrows, and on both sides and centre, in 10 to 70 fathoms (M.).

*Siphonodentalium (Pulsellum) lofotense. M. Sars.—Lower Loch Fyne (Robertson, in *Trans. Nat. Hist. Soc. Glasgow,* vol. ii., N.S., p. 151).

PELECYPODA.

Tetrabranchia.

Anomia ephippium, Linné,—Common in the Loch Tarbert district (B. & S.). In Upper Loch Fyne at Minard Narrows, and on both sides and centre, in 10 to 65 fathoms—the variety *striata* at Minard and on the east side, in 12 to 20 fathoms (M.).

Anomia patelliformis, Linné.—In Upper Loch Fyne at Minard Narrows, and on the east side in 10 to 20 fathoms (M.).

Ostrea edulis, Linné.—Generally distributed in the Tarbert district, but small and scarce, the variety *deformis,* Jeff., between tide-marks East Loch Tarbert (B. & S.). Upper Loch Fyne, between tide-marks, both the ordinary form and the variety *parasitica,* Turt. (M.). Loch Fyne near Tarbert (Alf. Brown).

Lima hians (Gmelin).—Off Battle Island, in 40 fathoms (B. & S.). In Upper Loch Fyne, near the shore on both sides, in about 15 fathoms (M.).

Lima subauriculata (Montagu).—Off Battle Island in 40 fathoms—dead shells (B. & S.). Off Silvercraigs, Loch Fyne; common in 10 fathoms in muddy sand and shells (Alf. Brown).

Lima Loscombi, G. B. Sowerby.—Rather rare in Upper Loch Fyne, in 10 to 15 fathoms on the west side (M.).

Pecten maximus (Linné).—Off Skate Island in 14 fathoms, and in other localities, but not very plentiful (B. & S.). Living specimens have been taken in Upper Loch Fyne (M.). Off Moll Dhu and Silvercraigs, Loch Fyne (Alf. Brown). Adult living specimens have been taken in East Loch Tarbert (Mihi).

Pecten varius, Linné.—East Loch Tarbert (B. & S.). Found abund-

* Dr Robertson (*op. cit.*, p. 152) also records the discovery of *Siphonodentalium (Pulsellum) affine,* M. Sars, 'off Skate Island at the mouth of Loch Fyne, in 90 to 100 fathoms.'

antly at extreme low water in Upper Loch Fyne (M.). In 10 fathoms, off Silvercraigs, in stony and shelly ground—young live specimens only (Alf. Brown).

Pecten tigrinus, O. F. Müller.—Not uncommon off Battle Island in 40 fathoms, both the typical form and the variety *costatus*, Jeff. (B. & S.). In Upper Loch Fyne at Minard Narrows, and at both sides, in 10 to 30 fathoms (M.).

Pecten striatus, O. F. Müller.—Furlong Bay and Moll Dhu Point, not common (B. & S.). Upper Loch Fyne at Minard Narrows, at both sides and in the centre, and also near the head of the loch, in 10 to 30 fathoms (M.).

Pecten pusio (Linné).—A few off Skate Island in 14 fathoms, and in East Loch Tarbert (B. & S.). Upper Loch Fyne at Minard Narrows, and on the east side in 10 to 25 fathoms, as well as also between tide marks (M.).

Pecten pes-lutræ (Linné), (= *Pecten septemradiatus*, Müller).—Common in Upper Loch Fyne in the deepest water, where also the largest specimens are found. Very large and fine specimens were taken below Strachur—much finer than towards the head of the loch—the specimens procured towards the head of the loch in about 10 fathoms, and at Minard Narrows, being smaller than those from the deep water. All the specimens taken in a haul off Skate Island in 104 fathoms were dead. In deep water the shells are always associated with much manganese (M.). Plentiful in deep water off Skate Island (B. & S.).

The variety *albus*, Jeff., is not very rare in Loch Fyne ; perhaps one in every fifty belongs to this variety (M.).

The variety *Dumasii*, Payr., has been taken in Upper Loch Fyne in 70 fathoms (M.).

Pecten opercularis (Linné).—Common in the neighbourhood of Tarbert (B. & S.). Upper Loch Fyne at Minard Narrows ; at both sides and in the centre, and also near the head in 10 to 70 fathoms ; and also on the shore between tide marks (M.). At Moll Dhu and Silvercraigs (Alf. Brown).

Pecten similis, Laskey.—Dredged at Tarbert Bank, Loch Fyne (Mihi). A local and gregarious species, Moll Dhu, Loch Fyne (Alf. Brown). The fry of *Pecten maximus* are liable to be mistaken for this species by those who have not had an opportunity of seeing both forms.

Mytilus edulis, Linné.—Found at the heads of all the Clyde lochs. This is a widely distributed species both in the northern and southern hemispheres ; it has been recorded from Rio de la Plata, Falkland, and Kerguelen Islands, and from New Zealand (M.). There is an extensive bed of *Mytilus edulis* at Ardrishaig (Alf. Brown).

Modiola modiolus (Linné).—Common in the neighbourhood of East Loch Tarbert (B. & S.). Abundant at extreme low water in all the Clyde lochs, also at Minard Narrows, and at both sides of Upper Loch Fyne in 10 to 30 fathoms (M.).

Modiolaria marmorata (Forbes).—Common ; usually found embedded in the test of large Ascidians which are frequent on Tarbert Bank, in Loch Fyne, and common at extreme low water in East Loch Tarbert (B. & S.). Upper Loch Fyne at Minard Narrows, in 12 to 20 fathoms (M.). Moll Dhu, Loch Fyne (Alf. Brown).

Modiolaria discors (Linné).—Attached to the roots of *Laminaria* and other algæ in East Loch Tarbert, and in neighbouring bays in Loch Fyne (B. & S.).

Nucula nucleus.—(Linné).—Upper Loch Fyne at Minard Narrows, in 12 to 20 fathoms (M.).

Nucula sulcata, Brown.—Upper Loch Fyne at Minard Narrows, in 12 to 15 fathoms (M.). Tarbert, Loch Fyne (Barlee).

Nucula nitida, G. B. Sowerby.—Battle Island ; not common (B. & S.). Upper Loch Fyne at Minard Narrows, and on the west and east sides and in the centre, in 10 to 70 fathoms (M.).

Nucula tenuis (Montagu).—Occasionally off Battle Island (B. & S.). On both sides and in the centre of Loch Fyne, in 10 to 70 fathoms (M.). Locally plentiful opposite Otter, Loch Fyne (Alf. Brown).

Leda minuta (O. F. Müller).—Off Battle Island (B. & S.). Abundant in Lower Loch Fyne in 100 to 105 fathoms (M.).

Astarte sulcata (Da Costa).—Laggan Bay and East Loch Tarbert (B. & S.). Upper Loch Fyne at Minard Narrows, and on both sides and in the centre, in 11 to 70 fathoms (M.). Off Otter Spit and Moll Dhu (Alf. Brown).

Astarte elliptica (Brown).—Laggan Bay (B. & S.). Upper Loch Fyne at Minard Narrows, and on the east side in 10 to 20 fathoms ; also in Lower Loch Fyne in 12 fathoms (M.). This, which is sometimes described as a variety of *Astarte sulcata*, is reinstalled to specific rank in Dr Norman's catalogue.

Astarte compressa (Montagu).—Upper Loch Fyne at Minard Narrows, in 11 to 20 fathoms, abundant (M.). In sand and nullipore, in 10 to 12 fathoms, off Silvercraigs, Loch Fyne; not uncommon (Alf. Brown). The variety *striata* (Leach), has been dredged at Tarbert Bank (Mihi).

Astarte triangularis (Montagu).— Off Silvercraigs, Loch Fyne—dead shells (Alf. Brown).

Turtonia minuta (Fabricius).—White Shore, East Loch Tarbert (B. & S.).

Lasœa rubra (Montagu).—East Loch Tarbert (B. & S.); var. *pallida*, Jeff., Loch Fyne (Alf. Brown).

Montacuta substriata (Montagu).—Off Otter, Loch Fyne (Alf. Brown). This species is frequently found on the spines of living *Spatangus purpureus*.

Montacuta bidentata (Montagu).—East Loch Tarbert (B. & S.). Off the pier at Ardrishaig—living (Alf. Brown).

Montacuta tumidula, Jeffreys.—Off Tarbert in 25 fathoms; Loch Fyne in 45 to 56 fathoms. (Somerville and J. T. Marshall in *Journal of Conchology*, vol. viii. p. 349, Jan. 1897.)

Decipula ferruginosa (Montagu).—Off the pier at Ardrishaig in rather less than 6 fathoms—fine specimens, living (Alf. Brown).

Lepton nitidum, Turton.—Off Silvercraigs, and at Otter Spit, Loch Fyne; living specimens (Alf. Brown).

Cardium echinatum, Linné.—Generally distributed, East Loch Tarbert and Loch Fyne (B. & S.). Upper Loch Fyne at Minard Narrows, and from the shore to 60 fathoms (M.). Loch Fyne (Alf. Brown).

Cardium papillosum, Poli.—Upper Loch Fyne (M.). This is an addition to Alf. Brown's Catalogue.

Cardium edule, Linné.—More common in West than in East Loch Tarbert (B. & S.). Found on shores, in bays, and in all the lochs of the Clyde district (M.). Loch Fyne (Alf. Brown).

Cardium exiguum, Gmelin.—East Loch Tarbert (B. & S.). Off Otter, Loch Fyne (Alf. Brown).

Cardium nodosum, Turton.—Has been taken in Upper Loch Fyne, on the east side and in the centre, in 30 to 70 fathoms (M.).

Cardium fasciatum, Montagu.—East Loch Tarbert (B. & S.). Upper Loch Fyne at Minard Narrows, and on east side in 10 to 30 fathoms (M.). Off Silvercraigs and off Otter, Loch Fyne (Alf. Brown).

Lævicardium norvegicum (Spengler).—East Loch Tarbert, and in Loch Fyne—single valves (B. & S.). At Minard Narrows, Upper Loch Fyne (M.).

Cyprina islandica (Linné).—Dead, but fresh, shells, near Laggan in 30 fathoms (B. & S.). At Minard Narrows; and on both sides and in the centre of Upper Loch Fyne, as well as near the head of the loch, in 5 to 65 fathoms; has also been obtained on the shore (M.).

Tapes pullastra (Montagu).—Common at White Shore, East Loch Tarbert (B. & S.). Upper Loch Fyne, on the shore—very plentiful on the Spit at Minard Narrows (M.).

Tapes virginea (Linné).—Has been observed in Upper Loch Fyne at Minard Narrows in 11 to 20 fathoms, and also at low water (M.).

Circe minima (Montagu).—Valves and dead, but fresh, shells in East Loch Tarbert and Loch Fyne (B. & S.). Upper Loch Fyne at Minard Narrows in 11 to 25 fathoms (M.). Sivercraigs, Loch Fyne (Alf. Brown).

Dosinia exoleta (Linné).—Upper Loch Fyne at Minard Narrows, also on the west side from the shore to 25 fathoms (M.).

Dosinia lincta (Pulteney).—Upper Loch Fyne, from the shore to 70 fathoms (M.).

Venus casina, Linné.—Upper Loch Fyne at Minard Narrows, in 12 to 20 fathoms, and in Lower Loch Fyne in 104 fathoms (M.). Off Silvercraigs, Loch Fyne (Alf. Brown).

Venus gallina, Linné.—Common along all the shores and in all the lochs of the Clyde district (M., Alf. Brown, and others).

Venus ovata, Pennant.—Off Battle Island, and in East Loch Tarbert (B. & S.). Common in all the Clyde lochs in depths of 10 to 70 fathoms (M.).

Venus fasciata (Da Costa).—Common in East Loch Tarbert and Loch Fyne (B. & S.). Minard Narrows, Upper Loch Fyne, in 12 to 20 fathoms (M.). Colour and sculpture variable.

Lucinopsis undata (Pennant).—Off Otter Spit, Loch Fyne; generally in about 8 to 12 fathoms (Alf. Brown).

Axinus flexuosus (Montagu). East Loch Tarbert (B. & S.). In all the upper lochs of the Clyde district in depths varying from 5 to 70 fathoms (M.).

Axinus ferruginosus (Forbes).—Loch Fyne, off Tarbert, in 16 to 18 fathoms (B. & S.). Upper Loch Fyne in 10 to 35 fathoms, and in Lower Loch Fyne in 20 to 100 fathoms (M.). Alive at Moll Dhu, Loch Fyne, on a muddy bottom (Alf. Brown).

Axinus croulinensis, Jeffreys.—Lower Loch Fyne (Robertson, *Trans. Nat. Hist. Soc. of Glasgow*, vol. ii., N.S., p. 152).

Psammobia ferröensis (Chemnitz).—At Minard Narrows, Upper Loch Fyne, in 10 to 30 fathoms, and also on the west and east sides (M.).

Solen siliqua, Linné.—East Loch Tarbert—variety *arcuata*, Jeff. (B. & S.). Large and fine specimens are occasionally obtained at extreme low water near Strachur in Upper Loch Fyne (M.).

* *Solen ensis*, Linné.—Taken in all the lochs of the Clyde district (M.).

Solen pellucidus, Pennant.—Loch Fyne, off Silvercraigs, and off Otter (Alf. Brown). Not very uncommon, but very fragile.

Mactra subtruncata (Da Costa).—Frequent in East Loch Tarbert (B. & S.). Upper Loch Fyne at Minard Narrows, in 12 to 20 fathoms, and also on the shore (M.).

Mactra elliptica, Brown.—Off the Otter Spit, Loch Fyne (Alf. Brown). Upper Loch Fyne, at Minard Narrows in 11 to 25 fathoms (M.). Jeffreys in *British Conchology* and Brown in *Mollusca of the Firth of Clyde*, include *M. elliptica*, Brown, under *M. solida*, Linné, as a variety of that species, but in Dr Norman's catalogue it is entered as a separate species.

Lutraria elliptica, Lamarck.—Upper Loch Fyne at Minard Narrows, in 12 to 20 fathoms (M.).

* *Solen siliqua* var. *arcuata* has sometimes been mistaken for *Solen ensis*; the latter, however, is usually considerably smaller than the variety *arcuata* of *S. siliqua*.

Mya arenaria, Linné.—Upper Loch Fyne, between tide marks (M.). Between tide marks East Loch Tarbert (Mihi).

Mya truncata, Linné.—East Loch Tarbert—dead shells (B. & S.). Between tide marks throughout the Clyde district where the conditions are favourable (M.).

Corbula gibba (Olivi).—Loch Fyne and East Loch Tarbert, not common (B. & S.). At Minard Narrows, and on both sides and in the centre, as well as near the head of Upper Loch Fyne, in 10 to 70 fathoms (M.). Off Silvercraigs, Loch Fyne (Alf. Brown).

Saxicavella plicata (Montagu).—A living specimen was obtained in muddy sand and nullipore at the mouth of Loch Gilp (Alf. Brown).

Saxicava rugosa (Linné).—Shores of East Loch Tarbert (B. & S.). Minard Narrows, and on both sides and in the centre of Upper Loch Fyne, in 10 to 36 fathoms (M.). The variety *artica* (Linné) has also been obtained at Minard (M.).

Xylophaga dorsalis, Turton.—A considerable number of specimens were obtained in a piece of partially rotten wood dredged in East Loch Tarbert; the species was also obtained under similar conditions off Skate Island, Loch Fyne (B. & S.). Also from a piece of old wood obtained off Inveraray (G.).

Teredo, sp. (?).—Specimens of a *Teredo* were observed in a piece of wood obtained on the east side of Upper Loch Fyne (M.).

Dibranchia.

Lucina borealis (Linné).—Dead shells frequent on the shore (B. & S.). On the shores of Upper Loch Fyne (M.). Off Silvercraigs, Loch Fyne (Alf. Brown).

Lucina spinifera (Montagu).—Taken on one occasion in Upper Loch Fyne, in 60 fathoms (M.). Off Otter and Moll Dhu, Loch Fyne; scarce, and not full grown (Alf. Brown).

Tellina crassa, Pennant.—One specimen off Silvercraigs, Loch Fyne (Alf. Brown). Single valves are frequent.

Tellina tenuis, Da Costa.—On the shore, Upper Loch Fyne (M.). East Loch Tarbert and neighbourhood; common (B. & S.).

Tellina fabula, Gronovius.—Off the pier at Ardrishaig (Alf. Brown).

Tellina balthica, Linné.—On the shore of Upper Loch Fyne (M.). Plentiful and fine at Lochgilphead; the variety *attenuata* (Jeff.) also occurs here (Alf. Brown).

Abra alba (S. Wood).—Off Battle Island, in muddy sand (B. & S.). In Upper Loch Fyne, on the east side and in the centre, and also near the head, in 25 to 70 fathoms (M.). Plentiful in Loch Gilp, in 5 to 7 fathoms, at the mouth of the loch (Alf. Brown).

Abra prismatica (Montague).—In muddy sand, in 6 fathoms, off Ardrishaig Pier; rather common (Alf. Brown).

Cuspidaria cuspidata (Olivi).—Off Battle Island; rare (B. & S.). At Minard Narrows, and on both sides and in the centre of Upper Loch Fyne, in 10 to 70 fathoms (M.). Loch Fyne (Robertson, see Brown's *Mollusca of the Firth of Clyde*, p. 41).

Cuspidaria costellata (Deshayes).—Tarbert Bank, Loch Fyne, and off Battle Island, in 40 fathoms (B. & S.). Minard Narrows, on the east and west sides; and off Furnace, in 10 to 25 fathoms (M.). Loch Fyne (M'Andrew, Forbes, and Barlee).

Cuspidaria abbreviata, Forbes.—In the deeper portions of Loch Fyne, rare (B. & S.). Minard Narrows, east side and centre, in 15 to 70 fathoms (M.). Loch Fyne (M'Andrew, Barlee, A. M. Norman).

Lyonsia norvegica (Chemnitz).—Loch Fyne; not common (B. & S.). Upper Loch Fyne, at Minard Narrows, and along both sides, in 8 to 20 fathoms (M.).

Cochlodesma prætenue (Pulteney).—Upper Loch Fyne, in depths of 5, 36, and 70 fathoms, as well as at low-water (M.).

Thracia papyracea (Poli).—Upper Loch Fyne, on the shore (M.).

Thracia villosiuscula (Macgillivray), variety *distorta* (Montagu).—Off Silvercraigs, Loch Fyne (Alf. Brown).

BRACHIOPODA OF LOCH FYNE.

Terebratulina caput-serpentis (Linné).—Fairly common of Battle Island, in 40 fathoms (B. & S.). At Minard, west and east sides and centre, in 10 to 65 fathoms (M.). Off Moll Dhu and Silvercraigs (Alf. Brown).

Crania anomala (O. F. Müller).—Abundant and large near Moll Dhu Point; frequent also in other parts of the loch; usually at a depth of less than 20 fathoms (B. & S.). Upper Loch Fyne, at Minard Narrows, in 11 to 20 fathoms, and towards the east side in 45 fathoms (M.).

CRUSTACEA OF LOCH FYNE.

A History of the Crustacea, by the Rev. T. R. R. Stebbing, has been followed as to the nomenclature of the MALACOSTRACA (the AMPHIPODA excepted).

BRACHYURA.

Inachus dorsettensis (Pennant).—Frequent in 10 to 20 fathoms in Loch Fyne (B. & S.). Upper Loch Fyne at Minard, and on both sides, in 10 to 30 fathoms (M.).

Hyas araneus (Linné).—Common between tide-marks (B. & S.). Upper Loch Fyne at Minard, on both sides and centre, and near the head of the loch, in 10 to 30 fathoms, and on the shore between tide-marks (M.).

I

Hyas coarctatus, Leach.—Common in the off-shore waters of Loch Fyne (B. & S.). Upper Loch Fyne at Minard, and on both sides, in 10 to 30 fathoms (M.).

Macropodia rostrata (Linné).—Upper Loch Fyne at Minard, and west side, in 10 to 25 fathoms (M.).

Macropodia longirostris (Fabricius).—Not common in Loch Fyne (B. & S.).

Achæus Cranchii, Leach.—Upper Loch Fyne, east side, in 10 fathoms (M.).

Cancer pagurus, Linné.—Frequent on the rocky shores of Loch Fyne (B. & S.).

Carcinus mænas (Linné).—Common (B. & S.). Both sides of Upper Loch Fyne, in 10 to 30 fathoms, and on the shore (M.). A specimen of this crab was obtained having a *Sacculina carcini* adhering to its abdomen (G.).

Portunus puber (Linné).—Frequent in 10 to 15 fathoms or more in Loch Fyne (B. & S.). Taken in fair numbers in 1885, but not observed during 1892 (M.).

Portunus depurator (Linné).—Frequent in Loch Fyne (B. & S.). At Minard and east side of Upper Loch Fyne, in 5 to 30 fathoms ; one of the commonest species of *Portunus* in the Clyde district (M.).

Portunus marmoreus, Leach.—Frequent in Loch Fyne (B. & S.).

Portunus pusillus, Leach.—Frequent in Loch Fyne (B. & S.). Upper Loch Fyne, east side, in 5 fathoms (M.).

Ebalia tuberosa (Pennant).—Tarbert Bank, Loch Fyne (Mihi).

ANOMURA.

Porcellana longicornis (Pennant).—Between tide-marks ; not very plentiful (B. & S.).

Lithodes maia (Linné).—Common in Loch Fyne, but mostly of medium size (B. & S.). Vicinity of Furnace (G.). Loch Fyne (Dr Scoular *).

Eupagurus bernhardus (Linné).—Common in Loch Fyne (B. & S.). Upper Loch Fyne at Minard, on the west and east sides, and in the centre, in 10 to 70 fathoms, and also near the head of the loch (M.). (' *Pagurus ulidianus* ' is a synonym of this species.)

Eupagurus Prideaux (Leach).—Common in Loch Fyne, and usually associated with *Adamsia palliata* (B. & S.). Upper Loch Fyne at Minard, on west and east sides and in the centre, in 10 to 70 fathoms (M.).

Eupagurus sculptimanus, Lucas.—Upper Loch Fyne at Minard, on the west side, in 10 to 25 fathoms (M., as *Pagurus Forbesii*). *Eupagurus Forbesii* is a synonym of *Eupagurus sculptimanus*.

---

* *Proceedings of Nat. Hist. Soc. of Glasgow*, First Series, vol. i. p. 8.

Eupagurus pubescens (Kroyer).—East Loch Tarbert (Mihi). Upper Loch Fyne at Minard, on the west side and in the centre, in 10 to 20 fathoms, also off Dunderave (M.). This species is usually surrounded more or less by a sponge—*Suberites suberea.*

Anapagurus Hyndmanni (Thompson).—East Loch Tarbert (Mihi).

Anapagurus lævis (Thompson).—Frequent in East Loch Tarbert and Loch Fyne (B. & S.). Upper Loch Fyne at Minard, on the west and east sides and in the centre, off Dunderave, in 10 to 36 fathoms (M.). A female with ova dredged in 105 fathoms in Loch Fyne in August (Henderson).

Anapagurus chiroacanthus, Lilljeborg (= *Eupagurus ferrugineus,* Norman).—East Loch Tarbert (B. & S.). Upper Loch Fyne, west side, in 10 to 15 fathoms (M.).

Munida rugosa (Fabricius).—Upper Loch Fyne at Minard, and on east side, in 12 to 20 fathoms (M.). Loch Fyne ; common (Dr Scouler, *op. cit.*).

Galathea squamifera, Leach.—Common in East Loch Tarbert and Loch Fyne (B. & S.). Upper Loch Fyne, west side, in 10 to 15 fathoms (M.).

Galathea nexa, Embleton.—Upper Loch Fyne at Minard, and in the centre, near the head of the loch, in 12 to 20 fathoms (M.). Loch Fyne (Dr Scouler, *op. cit.*).

Galathea dispersa, Spence Bate.—Upper Loch Fyne at Minard, and on both sides, and in the centre near the head, in 10 to 30 fathoms (M.).

Galathea intermedia, Lilljeborg.—East Loch Tarbert and Buck Bay (B. & S.). Loch Fyne at Minard, and on both sides, in 10 to 20 fathoms (M.).

MACRURA.

Calocaris Macandreæ (Bell).—Occurs sparingly in the deeper portions of Loch Fyne, in 60 to 90 fathoms (B. & S.). Upper Loch Fyne, in the centre, in 60 to 65 fathoms (M.). Loch Fyne (M'Andrew). Loch Fyne, in 40 to 105 fathoms (Henderson).

Palinurus vulgaris, Latr.—A single specimen taken in herring-nets by Ardrishaig fishermen in the spring (B. & S.).

Astacus Gammarus (Linné).—The common lobster is generally distributed in Lower Loch Fyne, where the shores are rocky. There is a small summer lobster fishing carried on in Buck Bay (B. & S.). East Loch Tarbert (Mihi). The fresh-water lobster of the English rivers is not an *Astacus,* but belongs to the genus *Potamobius* of Leach (see Stebbing's *History of Crustacea,* p. 207).

Pontophilus spinosus (Leach).—Upper Loch Fyne at Minard, in 11 to 25 fathoms (M.). East Loch Tarbert ; rare (B. & S.).

Crangon vulgaris, Fabricius.—Obtained in East Loch Tarbert, but not plentiful (B. & S.). Inshore, Upper Loch Fyne (M.).

Crangon Allmani, Kinahan.—Frequent in the offshore water (B. & S.). Upper Loch Fyne, both sides and centre, and near the head, in 10 to 70 fathoms (M.).   Off Inveraray (G.).

Cheraphilus echinulatus, M. Sars.—A single specimen off Skate Island, Loch Fyne, in 105 fathoms—mud (Henderson).

Cheraphilus neglectus, G. O. Sars.—Obtained occasionally in East Loch Tarbert and in neighbouring parts of Loch Fyne (Mihi).

Nika edulis, Risso.—Several specimens off Skate Island, Loch Fyne, in 105 fathoms (Henderson).

Spirontocaris Cranchii (Leach).—Loch Fyne (Forbes and M'Andrew. East Loch Tarbert (Mihi).

Spirontocaris Gaimardii (M. Edwards).—Loch Fyne (Forbes and M'Andrew), East Loch Tarbert as *Hippolyte pandaliformis* (B. & S.). Upper Loch Fyne, both sides, and in the centre, in 10 to 60 fathoms (M.).

Spirontocaris securifrons (Norman).—Common in 20 to 40 fathoms in Loch Fyne (B. & S.).   Upper Loch Fyne at Minard, east side, centre, and near the head, in 10 to 70 fathoms (M.).

Hippolyte varians, Leach.—*Zostera* bed in East Loch Tarbert, abundant (B. & S.).

Hippolyte fasciger, Gosse.—East Loch Tarbert, amongst *Zostera*, frequent (Mihi).

Caridion Gordoni (Spence Bate).—Upper Loch Fyne, on both sides, in 10 to 20 fathoms (M.).

Pandalus Montagui, Leach, (= *P. annulicornis*).—Common, East Loch Tarbert and Loch Fyne (B. & S.).   Abundant and of large size in Upper Loch Fyne in 70 fathoms.   The species is generally found everywhere in depths greater than 40 fathoms; but is also obtained in comparatively shallow water (M.).

Pandalus brevirostris, Rathke.—East Loch Tarbert (B. & S.).   West side of Upper Loch Fyne, in 10 to 15 fathoms (M.).   Loch Gair and Strachur Bay (G.).

Pasiphœa sivado (Risso).—Four specimens taken off Skate Island in 105 fathoms; one measures 4 inches in length, and two others are scarcely inferior in size (Henderson).   Off Strachur, in Upper Loch Fyne, in 70 fathoms (M.).   A specimen of *Pasiphœa* was obtained in the large midwater-net of the ' Garland,' at 5 fathoms below the surface, between Tarbert and Avidh Island, Loch Fyne.

SCHIZOPODA.

Euphausiidæ.

Nyctiphanes norvegica (M. Sars).—Loch Fyne; found also in the stomachs of herring and *Acanthias* (B. & S.).   In great abundance in

Upper Loch Fyne just above the mud in the deepest water; also in
Lower Loch Fyne, in 80 to 100 fathoms. The young are found in great
profusion at all seasons of the year, in from 5 to 20 fathoms above the
mud in the deepest water in Loch Fyne (M.).

Boreophausia Raschii (M. Sars).—Loch Fyne (B. & S.). Upper Loch
Fyne, between Lowburn and Dunderave, taken in the bottom tow-net
(G.).

Mysidæ.

Mysidopsis gibbosa, G. O. Sars.—East Loch Tarbert (B. & S.). In
Barmore Bay and other parts of Loch Fyne. Head of Upper Loch Fyne
(G.).

Mysidopsis didelphys (Norman).—Near the head of Loch Fyne, May
1896 (G.).

Mysidopsis augusta, G. O. Sars.—Barmore Bay, Loch Fyne, in 4
fathoms; rare (B. & S.)

Leptomysis linguara, G. O. Sars.—East Loch Tarbert; moderately frequent (B. & S.).

Leptomysis gracilis, G. O. Sars.—Between Loch Gair and Largymore,
and between Lowburn and Dunderave Castle, Upper Loch Fyne (G.).

Macromysis flexuosa (O. F. Müller).—East Loch Tarbert, among *Zostera*; frequent (B. & S.). Upper Loch Fyne, inshore (M.).

*Praunus inermis (Rathke).—East Loch Tarbert, among *Zostera*
(B. & S.). Upper Loch Fyne, between Inverae and Furnace (G.).

*Praunus neglectus (G. O. Sars).—Mouth of Loch Fyne, 60 fathoms,
mud; a single specimen from a tow-net attached to the trawl (Henderson).

Hemimysis Lamornæ (R. Q. Couch). — East Loch Tarbert, among
zostera, not very rare (B. & S.). Tarbert Bank, in 20 to 25 fathoms (G.).
When alive, the species is of a bright red or scarlet colour.

Neomysis vulgaris (J. V. Thompson).—Loch Dhu, near Inveraray (J.
Pringle). Prof. Bell, in *British Stalk-eyed Crustacea*, in describing this
species, refers to a *Mysis* obtained by Dr Leach at Loch Ranza, Arran,
and described by him as *M. integer*. The description of *M. integer* by Dr
Leach is imperfect, but Prof. Bell is of the opinion that it is identical with
M. vulgaris, J. V. Thompson. Dr Leach's record appears to be the only
one for the Clyde district, hitherto—it is the only one referred to by Dr
Henderson, in his *Decapod and Schizopod Crustacea of the Clyde.*
It is of interest, therefore, to have this confirmation of the fact that *Mysis*
(now *Neomysis*) *vulgaris* is a member of the Clyde fauna. Loch Dhu is
a small brackish water loch at the mouth of the river Shira, near Inveraray, into which the tide flows and ebbs. Mr Pringle, of H.M. Ordnance
Survey, who obtained the species in Loch Dhu, kindly handed his specimens over to me for examination, when I found them to belong to the
Mysis referred to. This *Mysis* is not an uncommon species in Scotland

* See *History of Crustacea*, by the Rev. T. R. R. Stebbing, p. 227, as to the
priority of *Praunus*.

—it is sometimes frequent in the Forth ; I have also obtained it in Loch Wester, in Caithness ; in a small loch on the Island of Barra, Outer Hebrides ; and in a gathering from Loch Belmont in Shetland.

Schistomysis ornatus (G. O. Sars).—Near the head of Loch Fyne, among trawl refuse (G.).

Schistomysis arenosa (G. O. Sars).—East Loch Tarbert, among *Zostera* (B. & S.).

Siriella armata (M. Edwards).—East Loch Tarbert and neighbouring parts of Loch Fyne (B. & S.).

Siriella Brooki, Norman.—East Loch Tarbert, among *Zostera* (B. & S.). This requires careful examination to distinguish it from the next species.

Siriella Clausii (G. O Sars).—East Loch Tarbert, among *Zostera ;* a somewhat rare species (B. & S.).

Erythrops elegans (G. O. Sars).—East Loch Tarbert, and in the vicinity of Barmore, Loch Fyne (Mihi).

CUMACEA.

Iphinoë trispinosa (Goodsir).—At the head of Upper Loch Fyne, in the bottom tow-net (G.).

Vaunthompsonia cristata, Spence Bate.—East Loch Tarbert, not common (B. & S.).   Off Largabruach, Upper Loch Fyne ; dredged (G.).

Lamprops fasciata, G. O. Sars.—East Loch Tarbert, Loch Fyne, frequent near low water, at White Shore (B. & S.).

Hemilamprops rosea, Norman.—A single specimen of this Cumacean was taken in the bottom tow-net between Inverae and Furnace, Upper Loch Fyne (G.).   In this specimen the telson was furnished with eight terminal spines.

Hemilamprops uniplicata, G. O. Sars.—East Loch Tarbert (B. & S.).

Leucon nasicus, Kroyer.—Between Lowburn and Dunderave, at the head of Upper Loch Fyne (G.).

Eudorella truncatula (Spence Bate).—Upper Loch Fyne, between Lowburn and Dunderave Castle, in the bottom tow-net (G.).

Diastylis rugosa, G. O. Sars.—East Loch Tarbert ; not very common (B. & S.).

Diastylis Rathkii, Kroyer.—East Loch Tarbert ; not common (Mihi).

Diastylus biplicata (Sars).—Head of Loch Fyne, in bottom tow-net (G.).

Pseudocuma cercaria, Van Beneden.—East Loch Tarbert ; frequent (Mihi).   Upper Loch Fyne, off Largabruach (G.).

Campylaspis costata, G. O. Sars.—Upper Loch Fyne, between Lowburn and Dunderave, in the bottom tow-net (G.).

Campylaspis rubicunda, Lilljeborg.—Upper Loch Fyne, between Lowburn and Dunderave, in the bottom tow-net (G.).

Cumella pygmœa, G. O. Sars.—East Loch Tarbert (Mihi). Tarbert Bank, Lower Loch Fyne, in 20 to 25 fathoms ; and at Largabruach, Upper Loch Fyne (G.).

Nannasticus unguiculatus, Spence Bate.—Upper Loch Fyne, between Lowburn and Dunderave, on the bottom tow-net, and at Tarbert Bank, Lower Loch Fyne, dredged (G.).

ISOPODA.

Tanais tomentosus, Kroyer [= *T. vitatus*, (Rathke)].—Loch Fyne, near the mouth of East Loch Tarbert, on rocks which are situated between tidemark, and more or less covered with barnacles ; frequent (Mihi).

Tanaopsis laticaudatus, G. O. Sars (= *Leptognathia laticaudata*, G. O. Sars).—Moderately frequent in Loch Gair, and in dredged material collected off Largabruach, Upper Loch Fyne (G.).

Eurycopa phalangium, G. O. Sars.—This curious little Isopod was obtained in the bottom tow-net, between Lowburn and Dunderave Castle, Upper Loch Fyne ; very rare (G.). I am indebted to the Rev. Mr Stebbing for the identification of this species.

Anceus maxillaris (Montagu).—Upper Loch Fyne (M.). Off Inverary (G.).

Æga bicarinata, Leach.—Dredged in Loch Fyne in 15 fathoms ; bottom, mud, shells and gravel (Robertson).

Cirolana spinipes, Bate and Westwood.—Dredged at the mouth of Loch Fyne, in 37 fathoms ; bottom, mud and small gravel (Robertson). Loch Fyne, in 30 fathoms (Mihi). This last was a female with ova.

Eurydice pulchra, Leach.—Upper Loch Fyne (M.). East Loch Tarbert, not unfrequent (Mihi).

Sphæroma curtum, Leach.—East Loch Tarbert, not common (Mihi).

Limnoria lignorum (Rathke).—East Loch Tarbert, on old wood ; generally distributed throughout Loch Fyne (Mihi).

Idotea tricuspidata (Desmarest).—Upper Loch Fyne, at Minard, and on the west side, in 10 to 25 fathoms (M.). East Loch Tarbert (Mihi).

Idotea pelagica Leach.—Upper Loch Fyne (M.). East Loch Tarbert, not very common (Mihi). *Idotea pelagica* is a much smaller Isopod than *I. tricuspidata*, as well as being of a different habit, and it is also proportionally stouter. I have a specimen of the former (*with ova*) not more than

---

* Professor Sars, in his new work on the *Crustacea of Norway* (vol. ii.), has instituted a new genus for this Isopod, viz. *Tanaopsis*.

7 mm. in length. (Notwithstanding the marked difference between ova-bearing specimens of *I. tricuspidata* and *I. pelagica*, it is now customary to consider both as belonging to the one species, viz. *Idotea marina*, Linné.)

Astacilla longicornis (Sowerby).—Upper Loch Fyne, in the centre, in 36 to 70 fathoms; also off Dunderave (M.). East Loch Tarbert (Mihi). Off Skate Island, in 105 fathoms (Robertson).

Janira maculosa, Leach.—East Loch Tarbert, under stones at low water, and dredged; generally distributed (Mihi).

Jæra (?) Nordmanni, Rathke.—Upper Loch Fyne, between Lowburn and Dunderave Castle, in a bottom tow-net gathering (G.). East Loch Tarbert (Mihi).

Leptaspidia brevipes, Bate and Westwood.—East Loch Tarbert, not common (Mihi).

Cryptothir balani (Spence Bate).—East Loch Tarbert, associated with *Balanus balanoides*. While examining a number of specimens of *Balanus balanoides* at Tarbert Laboratory, a specimen of *Cryptothir*, a female, was obtained in the sixth *Balanus* examined, but though several hundreds of the same species of *Balanus* were afterwards carefully dissected, no more *Cryptothir balani* were found (Mihi).

Athelges paguri (Rathke).—Occasionally obtained on *Eupagurus bernhardus* in East Loch Tarbert and neighbouring parts of Loch Fyne.

Pseudione Hyndmanni (Bate and Westwood).—This species (both the form known as "*Phryxus Hyndmanni*" and that described as "*Phryxus fusticaudatus*"),* was obtained in Loch Fyne, near Tarbert. "*Phryxus Hyndmanni*" was taken from a specimen of *Hippolyta varians*, but "*Phryxus fusticaudatus*" was obtained from a specimen of *Eupagurus bernhardus*. As regards "*Phryxus fusticaudatus*" the following note was made at the time of its discovery—"It agrees very well with the figure given in Bate and Westwood's monograph; the clavate lobes on each side of pleon and the "spathulate" terminal segment are well developed." (This specimen was unfortunately lost.)

AMPHIPODA.

The Amphipoda of Norway, by Professor G. O. Sars, is followed in the classification and nomenclature of this group.

Hyperiidea.

Hyperia galba (Montagu).—East Loch Tarbert, associated with *Aurelia aurita* (B. & S.).

Hyperoche tauriformis (Spence Bate).—Loch Gair, near Quay Ferry, Upper Loch Fyne (G.). These two species of the *Hyperiidea*, though

---

* This specimen of *Phryxus fusticaudatus* was a mature female with numerous embryo young enclosed under the ovigerous plates: these embryos closely resemble the parasitic Epicaride known as *Microniscus*, and which is found sometimes on *Calanus*.

occasionally observed within the Clyde area, are somewhat uncommon; very few specimens of either have, so far as known, been obtained in Loch Fyne. Both species are moderately common on the east coast of Scotland.

Gammaridea.

Talitrus locusta (Pallas).—Upper Loch Fyne, off Inveraray (G.)

Orchestia littorea (Montagu), [= *O. Gammarellus* (Pallas)].—East Loch Tarbert. This is a species of frequent occurrence wherever the conditions are suitable.

Hyale Lubbockiana (Spence Bate), [= *Allorchestes imbricates*, Spence Bate, and *Hyale Nelssoni*, Boeck (in part)].—East Loch Tarbert, moderately frequent.

Lysianax Costæ (Milne Edwards).—Upper Loch Fyne (M.). East Loch Tarbert (Mihi).

Perrierella Audouiniana (Spence Bate), (= *Lysianassa Audouiniana*, Spence Bate, and *Pararistias Audouiniana*, Robertson).—Upper Loch Fyne, at Minard, in 11 to 25 fathoms (M.). Tarbert Bank, Lower Loch Fyne (G.).

Calisoma crenata (Spence Bate).—Loch Fyne, in 80 fathoms (Robertson). Upper Loch Fyne in the centre in 36 to 70 fathoms—inside a dead *Brissopsis lyrifera* (M.). Between Lowburn and Dunderave, rare (G.).

Hippomedon denticulatus, Spence Bate (= *Hippomedon Holbölli*, Boeck).—Loch Fyne in 80 fathoms, bottom mud (Robertson). East Loch Tarbert (Mihi). Tarbert Bank, Lower Loch Fyne, in 20 to 25 fathoms (G.).

Sophrosyne Robertsoni, Stebbing and Robertson.—(?) Upper Loch Fyne (M.).

Orchomene Batei, G. O. Sars.—Loch Fyne, in 10 to 12 fathoms (Robertson). Tarbert Bank, Loch Fyne (G.).—Upper Loch Fyne, west and east sides and centre, in 10 to 70 fathoms (M.)

*Orchomenella minuta (Kroyer).—This species is reported from Minard, where it was obtained in 11 to 25 fathoms (M.).

Tryphosa Sarsi (Bonnier).—Upper Loch Fyne, near Largabruach, and off Inveraray (G.).

Tryphosites longipes (Spence Bate).—East Loch Tarbert (Mihi). Upper Loch Fyne (M.).

Hoplonyx cicada (Fabricius), [= *Anonyx Holbölli*, Spence Bate (not Kroyer)].—Tarbert Bank, Lower Loch Fyne, in 20 to 25 fathoms (G.).

Bathyporeia Robertsonii, Spence Bate.—East Loch Tarbert. (Prof.

---

*" Orchomene propinquus " is also recorded as having been obtained on the west side of Upper Loch Fyne, in 10 to 25 fathoms (M.). I do not find this species in Robertson's catalogue, or in Sars' monograph.

G. O. Sars seems to have no doubt as to *B. Robertsoni* being a ' good ' species).

Urothoe marina, Spence Bate.—Near Barmore, Loch Fyne (Mihi). Upper Loch Fyne (M.).

Urothoe norvegica, Boeck.—Upper Loch Fyne, west and east sides and centre, in 10 to 70 fathoms (M.).

Phoxocephalus Holbölli (Kroyer).—Upper Loch Fyne, on the east side and in the centre, in 10 to 70 fathoms (M.). *Phoxus simplex*, Spence Bate, is probably synonymous with this species.

Harpinea neglecta G. O. Sars.—East Loch Tarbert (Mihi). Loch Fyne, in 80 fathoms (Robertson). Loch Gair, frequent (G.).

Ampelisca typica (Spence Bate).—East Loch Tarbert (Mihi). Upper Loch Fyne, west side, in 10 to 25 fathoms (M.).

Ampelisca lævigata, Lilljeborg.—Taken by the sieve in pure sand, at low water, at Crarae, Loch Fyne (R.). Upper Loch Fyne, in the centre, in 36 to 70 fathoms (M.).

Ampelisca tenuicornis, Lilljeborg.—Dredged in Loch Fyne, in 80 fathoms, bottom soft mud (R.). Upper Loch Fyne, at Minard, west and east sides and centre, in 10 to 70 fathoms (M.). Loch Gair (G.).

Ampelisca spinipes, Boeck.—Loch Fyne, near Skate Island, in 100 fathoms and also in 80 fathoms (R.).

Ampelisca Eschrichtii, Kroyer.—Upper Loch Fyne, at Minard, in 11 to 25 fathoms (M.).

Haploops setosa, A. Boeck.—Dredged off Skate Island, Loch Fyne, in 100 fathoms ; two only were met with (Robertson).

Stegocephaloides christianiensis, Boeck.—Loch Fyne, in 40 to 70 fathoms. Upper Loch Fyne (M.).

[*Stegocephaloides aurates*, G. O. Sars, is inserted among the MS. records of the steam yacht ' Medusa ' for Upper Loch Fyne, east side and centre, in 15 to 70 fathoms, but this species is not recorded in the catalogue of the late Dr Robertson.]

Amphilochus manudens, Spence Bate.—Upper Loch Fyne, between Lowburn and Dunderave, in the bottom tow-net (G.).

Cyproidea damnoniensis, Stebbing.—Upper Loch Fyne (M.).

Stenothoe marina (Spence Bate).—Upper Loch Fyne (M.). At Tarbert Bank, Lower Loch Fyne, in 20 to 25 fathoms (G.),

Stenothoe monoculodes (Montagu).—Upper Loch Fyne, west and east sides and centre, in 10 to 70 fathoms (M.).

Metopa, sp.—Upper Loch Fyne, between Lowburn and Dunderave (G.).

Cressa dubia (Spence Bate).—Tarbert Bank, Loch Fyne, dredged (G.).

Leucothoë spinicarpa (Abildgaard).—Taken in Loch Fyne, in 92 fathoms (R.). In the branchial chamber of large Ascidians (*Ascidia mentula*) collected at low water in East Loch Tarbert ; also in Ascidians dredged on Tarbert Bank, Loch Fyne (G.).

* *Leucothoë Lilljeborgii*, Boeck.—Loch Gair, and between Lowburn and Dunderave, Upper Loch Fyne (G.).

Monoculodes carinatus, Spence Bate.—East Loch Tarbert (Mihi). Upper Loch Fyne (M.).

Monoculodes Packardi, Boeck.—Upper Loch Fyne, between Lowburn and Dunderave in the bottom tow-net (G.).

Perioculodes longimanus (Spence Bate).—Upper Loch Fyne, between Lowburn and Dunderave (G.).

Pontocrates altamarinus (Spence Bate).—Upper Loch Fyne (M.).

Synchelidium brevicarpum (Spence Bate).—Upper Loch Fyne (M.). Tarbert Bank, Lower Loch Fyne (G.).

Halimedon parvimanus (Spence Bate †).—Upper Loch Fyne, west and east sides and centre, in 10 to 70 fathoms (M.). Near Largabruach, dredged, and between Lowburn and Dunderave, in the bottom tow-net (G.). This species, which is the *Œdiceros parvimanus* of Spence Bate, is not uncommon in Loch Fyne.

Epimeria cornigera (Fabricius).—Upper Loch Fyne, at Minard, in 11 to 25 fathoms (M.).

Iphimedia obesa, Rathke.—Upper Loch Fyne, on both the west and east sides and in the centre, in 10 to 70 fathoms (M.). Near Largabruach, dredged ; and in the vicinity of Carndow, near the head of the loch (G.).

Iphimedia minuta, G. O. Sars.—Upper Loch Fyne, between Lowburn and Dunderave, in the bottom tow-net ; also taken with the dredge at Tarbert Bank, Lower Loch Fyne, in 20 to 25 fathoms (G.).

Eusirus longipes, Boeck.—Upper Loch Fyne, at Minard, and on the west side, in 10 to 25 fathoms (M.). Between Lowburn and Dunderave, in the bottom tow-net (G.).

Apherusa bispinosa (Spence Bate).—Upper Loch Fyne, at Minard, west and east sides and centre, in 10 to 70 fathoms (M.). Loch Gair and near Largabruach (G.).

---

* I have a *Leucothoë* from East Loch Tarbert (Loch Fyne) that differs from *L. Lilljeborgii* in having the penultimate as well as the last pair of epimeral plates strongly toothed at the lateral corners—the last pair having the corner tooth defined above by a sinus as in *L.* Lilljeborgii ; the palm of the second gnathopode of this Tarbert specimen differs also in its general outline. This is probably the form described by Dr Robertson as *Leucothoë incisa* (see Dr Robertson's *Amphipoda of the Clyde*, Part ii., p. 23.

† See 'The Amphipoda of Bate and Westwood's "British Sessile-eyed Crustacea,"' by A. O. Walker (*Ann. and Mag. Nat. Hist.*, Sixth Series, vol. xv. p. 466, 1895).

Paratylus Swammerdami (Milne Edwards).—East Loch Tarbert (Mihi).

Paratylus vedlomensis (Spence Bate)—Upper Loch Fyne, in the middle of the loch, in 36 to 70 fathoms (M.).

Dexamine spinosa (Montagu).—Upper Loch Fyne, west and east sides and centre, in 10 to 70 fathoms (M.). Off Largabruach, dredged (G.).

Dexamine Thea, Boeck.—Upper Loch Fyne (M.). A small species, and easily overlooked.

Amathilla homari (Fabricius).—East Loch Tarbert, at extreme low water. This was one of the largest specimens I have seen (Mihi).

Gammarus marinus, Leach.—East Loch Tarbert (Mihi).

Gammarus locusta (Linné).—Upper Loch Fyne, at Minard, in 11 to 25 fathoms (M.). Off Inveraray (G.). East Loch Tarbert (Mihi).

Melita obtusata (Montagu).—Upper Loch Fyne, west side, in 10 to 25 fathoms (M.). Off Inveraray, and near Carndow at the head of the loch (G.).

Mœra Othonis (Milne Edwards).—Upper Loch Fyne, at Minard, on the west and east sides and in the centre, in 10 to 70 fathoms (M.). Loch Fyne in 90 fathoms [Robertson, as *Mœra longimana* (Thompson)]. Tarbert Bank, Loch Fyne, dredged (G.). *Mœra longimana* (Thompson) is considered to be the male of *M. Othonis*, but Spence Bate's figure of the hand of the second *gnathopods* of *Mœra (Megamœra) longimana* is quite different from that of Sars' figure of the hand of the same gnathopods of the male of *Mœra Othonis*.

Cheirocrates Sundewalli (Rathke).—Upper Loch Fyne, at Minard Narrows and in the centre of the loch, in 11 to 70 fathoms (M.). Dredged near Largabruach, and also at Tarbert Bank (G.).

Cheirocrates intermedius, G. O. Sars.—Dredged in Loch Gair in Upper Loch Fyne, rare (G.). East Loch Tarbert, 1886 (Mihi).

Cheirocrates assimilis (Lilljeborg).—Lower Loch Fyne, in 104 fathoms (Robertson). Tarbert Bank, in 20 to 25 fathoms (G.).

Microdeutopus anomalus (Rathke).—Upper Loch Fyne, on the east side, in 15 to 30 fathoms (M.). East Loch Tarbert (Mihi).

Microdeutopus danmoniensis (Spence Bate).—Upper Loch Fyne, at Minard, on the east side and in the centre, in 11 to 70 fathoms (M.). According to Sars this is the *Microdeutopus gryllotalpa* of Spence Bate, but not of Costa.

Aora gracilis (Spence Bate).—Loch Gair, Upper Loch Fyne, dredged ; not common (G.).

Leptocheirus pilosus, Zaddach.—Tarbert Bank, Lower Loch Fyne, in 20 to 25 fathoms (G.). The antennal appendages are two-jointed in these Tarbert Bank specimens.

Gammaropsis erythrophthalma, Lilljeborg.—Upper Loch Fyne, at Minard, on the east side and in the centre, in 11 to 70 fathoms (M.).

Podoceropsis Sophiæ, Boeck.—Upper Loch Fyne, at Minard, west and east sides, in 10 to 30 fathoms (M.).

Podoceropsis excavata (Spence Bate).—Upper Loch Fyne (M.).

Amphithoe rubricata (Montagu).—Upper Loch Fyne (M.). East Loch Tarbert (Mihi).

Pleonexes gammarodes, Spence Bate.—East Loch Tarbert, at low-water, and dredged (Mihi).

Podocerus pusillus, G. O. Sars.—Tarbert Bank, Loch Fyne, dredged (G.).

Erichthonius abditus (Templeton).—Upper Loch Fyne, on the east side, in 15 to 30 fathoms (M.). East Loch Tarbert, both ♂ and ♀ (Mihi). Loch Gair (G.).

Corophium Bonellii, Milne Edwards.—Off Inveraray, Loch Fyne, also near Carndow, and in Loch Gair (G.).

Corophium crassicorne, Bruzelius.—East Loch Tarbert (Mihi).

Dulichia falcata (Spence Bate).—Tarbert Bank, Lower Loch Fyne (G.).

CAPRELLIDEA.

Phtisica marina, Slabber.—Upper Loch Fyne (M.). Between Low-burn and Dunderave, and off Ardno (G.). East Loch Tarbert (Mihi). Tarbert Bank, in 20 to 25 fathoms (G.).

Pariambus typicus (Kroyer).—East Loch Tarbert, on the common starfish *(Asterias rubens)* (Mihi).

Caprella linearis (Linné).—East Loch Tarbert, amongst *Zostera* (Mihi).

Caprella acanthifera, Leach.—East Loch Tarbert, amongst *Zostera* (Mihi).

PHYLLOCARIDA (PHYLLOPODA).

Nebalia bipes, Milne Edwards.—Upper Loch Fyne, off Inveraray, and off Largabruach (G.). East Loch Tarbert (Mihi).

CLADOCERA.

Evadne Nordmanni, Lovén.—Upper Loch Fyne, between Carndow and Dunderave, off Inveraray and off Furnace (G.). A generally distributed species, but sometimes it may occur in abundance ; at other times it may be very scarce.

Podon polyphemoides, Leuckart.—Near the head of Loch Fyne ; this species was found in considerable abundance about 12 inches or so below the surface of the water.

(?) *Podon intermedius*, Lilljeborg.—Upper Loch Fyne.

OSTRACODA.

PODOCOPA.

Paracypris polita, G. O. Sars.—Loch Fyne, off Tarbert, rare (Norman).

Pontocypris mytiloides, Norman.—East Loch Tarbert ; not uncommon (B. & S.).

Pontocypris trigonella, G. O. Sars.—East Loch Tarbert (B. & S.). Loch Gair, and off Inveraray ; not uncommon (G.).

Argillœcia cylindrica, G. O. Sars.—Off Tarbert, Loch Fyne, in 25 fathoms (Norman).

Bairdia complanata, G. S. Brady.—Loch Fyne (Norman).

Cythere lutea, Müller.—East Loch Tarbert (B. & S.).   A moderately common species, especially in shallow water.

Cythere confusa, Brady and Norman.—East Loch Tarbert (B. & S.). Loch Gair, Upper Loch Fyne (G.).   Generally distributed throughout the district.

Cythere porcellanea, G. S. Brady.—Off Inveraray, Upper Loch Fyne (G.).

Cythere (?) *semipunctata*, G. S. Brady.—Off Tarbert, Loch Fyne (Norman).   Loch Fyne (Mihi).

Cythere crispata, G. S. Brady.—East Loch Tarbert (B. & S.).

Cythere gibbosa (Brady and Robertson).—Loch Gilp (Loch Fyne) (Brady and Robertson).

Cythere albo-maculata, Baird.—East Loch Tarbert (B. & S.).   Generally distributed.

Cythere Robertsoni, G. S. Brady.—Loch Fyne, off East Loch Tarbert (Mihi).   Not very uncommon.

Cythere convexa, Baird.—East Loch Tarbert (B. & S.).   A moderately common species.

Cythere marginata, Norman.—Off Tarbert, Loch Fyne, in 25 fathoms (A. M. Norman).

Cythere cluthæ, Brady, Crosskey, and Robertson.—Loch Fyne, in 20 fathoms (Mihi—see *Monograph of the Marine and Fresh-Water Ostracoda of the North Atlantic and North-Western Europe*, Part i. p. 145).

Cythere villosa (G. O. Sars).—Off Inveraray (G.). East Loch Tarbert (B. & S.). Common and generally distributed.

Cythere tuberculata (G. O. Sars).—East Loch Tarbert (B. & S.).

Cythere concinna, Rupert Jones.—Dredged in Loch Fyne (Brady and Robertson). East Loch Tarbert (B. & S.).

Cythere angulata (G. O. Sars).—Off Tarbert Loch Fyne (A. M. Norman). East Loch Tarbert (B. & S.).

Cythere antiquata (Baird).—East Loch Tarbert (B. & S.).

Cythere Jonesii (Baird).—Loch Fyne, off Tarbert, in 25 fathoms, and off Skipness in 41 fathoms (A. M. Norman). Loch Fyne (B. & S.).

Cytheridea papillosa, Bosquet.—Loch Fyne (B. & S.). More or less generally distributed.

Cytheridea punctillata, Brady.—Loch Fyne, off Inveraray, off Tarbert, and off Skipness (A. M. Norman). Loch Fyne (B. & S.). Upper Loch Fyne (G.).

Cytheridea (?) subflavescens, G. S. Brady.—Loch Fyne, in 40 fathoms ; off Skipness, off Tarbert, in 25 fathoms (A. M. Norman). Loch Fyne (G. S. Brady and D. Robertson).

Eucythere declivis (Norman) = *Eucythere argus* (G. O. Sars).—Loch Fyne (B. & S.). A moderately common species.

Krithe bartonensis (Jones).—Loch Fyne, off Inveraray and off Tarbert, in 25 fathoms (A. M. Norman). Off Inveraray, and between Carndow and Ardno, north side of the loch (G.).

Loxoconcha impressa (Baird).—East Loch Tarbert (B. & S.). Loch Gair and off Largabruach (G.). A moderately common species.

Loxoconcha guttata (Norman).—Loch Fyne, at Inveraray, and off Skipness (A. M. Norman).

Loxoconcha multifora (Norman).—East Loch Tarbert (B. & S.). Loch Fyne, as *Cythere multifora* (A. M. Norman).

Loxoconcha tamarindus (Jones).—East Loch Tarbert (B. & S.). Moderately common everywhere.

Xestoleberis aurantia (Baird).—East Loch Tarbert (B. & S.). Not very rare in Loch Fyne.

Xestoleberis depressa (G. O. Sars).—Loch Fyne (Mihi). Loch Gair and off Inveraray (G.).

Cytherura gibba (Müller).—East Loch Tarbert (B. & S.).

Cytherura cornuta (G. S. Brady).—East Loch Tarbert (B. & S.). Off Inveraray, Upper Loch Fyne (A. M. Norman). Loch Fyne (G. S. Brady and D. Robertson).

Cytherura sella, G. O. Sars.—East Loch Tarbert, as *Cy. flavescens* (B. & S.). A moderately common species throughout the district.

Cytherura acuticostata, G. O. Sars.—Loch Fyne (A. M. Norman).

Cytherura striata, G. O. Sars.—East Loch Tarbert (B. & S.). A common and generally distributed species.

Cytherura angulata (G. S. Brady).—East Loch Tarbert (B. & S.).

Cytherura undata, G. O. Sars.—East Loch Tarbert (B. & S.). This is a small species and easily overlooked.

Cytherura producta, G. O. Brady.—Loch Fyne, off Tarbert, in 25 fathoms (A. M. Norman).

Cytherura nigrescens (Baird).—East Loch Tarbert (B. & S.). There is scarcely a haul made inshore with the dredge in which this species does not occur.

Cytherura similis, G. O. Sars.—Loch Fyne, off Skipness (A. M. Norman). Generally distributed throughout the Clyde area.

Cytherura cellulosa, Norman.—East Loch Tarbert (B. & S.). This is a very small species and easily overlooked.

Cytheropteron latissimum (Norman).—East Loch Tarbert (B. & S.). Loch Fyne (A. M. Norman).

Cytheropteron nodosum, G. S. Brady.—East Loch Tarbert (B. & S.). Loch Fyne, off Tarbert (A. M. Norman).

Cytheropteron inflatum, Brady, Crosskey, and Robertson.—Loch Fyne (A. M. Norman). This is the only British record for *Cytheropteron inflatum* as a recent species.

Cytheropteron punctatum, G. S. Brady.—Loch Fyne, off Tarbert, in 25 fathoms (A. M. Norman). East Loch Tarbert (Mihi).

Cytheropteron alatum, G. O. Sars.—Loch Fyne, off Tarbert, in 25 fathoms (A. M. Norman). East Loch Tarbert, as *Cytheropteron arcuatum* (B. & S.).

Cytheropteron angulatum, Brady and Robertson.—East Coast Tarbert (B. & S.). Loch Fyne, off Tarbert, in 25 fathoms (A. M. Norman).

Cytheropteron humile, Brady and Norman.—Off Inveraray, Upper Loch Fyne (G.). Hitherto this Ostracod has almost invariably been obtained from pieces of partially decayed wood brought up in the dredge or trawl—wood that has been more or less perforated by boring Mollusks or Crustacea. In such situations I have usually found this Ostracod associated with the Copepod *Laophonte simulans,* T. Scott.

Bythocythere constricta, G O. Sars.—Loch Fyne (G. S. Brady and D. Robertson).

Bythocythere turgida, G. O. Sars.—Furlong Bay, Loch Fyne (B. & S.).

Bythocythere simplex (Norman).—East Loch Tarbert (B. & S.). This is a moderately common species in Loch Fyne and in the Clyde generally.

Pseudocythere caudata, G. O. Sars.—Loch Fyne (B. & S.). This Ostracod is not very rare, but is easily overlooked.

Sclerochilus contortus (Norman).—Loch Fyne (B. & S.). A moderately common and widely distributed species.

Paradoxostoma variable (Baird).— East Loch Tarbert, Loch Fyne (B. & S.). A common and generally distributed species.

Paradoxostoma pulchellum, G. O. Sars.—Furlong Bay, Loch Fyne (B. & S.).

Paradoxostoma Hodgei, G. S. Brady.—Off Tarbert, Loch Fyne, in 25 fathoms (A. M. Norman).

Paradoxostoma flexuosum, G. S. Brady.—Loch Fyne, off Inveraray, in 25 to 40 fathoms, and also off Skipness (A. M. Norman).

Paradoxostoma affine, T. Scott.—Off Inveraray, Upper Loch Fyne, in 25 to 40 fathoms (A. M. Norman).

Machœrina tenuissima (Norman).—Loch Fyne as *Xiphichilus tenuissima* (B. & S.).

MYODOCOPA.

Asterope mariœ (Baird).—East Loch Tarbert (B. & S.). Off Tarbert, Loch Fyne (A. M. Norman). Off Largabruach, Upper Loch Fyne (G.).

Asterope teres (Norman).—East Loch Tarbert (B. & S.). This species is much less common in Loch Fyne than the previous one.

Philomedes interpuncta (Baird).—East Loch Tarbert (B. & S.). Off Largabruach, Upper Loch Fyne (G.).

CLADOCOPA.

Polycope orbicularis, G. O. Sars.—East Loch Tarbert (B. & S.). Loch Fyne (Brady and Norman).

Polycope punctata, G. O. Sars.—Some specimens dredged in Loch Fyne are doubtfully referable to this species (Brady and Norman).

THE COPEPODA OF LOCH FYNE.

In the preparation of this list, *The Monograph of the British Copepoda*, by Prof. G. S. Brady; *The Copepoda of the Bay of Naples*, by Dr Giesbrecht; Dr Canu's Monograph *Les Copepodes de Boulonnais*, besides many separate papers, have been consulted.

GNATHOSTOMATA.

Calans finmarchicus (Gunner).—Very abundant in Upper Loch Fyne in deep water, where they are found all the year through; less abundant towards the lower end of the loch.

K

Pseudocalanus elongatus, Boeck.—Generally distributed, and more or less frequent all over the loch. The following are a few of the localities where the species has been obtained : In surface and bottom tow-nets at the head of Loch Fyne, off Inveraray in a tow-net at 3 fathoms from the surface ; in surface and bottom tow-nets, between Loch Gair and Largymore ; and in surface tow-net between Tarbert and Avidh Island (G.).

× Bradypolius,   (Brady).

(?) ~~Pseudocalanus~~ *armatus,* ~~Boeck~~.—In bottom tow-net gatherings from the vicinity of Largymore and Furnace, and between Lowburn and Dunderave Castle (G.).

Stephos gyrans (Giesbrecht).—(Pl. II. fig. 9 ; Pl. III. figs. 17, 18). A single specimen—a female—of this interesting species was obtained in some dredged material from Loch Gair, Upper Loch Fyne. A careful comparison of the various appendages of this Loch Gair specimen with Dr Giesbrecht's description and figures, leaves no doubt as to its identity with the specimens obtained at Naples. Dr Giesbrecht discovered his specimens in the tanks of the Zoological Station at Naples. The species appears to be quite distinct from *Stephos minor* (T. Scott) from the Firth of Forth. The Loch Gair specimen measures ·8 mm. in length (about $\frac{1}{30}$th of an inch).

Euchæta norvegica, Boeck.—Common in the deep water of Upper Loch Fyne, and also obtained occasionally at the surface (G.). Upper Loch Fyne in 60 to 70 fathoms ; taken at all seasons of the year in great abundance in Upper Loch Fyne, at from 5 to 15 fathoms above the mud in the deepest water. The females seem to carry a succession of ova, which are of a bluish colour, all through the year (M.).

Scolecithrix hibernica, A. Scott.[*]—Taken in a bottom tow-net gathering between Carndow and Dunderave Castle, and between Loch Gair and Minard Castle (G.). This species was first obtained in deep water off the coast of county Down, Ireland, by my son, Mr Andrew Scott.

Centropages typicus, Kröyer.—East Loch Tarbert and neighbouring parts of Loch Fyne (Calderwood).

Centropages hamatus (Lilljeborg).—East Loch Tarbert and Loch Fyne (Calderwood). Off Inveraray, between Carndow and Dunderave Castle, and in Loch Gair (G.). This species is moderately common all over the district ; Loch Fyne (M.).

Temora longicornis (Müller).—This also is a moderately common and widely distributed species. It has been obtained off Carndow, off Inveraray, off Minard, and in Loch Gair (G.). In East Loch Tarbert and Loch Fyne (Calderwood).

Metridia hibernica (Brady and Robertson).—East Loch Tarbert and Loch Fyne (Calderwood). Off Largabruach, Upper Loch Fyne (G.). Dr W. Giesbrecht gives the following synonymy for this species.[†]

---

[*] *Ann. and Mag. Nat. Hist.*, Nov. 1896, p. 362, Pls. XVII. and XVIII.
[†] *Pelagischen Copepoden des Golfes von Neapel*, p. 340.

Metridia hibernica (Brady and Robertson).

1873 *Paracalanus hibernicus,* Brady and Robertson (*Ann. and Mag. Nat. Hist.,* S. 4, vol. xii. p. 126, Pl. VIII. figs. 1–3).

1878 *Metridia armata,* Brady (*Mon. Brit. Copep.,* vol. i. p. 42). (not *Metridia armata,* Boeck.)

1887 *? Pleuromma armatum,* Pouchet and de Guerne (*Compt. Rend. Acad. Paris,* T. 104, pp. 712–715).

Anomalocera Patersonii, Templeton.—East Loch Tarbert and Loch Fyne (Calderwood). This copepod is at times moderately common in Lower Loch Fyne.

Parapontella brevicornis (Lubbock). East Loch Tarbert and Loch Fyne (Calderwood).

Acartia Clausii, Giesbrecht.—Head of Loch Fyne, between Carndow and Ardno. Off Inveraray, and in the vicinity of Minard (G.). Upper Loch Fyne, in 30 to 70 fathoms, as *Dias longiremis* (M.). ? East Loch Tarbert as *Dias longiremis* (Calderwood). I have examined specimens of *Acartia* from various parts of Upper and Lower Loch Fyne, and *A. Clausii* is the only species that has yet been observed in this part of the Clyde area.

Paramisophria, nov. gen.—Somewhat like *Misophria* in general appearance. Antennules short, and composed of about twenty-one joints. Antennæ, with the primary branch short, three-jointed; secondary branches longer than the primary, and two-jointed; mouth organs as in the *Calanidæ.* Swimming feet nearly as in *Pseudocyclops*; both branches three-jointed. Each part of the fifth pair consists of a more or less simple two-jointed branch.

This genus, though apparently a true member of the family *Misophriadæ,* differs from the two genera *Misophria* and *Pseudocyclops,* most closely allied to it, in the structure of the antennæ and of the fifth pair of thoracic feet.

Paramisophria cluthæ, nov. spec. (Pl. II., figs. 3–8; Pl. III., figs. 13–16). *Description of the female*—length, 1·4 mm. ($\frac{1}{18}$th of an inch). Body, robust; abdomen, short; only about one-fourth of the length of the cephalothorax (fig. 13, Pl. III.). Antennules short; twenty-one jointed; joints very short (fig. 14, Pl. III.). The formula shows approximately the proportionate lengths of all the joints :—

$$\frac{14\cdot 5\cdot 4\cdot 2\cdot 2\cdot 2\cdot 2\cdot 2\cdot 2\cdot 2\cdot 2\cdot 2\cdot 3\cdot 4\cdot 4\cdot 6\cdot 8\cdot 9\cdot 6\cdot 6\cdot 11}{1\cdot 2\cdot 3\cdot 4\cdot 5\cdot 6\cdot 7\cdot 8\cdot 9\cdot 10\cdot 11\cdot 12\cdot 13\cdot 14\cdot 15\cdot 16\cdot 17\cdot 18\cdot 19\cdot 20\cdot 21}.$$

Antennæ short, three-jointed, end joint small; secondary branches considerably longer than the primary branches, and composed of two elongate sub-equal joints (fig. 3, Pl. III.). Anterior foot-jaws four-jointed, stout, the second and last joints short; the first and second joints, with a few small papilliform and setiferous appendages on the inner aspect; the end joint is furnished with a number of long plumose setæ (fig. 4, Pl. II.). Posterior foot-jaws elongate, moderately stout, (?) seven-jointed, first two joints large and sub-equal, the third, fifth, and sixth small, and of nearly equal length, fourth joint about one and a half times the length of the preceding joint, the (?) seventh is very minute; the three last are strongly setiferous (fig. 5, Pl. II.). Swimming feet are somewhat like those of *Pseudocyclops*, both branches are three-jointed and furnished with numerous plumose setæ on the inner margins; in the first pair the marginal spines are slender (fig. 6, Pl. II.). In the fourth pair the spines are short, stout, and sabre-like (fig. 7, Pl. II.). The fifth pair consists each of a single two-jointed branch, the first joint is short, but produced interiorly at the distal end into a cylindrical process about as long, and half as broad as the joint itself, and provided at the apex with a small spine and an elongate plumose seta; second joint sub-cylindrical, and fully three times the length of the first joint, and armed with five stout spines, arranged along the outer margin and apex (fig. 8, Pl. II.). Abdomen composed of four segments, the last two being together scarcely equal in length to the preceding segment; caudal stylets rather longer than the entire length of the last two abdominal segments (fig. 16, Pl. III.). Male unknown. *Habitat*—Off Largabruach, Upper Loch Fyne; dredged.

Misophria pallida, Boeck.—Dredged near Largabruach, Upper Loch Fyne (G.).

Thorellia brunnea, Boeck.—Head of Loch Fyne between Carndow and Ardno; off Largabruach and Loch Gair, Upper Loch Fyne (G.). This is not an uncommon copepod in the Clyde in dredged material.

Cyclopina littoralis, G. S. Brady.—In the vicinity of Carndow, near the head of Upper Loch Fyne; and also near Largabruach; in dredged material (G.).

Cyclopina gracilis, Claus.—This species has been obtained in Loch Gair; and also off Inveraray, Upper Loch Fyne (G.).

Oithona (?) similis, Claus.—East Loch Tarbert and Loch Fyne (Calderwood). Off Inveraray; and between Carndow and Ardno (G.). Upper Loch Fyne, in 30 to 70 fathoms (M.).

Notodelphys Allmani, Thorell.—East Loch Tarbert, and Tarbert Bank, Loch Fyne, in 20 to 25 fathoms; in branchial chamber of large Ascidians (Mihi).

Notodelphys agilis, Thorell.—Also found in the branchial chamber of large Ascidians collected at extreme low-water in East Loch Tarbert.

Notodelphys prasina, Thorell.—With the others in the branchial chamber of large Ascidians (Mihi). This species has short caudal stylets.

Doropygus porcicauda, Brady.—A copepod, belonging apparently to this species, was obtained in the branchial chamber of Ascidians from East Loch Tarbert, and also from Largabruach, Upper Loch Fyne (G.). Though slightly imperfect, the Largabruach specimen resembles this, and no other described British species.

Botachus cylindratus, Thorell.—Obtained in the branchial chamber of Ascidians collected in East Loch Tarbert.

Notopterophorus papilio, Hesse.—This curious copepod was of frequent occurrence in the branchial chamber of large Ascidians from East Loch Tarbert (Mihi).

Ascidicola rosea, Thorell.—Upper Loch Fyne, between Carndow and Ardno ; from an Ascidian brought up in the trawl-net (G.).

Longipedia coronata, Claus.—Loch Gair, and off Largabruach, Upper Loch Fyne ; in dredged material (G.).

Longipedia minor, T. and A. Scott.—In material dredged off Largabruach, Upper Loch Fyne (G.). This is not much more than half the size of the other ; both forms were obtained off Largabruach, and specimens of both carried ova.

Ectinosoma Sarsi, Boeck (*E. spinipes,* Brady).—In material dredged off Largabruach, Upper Loch Fyne (G.).

Ectinosoma melaniceps, Brady.—Head of Upper Loch Fyne, between Carndow and Ardno, and off Inveraray (G.).

Ectinosoma atlanticum, Brady and Robertson.—Upper Loch Fyne, in 30 to 64 fathoms (M.). Common in a tow-net gathering from Upper Loch Fyne. Collected by the late Mr Brook.

Ectinosoma curticorne, Boeck.—Dredged off Largabruach, Upper Loch Fyne ; rare (G.).

Ectinosoma Herdmani, T. & A. Scott.—Dredged off Largabruach and in Loch Gair, Upper Loch Fyne (G.).

Ectinosoma pygmœum, T. & A. Scott.—Dredged at Tarbert Bank, Lower Loch Fyne, in 20 to 25 fathoms (G.).

Bradya elegans, T. and A. Scott.—Loch Gair, and near Largabruach, Upper Loch Fyne ; dredged (G.).

Bradya similis, T. and A. Scott.—Upper Loch Fyne, near Largabruach, in dredged material (G.).

Zosima typica, Boeck.—In dredged material from Largabruach, Upper Loch Fyne (G.).

Amymone sphœrica, Claus.—East Loch Tarbert and Loch Fyne (Calderwood).

Stenhelia hispida G. S. Brady.—In material dredged in Loch Gair and near Largabruach (G.). This distinct and fine species was not very uncommon in the Largabruach gathering. In the spirit specimens, the last abdominal segment and caudal stylets were usually of a more or less dusky hue.

Stenhelia ima (G. S. Brady).—East Loch Tarbert, Loch Fyne (Mihi)(?), off Inveraray (G.).

Ameira longiremis, T. Scott.—Dredged in Loch Gair, Upper Loch Fyne (G.).

Ameira longicaudata, T. Scott.—This well marked species was obtained near the head of Upper Loch Fyne (G.).

Jonesiella spinulosa (Brady and Robertson).—Near Largabruach, Upper Loch Fyne, in dredged material (G.). *Jonesiella fusiformis*, which in some localities appears to be a more common species than the one recorded here, has not yet been observed in any of the collections made by the 'Garland' in the Clyde.

Delavalia robusta (Brady and Robertson).—Near Largabruach, Upper Loch Fyne, in dredged material; not very rare (G.).

Delavalia mimica, sp. n. (Pl. I. figs. 1–9.).

Description of the Female.—Length, ·65 mm. ($\frac{1}{13}$th of an inch). Body moderately stout, and somewhat like *Delavalia reflexa* in general appearance. Antennules eight-jointed, the penultimate joint is smaller than any of the others, while the last is rather more elongate than either of the preceding three or four joints (fig. 2). The antennæ and mouth organs are somewhat similar to those of *D. reflexa;* the principal seta of the end joint of the mandible-palp is moderately stout and curved, and of considerable length (fig. 3). The inner branches of the first pair of swimming feet are three-jointed, the first joint is rather longer than the entire length of the three-jointed outer branches, but the second and third are short, and together scarcely equal to half the length of the first joint (fig. 5). The second, third, and fourth pairs do not differ much from those of other species of *Delavalia* (fig. 6 shows the fourth pair). The fifth pair, which have a general resemblance to those of other species of *Delavalia*, differ in the following points : the armature of the basal joint consists of three moderately short and rather stout spines, situated on the apex of the slightly produced inner portion, in addition to two spiniform setæ, one of which is exterior and the other interior to the spines; the innermost of the three spines, which is also the largest, bears two minute marginal hairs near its extremity—one on either side. The secondary joint is lamelliform, moderately broad, and subcylindrical. The margins, which, in the middle of the joint are somewhat parallel, converge towards both ends, and the distal end, which for this reason assumes a triangular form, is provided with five setæ and a short stout spine ; this spine is situated at the beginning of the exterior distal slope, while the five setæ are arranged, two on each sloping distal margin, and one at the apex, as shown in the drawing (fig. 7).

Description of the Male.—The male differs little from the female except in the form and armature of the fifth pair of feet. The basal joint of the fifth pair bears interiorly a single, and somewhat peculiar, stout, and moderately large spine, and a small spiniform seta ; the secondary joint is small, subovate, and armed with three spines on the oblique distal end of the exterior margin ; there is also a seta at the apex and another on the inner margin, as shown by the drawing (fig. 8). The second pair in the male were not modified, as is sometimes the case, but resembled those of the female.

Habitat.—Loch Gair, Upper Loch Fyne. It has also been obtained in other parts of the Clyde district, and in the Firth of Forth at Granton.

Remarks.—This species differs very markedly from any other *Delavalia* known to me in the structure of the first pair of swimming feet ; this pair are not very unlike those of some species of *Dactylopus* or *Ameira* ; but as the structure of the mandibles and of the fifth pair of feet is that of a true *Delavalia,* I prefer for the present to give this Loch Gair form a place in the genus to which it has so close an affinity.

Pontopolites typicus, T. Scott.—In dredged material from Loch Gair nd from near Largabruach, Upper Loch Fyne (G.).

Diosaccus tenuicornis (Claus).—East Loch Tarbert (Mihi). A moderately large and well marked species.

Laophonte horrida (Norman).—East Loch Tarbert and Loch Fyne (Calderwood). Loch Gair, Largabruach, and near the head of Upper Loch Fyne (G.).

Laophonte thoracica, Boeck.—East Loch Tarbert (Mihi). Near Largabruach and off Inveraray, Upper Loch Fyne (G.).

Laophonte hispida (Brady and Robertson). Loch Gair, and near Largabruach Upper Loch Fyne, in dredged material (G.).

Laophonte simulans, T. Scott.—Off Inveraray. Several specimens from a piece of partially decayed wood brought up in the trawl-net (G.). This was associated with *Cytheropteron humile* (B. & N.).

Laophonte similis (Claus).—Taken in East Loch Tarbert and in Loch Fyne (Calderwood).

Laophonte lamellifera (Claus).—Also taken in East Loch Tarbert and Loch Fyne (Calderwood).

Laophonte serrata (Claus).—East Loch Tarbert (Mihi). This is quite a distinct and comparatively large species, and it is also one of the rarest species of the genus.

Laophonte depressa, T. Scott.—Loch Fyne, off Tarbert, in 20 to 25 fathoms, in dredged material.

Normanella dubia (Brady and Robertson).—Obtained in dredged material from Loch Gair and from near Largabruach (G.)

Cylindropsyllus laevis, G. S. Brady.—From pools between tide-marks in East Loch Tarbert.

Cletodes longicaudata (Brady and Robertson).—East Loch Tarbert, Largabruach, and Loch Gair, in dredged material (G.).

Cletodes linearis (Claus).—This somewhat rare species was obtained in material dredged in East Loch Tarbert, Loch Fyne.

Cletodes curvirostris, T. Scott.—In dredged material from Loch Gair and Largabruach (G.).

Itunella tenuiremis (T. Scott). = [Cletodes tenuiremis, T. Scott,* and Itunella (?) subsalsa, G. S. Brady†].—Dredged near Largabruach, Upper

* *Eleventh Annual Report of the Fishery Board for Scotland,* Part iii. p. 204, pl iii., figs. 21-28.
† *Nat. Hist. Trans. of Northumb., Durham, and Newcastle-upon-Tyne,* vol. xiii. p. 6, pl. i. Separate reprint.

Loch Fyne (G.). This copepod, when first described, was doubtfully referred to the genus *Cletodes* for reasons stated in the 'remarks' on the species. Sometime afterwards Dr Brady instituted the genus *Itunella* for what appears to be a closely allied form from the Solway. Having, during the last two or three years, examined many specimens of *Cletodes tenuiremis* from different localities, I now quite agree with Dr Brady in considering this copepod to be generically distinct from *Cletodes*, and have, therefore, adopted his generic appellation.

Enhydrosoma curvatum (Brady and Robertson).—This species was obtained in dredged material from Loch Gair and from near Largabruach (G.).

Dactylopus tisboides, Claus.—Loch Fyne and East Loch Tarbert (Calderwood). Dredged near Largabruach (G.).

Dactylopus similis, Claus.—East Loch Tarbert (Mihi). Near the head of Loch Fyne (G.). This is a distinct and moderately large species ; it does not appear to be common in the Loch Fyne district.

Dactylopus flavus, Claus.—Obtained by the ' Garland' near the head of Loch Fyne.

Dactylopus stromii (Baird).—East Loch Tarbert, Loch Fyne, in dredged material (Mihi).

Thalestris Clausii, Norman.—Loch Fyne and East Loch Tarbert (Calderwood). Dredged near Largabruach (G.).

Thalestris mysis, Claus.—This fine species has been recorded from East Loch Tarbert by Mr Calderwood. I also have found it there.

Thalestris longimana, Claus.—Loch Fyne and East Loch Tarbert (Calderwood). This is one of the more common and widely distributed species of *Thalestris*.

Thalestris forficuloides, T. and A. Scott.—Obtained off Inveraray and near Largabruach (G.).

Westwoodia nobilis (Baird).—Loch Fyne and East Loch Tarbert (Calderwood).

Harpacticus chelifer (Müller).—East Loch Tarbert (Calderwood). Near Largabruach and near the head of Upper Loch Fyne (G.).

Zaus spinatus, Goodsir.—Loch Fyne and East Loch Tarbert (Calderwood).

Alteutha depressa, Baird.—Loch Fyne and East Loch Tarbert (Calderwood, as *Peltidium depressum*). This copepod may usually be obtained where *Laminaria* is more or less common—as in the bays, where the water is shallow.

Alteutha interrupta (Goodsir).—Loch Fyne and East Loch Tarbert (Calderwood as *Peltidium interruptum*).

Peltidium purpureum, Philippi.—East Loch Tarbert (Mihi). (See a description of the species by Dr Brady in the *Fifth Annual Report of the Fishery Board for Scotland*.) Near Largabruach, Upper Loch Fyne (G.).

Porcellidium fimbriatum, Claus.—Loch Fyne and East Loch Tarbert (Calderwood). Upper Loch Fyne, near Largabruach, and near the head of the loch (G.).

Porcellidium subrotundum, Norman.—Upper Loch Fyne, in 30 fathoms (M.).

Idya furcata (Baird).—East Loch Tarbert (Calderwood). Upper Loch Fyne, near Largabruach, and off Inveraray (G.).

Idya longicornis, T. and A. Scott.—East Loch Tarbert, Loch Fyne (Mihi). This is a large and well-marked species.

Idya gracilis, T. Scott.—This was obtained in Loch Gair, off Inveraray, and near the head of Loch Fyne (G.).

Idya minor, T. and A. Scott.—Upper Loch Fyne, near Largabruach, and near the head of the loch (G.).

Scutellidium fasciatum (Boeck).—East Loch Tarbert (Calderwood). Obtained by washing the 'roots' of *Laminaria* and other large marine algæ.

Monstrilla (?) Danæ, Claparède.—A single representative of this curious group of Copepods was obtained in a bottom tow-net gathering collected between Dunderave and Ardno, near the head of Loch Fyne (G.). The specimen is a female, and has three abdominal segments. The first segment is about twice as long as the second, and is more tumid; it is also rather longer than the third segment; it bears two long 'genital setæ,' which are united at the base so as to form a very short but quite distinct basal part; the caudal setæ are three on each furca. The antennules are in length scarcely equal to one-third the length of the body; the first joint is about four and a half times the length of the second, the third is about one and a half times the length of the second, and the second and fourth joints are nearly equal in length; the antennules are four-jointed. The Loch Fyne specimen thus agrees very well with the brief description of *Monstrilla Danæ*, Claparède, in Mr Bourne's memoir on the *Monstrillidæ*.* The fifth feet, so far as I can make them out without dissection, resemble those of *Monstrilla gracilicauda*, Giesbrecht.

PARASITA.

Lichomolgus fucicolus, G. S. Brady.—Upper Loch Fyne, near Largabruach, in dredged material (G.).

Lichomolgus forficulus, Thorell.—In the branchial chamber of large Ascidians from Loch Fyne and East Loch Tarbert (Mihi).

Lichomolgus furcillatus, Thorell.—Off Inveraray, Upper Loch Fyne, from trawl refuse (G.).

* *The Quarterly Journal of Microscopical Science*, vol. xxx. (new series), p. 575 (February 1890). Reprint.

(?) *Lichomolgus maximus,* I. C. Thompson.—Obtained in specimens of the common *Pecten—Pecten opercularis*—from Loch Gair, off Inveraray, and from near the head of the loch (G.).   This is not a true *Lichomolgus,* and is only provisionally placed here.

Pseudanthessius liber (Brady and Robertson).—This species is from the same localities as the last (G.).

†*Dermatomyzon nigripes* (Brady and Robertson).—This fine species has been obtained in East Loch Tarbert (Calderwood) ; and near the head of Upper Loch Fyne (G.).

Artotrogus orbicularis, Boeck.—A single ♂ and ♀ specimen of this fine species were taken in material dredged at Tarbert Bank, Loch Fyne, in 20 to 25 fathoms (G.).

†*Bradypontius magniceps* (G. S. Brady).—Has been recorded for East Loch Tarbert (Calderwood) ; it has also been obtained near the head of Upper Loch Fyne among trawl refuse (G.).

Bradypontius Normani (B. and R.) (Pl. II. figs. 1 and 2 ; Pl. III. figs. 1–11).

Description of the female.—Length of the specimen figured, 1·5 mm. ($\frac{1}{16}$ of an inch).   In general appearance somewhat like *Bradypontius magniceps,* G. S. Brady, but the abdomen is more elongate and slender (fig. 1, Pl. III.).   Antennules short and nine-jointed ; the first joint is of moderate length, the third is fully three times the length of the preceding joint ; the next five joints are small, while the last is nearly twice the size of the penultimate joint (fig. 3, Pl. III.).   The formula shows approximately the proportionate lengths of the joints.

Proportionate lengths of the joints, .	13	· 5	· 17	· 7	· 5	· 6	· 5	· 7	· 13.
Number of the joints,	1	· 2	· 3	· 4	· 5	· 6	· 7	· 8	· 9.

The antennæ are somewhat like those of *Asterocheres Boecki,* G. S. Brady ; the secondary branch is small and uniarticulate, and bears a single apical seta (fig. 5, Pl. III.).   The mandibles are elongate and very slender (fig. 6, Pl. III.).   The maxillæ consist of two small branches, as shown by the figure (fig. 7, Pl. III.).   Both foot-jaws are robust and strongly clawed (figs. 8 and 9, Pl. III.).   The swimming feet, which are robust, have the inner margins of both branches furnished with numerous plumose setæ (figs. 1 and 2, Pl. II.).   The secondary joint of the fifth pair is small and sub-quadrangular, and furnished with three setæ ; a single seta also springs from the basal joint (fig. 10, Pl. III.).   The abdomen is composed of four segments—genital segment larger than the others.

Description of the male.—The male differs slightly in its general outline from the female ; the abdomen is five-jointed, and the genital segment is rather larger than the others.   The antennules are ten-jointed ; the first six are nearly as in the female ; the seventh joint is about equal to the combined lengths of the three preceding joints ; the eighth, ninth, and last are each rather smaller than the one that immediately precedes it ; the antennæ are hinged and adapted for grasping (fig. 4, Pl. III.).   The other appendages resemble those of the female.

† See Dr Giesbrecht on the family *Ascomyzontidæ,* Thorell.   (*Zoologischen Anzeiger,* Nos. 521, 522, 1897.)

Habitat.—Loch Gair, Upper Loch Fyne.

Remarks.—Living specimens of this handsome species are very prettily ornamented with usually eight brick-red blotches, six of which are submarginal ; and two central—one being at the anterior apex of the first body segment and one near the posterior portion of the thorax, as shown by figs. 1 and 2, Pl. III., which are reproduced from drawings of a living male and female specimen from Cromarty Firth. The markings on the cephalothoracic segment were similar in all the specimens examined, but one or other of the posterior marginal blotches were observed to be occasionally absent. The general colour of the dorsal surface was yellowish, tinged with brown. When examined under the microscope, the integument was also seen to be thickly besprinkled with minute circular markings. The secondary branches were quite distinct, though small. With the exception of the antennules, the various appendages were the same in the male as in the female.

†*Scottomyzon gibberum* (T. and A. Scott).—From specimens of the common star-fish (*Asterias rubens*) caught near the head of Loch Fyne (G.).

Caligus rapax, Milne Edwards.—East Loch Tarbert, on a coal-fish caught in Loch Fyne (Mihi).

Caligus diaphanus, Nordmann (with numerous specimens of *Udonella caligorum* adhering to the *Caligus*). From large Coalfish caught in Loch Fyne (Mihi).

Lepeoptheirus pectoralis (Müller).—From the pectoral fins of plaice (*Pleuronectes Platessa*), caught off Inveraray and in Ard-a-Eaolas Bay (G.).

Lernentoma cornuta (Müller).—Taken from the gills of long rough dabs (*Hippoglossoides limandoides*), caught off Inveraray (G.).

Lernentoma lophii (Johnston). From an Angler fish caught in Loch Fyne (Mihi).

Anchorella uncinata (Müller).—Found adhering to the inside of the mouth and on the gills of a young coal-fish caught in Loch Gair, Upper Loch Fyne (G.).

CIRRIPEDIA.

Balanus porcatus, Da Costa.—Upper Loch Fyne, at Minard, attached to Mytilus (M.).

Balanus Hameri, Ascanius.—Upper Loch Fyne, at Minard and on the west side, in 12 to 20 fathoms ; also on the shore (M.).

Balanus balanoides (Linné).—Upper Loch Fyne, at Minard and on the west side, in 12 to 20 fathoms ; also on the shore (M.).

Balanus crenatus (Brug.).—At Minard and on both sides of Upper Loch Fyne, in 10 to 20 fathoms, and between tide-marks (M.).

Verruca Strömia, O. F. Müller.—Off Inveraray, Upper Loch Fyne, in 4 to 10 fathoms (M.).

† See Dr Giesbrecht on the family *Ascomyzontidæ*, Thorell. (*Zoologischen Anzeiger*, Nos. 521, 522, 1897.)

Sacculina carcini (Thompson).—Parasitic on a specimen of *Carcinus maenas*, captured in the trawl-net near the head of Upper Loch Fyne (G.).

Peltogaster paguri (Rathke).—Parasitic on Eupagurus, East Loch Tarbert (Mihi).

POLYZOA OF LOCH FYNE.

The following list of Loch Fyne Polyzoa is compiled entirely from the MS. records of the steam-yacht 'Medusa.' The arrangement is that of the British Association *Report on the Marine Zoology, Botany, and Geology of the Irish Sea* (1896).

CHEILOSTOMATA.

Gemellaria loricata (Linné).—Upper Loch Fyne, at Minard Narrows, in 10 to 15 fathoms (M.).

Bugula turbinata, Alder.—Upper Loch Fyne, east side in 10 to 20 fathoms (M.).

Cellaria fistulosa (Linné).—Upper Loch Fyne, at Minard and at east side, in 10 to 20 fathoms (M.).

Membranipora pilosa (Linné).—Upper Loch Fyne, east side, in 15 fathoms (M.).

Membranipora Flemingii, Burk.—Upper Loch Fyne, at Minard, in 10 to 15 fathoms (M.).

Microporella impressa (Audouin).—Upper Loch Fyne, at Minard, in 10 to 20 fathoms (M.).

Schizoporella unicornis.—Upper Loch Fyne, at Minard and east side, in 10 to 30 fathoms (M.).

Hippothoa distans, MacGillivray (*Hippothoa flagellum* (Manzoni), (Hincks).—Upper Loch Fyne, at Minard, in 15 fathoms (M.).

Leprailia Pallasiana (Moll).—Upper Loch Fyne, at Minard (M.).

Leprailia cruenta, Norman.—Minard, Upper Loch Fyne (M.).

Porella compressa (Sowerby).—Upper Loch Fyne, east side, in 10 to 20 fathoms (M.).

Smittia reticulata (MacGillivray).—Upper Loch Fyne (M.).

Cellepora pumicosa (Linné).—Upper Loch Fyne, at Minard and east side, in 15 to 20 fathoms (M.).

Cellepora ramulosa (Linné).—East side of Upper Loch Fyne, in 15 to 20 fathoms (M.).

Cellepora avicularis, Hincks.—East side of Upper Loch Fyne, in 15 to 20 fathoms (M.).

CYCLOSTOMATA.

Crisia eburnea (Linné).—At Minard, Upper Loch Fyne, in 15 to 20 fathoms (M.).

Crisia denticulata (Lamarck).—Upper Loch Fyne, at Minard and east side, in 15 to 20 fathoms (M.).

Diastopora obelia (Johnston).—Upper Loch Fyne, east side, in 15 to 20 fathoms (M.).

Stomatopora granulata (Milne Edwards).—Upper Loch Fyne, at Minard (M.).

Lichenopora hispida (Fleming).—At Minard, Upper Loch Fyne, in 15 to 20 fathoms (M.).

CTENOSTOMATA.

Vesicularia spinosa (Linné).—Upper Loch Fyne, east side, in about 15 fathoms (M.).

Incerta sedis.

Escaroides rosacea.—Upper Loch Fyne, at Minard (M.).

(ARACHNIDA.)

(One or two species of *Pycnogons* and *Acarina* have been observed in Loch Gair and in other parts of Loch Fyne, but these have not yet been identified.)

THE VERMES OF LOCH FYNE.

CHÆTOPODA.

The following records have almost all been obtained from the MS. notes of the steam yacht ' Medusa.'

Tomopteris onisciformis, Eschscholtz.—Obtained in a surface tow-net gathering collected in Lower Loch Fyne by the Fishery steamer ' Garland.'

Filigrana implexa, Berkeley.—Upper Loch Fyne, near Minard, in 15 to 20 fathoms (M).

Spirorbis borealis, Mörch.—Upper Loch Fyne, between tide-marks (M.).

Serpula triquetra (Linné).—At Minard, and on both sides and in the centre of Upper Loch Fyne, in 10 to 36 fathoms, and also between tide-marks (M.).

Serpula vermicularis (Ellis).—At Minard, and on both sides and in the centre of Upper Loch Fyne, in 10 to 36 fathoms (M.).

Serpula contortuplicata (Linné).—On the east side of Upper Loch Fyne, in 15 to 30 fathoms ; and at Minard, in 11 to 25 fathoms (M.).

Dasychone argus.—This somewhat rare species was obtained on the east side of Upper Loch Fyne (M.).

Sabella pavonia (Savigny).—Taken in Upper Loch Fyne, on both sides and in the centre, in 10 to 70 fathoms (M.).

Sabella penicillus (Linné).—Taken in deep water, in the centre of Upper Loch Fyne and also near the head of the loch (M.).

Sabella, sp.—A form of *Sabella* that has not been identified was obtained on the west side of Upper Loch Fyne, in 10 to 25 fathoms (M.).

Polycirrus aurantiacus, Malmgren.—This annelide was obtained in the centre of Upper Loch Fyne, in 35 fathoms (M.).

Polymnia nasidensis, Chiaje.—Taken in Upper Loch Fyne, in 15 to 20 fathoms (M.).

Polymnia nebulosa.—This *Polymnia* was obtained in the same locality as the last, and it also occurred on the east side of Upper Loch Fyne (M.).

Notomastus, sp.—An annelide apparently belonging to this genus was obtained on the east side of Upper Loch Fyne (M.).

Thelepus circinatus, Fabricius.—Taken in 15 to 20 fathoms, in Upper Loch Fyne.

Terebella, sp.—A species of annelid was obtained in Upper Loch Fyne, in 15 to 20 fathoms, and also at Minard, that apparently belonged to this genus (M.).

Terebellides Stroemii, Sars.—Taken in Upper Loch Fyne, in 15 to 20 fathoms (M.).

Trophonia glauca, Malmgren.—This was obtained in the same locality as the last.

Pectenaria belgica (Pallas).—*Pectenaria* was taken at various depths from between tide-marks to the deep water in the centre of the loch ; and also in various localities from near the head of the loch downwards (M.). It has also been obtained in East Loch Tarbert in Lower Loch Fyne.

Maldane biceps, Sars.—This was taken in Upper Loch Fyne, in the centre of the Loch in 60 fathoms (M.).

Rhodine Lovéni.—This, like a few of the others, appears to be a deep-water form ; it occurred in the centre of the Loch in 70 fathoms (M.).

Clymene (Paxilla) gracilis.—This was obtained in Upper Loch Fyne, in 15 to 20 fathoms (M.).

Clymene amphistoma.—This also was procured in Upper Loch Fyne, in 15 to 20 fathoms (M.).

Cirratulus, sp.—A species of *Cirratulus* was obtained on the west side of Upper Loch Fyne, in 10 to 15 fathoms, and also between tide-marks (M.).

Arenicola piscatorum, Lamarck.—This was obtained between tide-marks in Upper Loch Fyne (M.).

Chætopterus variopedatus, Ren. [= *Chæt. insignis* (Baird)].*—Taken at Minard, and on both sides of Upper Loch Fyne, in 10 to 20 fathoms (M.).

Eumenia Jeffreysii.—Taken on the east side and in the centre of Upper Loch Fyne, in 20 to 70 fathoms, and also near the head of the Loch (M.).

Ammotrypane aulogaster, Rathke.—Upper Loch Fyne, in 15 to 20 fathoms (M.). In East Loch Tarbert, Lower Loch Fyne, dredged (Mihi).

Glycera tesselata, Grube ; variety *Macintoshii.*—This was taken in Upper Loch Fyne, at various depths from 15 to 20 fathoms, down to 60 fathoms in the centre of the Loch (M.).

Laetmonice filicornis, Kinberg, variety *Kinbergi,* was obtained in 70 fathoms in the centre of Upper Loch Fyne ; it also occurred on both sides of the Loch (M.).

Neomenia carinata.—This also was obtained in the centre of the loch in 70 fathoms (M.).

Eunice norvegica (Linné).—*Eunice norvegica* was procured on the east side of Upper Loch Fyne (M.).

Eunice, sp.—A species of this genus was obtained in 10 to 15 fathoms, in Upper Loch Fyne, but was not identified (M.).

Hyalinœcea tubicola (Müller).—This species, which appears to be widely distributed, was obtained at Minard, and also in the centre of the Loch, in 15 to 70 fathoms (M.).

[*Nothria tubicola* ((?) = *Hyalinœcea tubicola*) was obtained in somewhat similar localities as the last.]

Lumbriconereis nardonis, Grube.—Occurred on the east side of Upper Loch Fyne (M.).

Nereis Dumerilii (Aud. and M. Edw.) was obtained on the east side of Upper Loch Fyne in 10 to 15 fathoms (M.).

Nereis pelagica (Linné).—Taken at Minard, in the centre of Upper Loch Fyne, near the head, and also between tide-marks (M.). In east Loch Tarbert and neighbouring parts of Lower Loch Fyne.

Nereis, sp.—An unidentified species of *Nereis* was obtained at Minard, in the centre of the Loch, near the head, in 12 to 70 fathoms, and also between tide-marks (M.).

Nephthys Hombergi (Aud. and M. Edw.) was procured in Upper Loch Fyne, on the east side (M.).

* See J. Hornell, in *Tenth Annual Report of the L.M.B.C.,* p. 28 (1897).

Nephthys ciliata.—This form was obtained in Upper Loch Fyne, in 15 to 20 fathoms (M.).

Lepidonotus squamatus (Linné) was found on the east side of Upper Loch Fyne.   .

Halosydna gelatinosa, Sars.—This was taken on the east side of Upper Loch Fyne, and on the shore at low water (M.).

Polynoe squamata (Johnston).—At Minard ; on both sides of Upper Loch Fyne, in 10 to 30 fathoms ; and between tide-marks (M.).

Polynoe, sp.—A species of *Polynoë*, not identified, was obtained at Minard ; and on both sides, as well as in the centre and near the head of the Loch, in 10 to 35 fathoms ; it was also found on the shore between tide-marks (M.).

Aphrodite aculeata (Linné).—At Minard ; and on the east side and in the centre of Upper Loch Fyne, in 15 to 70 fathoms, as well as near the head of the Loch (M.).   East Loch Tarbert and adjacent parts of Lower Loch Fyne, not rare.

Hermione hystrix Savigny (= *Aphrodite hystrix*).—This species appears to be more restricted in its distribution than the last, and confined to deeper water ; it was obtained in Upper Loch Fyne, in the deep water of the centre—65 to 70 fathoms (M.).

Gephyrea.

Sipunculus bernhardus was obtained at Minard, in about 10 fathoms (M.).

(?) *Phascolosoma strombi* (Mont.).—A species of *Sipunculus*, which was probably a *Phascolosoma*, was obtained at Minard, and also in the centre of the Loch in depths ranging from 10 to 70 fathoms, in the dead shells of *Dentalium* (M.).   .

Chætognatha.

Sagitta bipunctata, Quoy and Gainard, appeared to be generally distributed all over the Loch.

Nemertea.

Lineus marinus (Mont.) is occasionally obtained at the roots of tangle and other sea-weeds, specimens many yards in length being sometimes observed.

Turbellaria.

Planaria, sp.—Specimens of a *Planaria* (probably *Planaria littoralis*) are occasionally observed in Loch Fyne ; but the *Planarians*, as well as the other groups of *Vermes*, require further study ; and when that is done considerable additions will no doubt be made to the preceding list.

THE ECHINODERMATA OF LOCH FYNE.

The *Catalogue of the British Echinoderms in the British Museum*, by Prof. Jeffrey Bell, is followed as to the arrangement and names in this list.

CRINOIDEA.

Antedon bifida (Pennant).—Common near the east shore of Loch Fyne (B. & S.).

ASTEROIDEA.

Porania pulvillus (O. F. Müller).—In Upper Loch Fyne at Minard in 11 to 25 fathoms (M.).

Stichaster roseus (O. F. Müller).—Taken at Minard, west side, in 10 to 25 fathoms, also on the shore at low water (M.) East Loch Tarbert (B. & S.).

Solaster papposus (Fabricius).—Minard, on both sides, in 10 to 30 fathoms, and also on the shore (M.). Frequent in Loch Fyne, smaller specimens between tide marks (B. & S.). Strachur Bay (G.).

Solaster endeca (Linné). At Minard, on both sides, in 10 to 25 fathoms (M.). Frequent some distance from shore, and also occurs at low water (B. & S.).

Henricia sanguinolenta (O. F. Müller).—On both sides of Loch Fyne, at Minard, in 10 to 40 fathoms (M.). East Loch Tarbert (B. & S.).

Asterias glacialis, Linné.—On the west side of Upper Loch Fyne, in 10 to 25 fathoms (M.). Frequent in 20 to 30 fathoms in Loch Fyne, sometimes of large size (B. & S.).

Asterias rubens, Linné.—Generally distributed, and more or less frequent all over the loch, and at all depths ; Minard, etc. (M.). East Loch Tarbert (B. & S.). Cairndow, Loch Gair, etc. (G.).

Asterias Murrayi, Bell.—Upper Loch Fyne, in 65 fathoms. (F. Jeffrey Bell, in *Catalogue of the Brit. Echin. in the British Museum*, p. 103, Pl. XII., figs. 1 and 2).

OPHIUROIDEA.

Ophiura ciliaris (Linné).—Frequent in East Loch Tarbert (B. & S.). Generally distributed in Upper Loch Fyne, in from 10 to 70 fathoms (M.).

Ophiura albida, Forbes.—Loch Fyne and East Loch Tarbert, common (B. & S.). Generally distributed in Upper Loch Fyne, in from 10 to 70 fathoms (M.).

Ophiura affinis, Lütken.—On both sides, and in the centre of Upper Loch Fyne, at Minard, in from 12 to 70 fathoms (M.).

Amphiura Chiajii, Forbes.—Off Inveraray (Robertson, in *Trans. N. H. S. Glasg.*, vol. i.). On both sides and centre of Upper Loch Fyne, and also near the head, in 15 to 70 fathoms (M.).

Amphiura filiformis (O. F. Müller). Off Buck Island, Loch Fyne (B. & S.). Both sides and centre of Upper Loch Fyne, in 10 to 70 fathoms (M.).

L

Amphiura elegans (Leach).—Frequent between tide-marks (B. & S.)

Ophiopholis aculeata (Linné).—Generally distributed, and more or less frequent in Loch Fyne, Upper and Lower (M., B. & S., G.).

Ophiocoma nigra (Abilgard).—Very abundant, in 15 to 20 fathoms (B. & S.). At Minard, Upper Loch Fyne, on both sides, in 10 to 30 fathoms (M). Abundant in Loch Gair (G.).

Ophiothrix fragilis (Abilgard).—Common, especially near the east shore of Loch Fyne (B. & S.) Upper Loch Fyne, on both sides, in from 10 to 30 fathoms, and also between tide-marks (M.).

ECHINOIDEA.

Echinus miliaris, Linné.—Between tide marks, East Loch Tarbert (B. & S.). Upper Loch Fyne, at Minard, on both sides and in the centre, in from 10 to 60 fathoms, and also between tide-marks (M.). Strachur Bay, Loch Gair, etc. (G.).

Echinus esculentus, Linné.—More or less common all over the loch, both upper and lower, usually on hard ground.

Spatangus purpureus (O. F. Müller).—Upper Loch Fyne, at Minard, and in the centre of the loch, in 12 to 70 fathoms (M.).

Echinocardium cordatum (Pennant).—A moderately common species in Loch Fyne, where the conditions are suitable, as in East Loch Tarbert (B. & S.). At Minard (M.), and Cairndow (G.).

Brissopsis lyrifera, Forbes.—In the centre of Upper Loch Fyne, in 55 to 60 fathoms (M.). Tarbert Bank, Lower Loch Fyne (Mihi).

HOLOTHURIOIDEA.

Synapta digitata (Montagu).—Loch Fyne (Dr Scouler in *Trans. N. H. S. Glasg.*, vol. i. p. 8).

Cucumaria Hyndmani (Thompson).—Obtained on the east side of Upper Loch Fyne, in 20 fathoms (M.).

Cucumaria pentactes (?) (Linné).—Upper Loch Fyne, on the east side, in about 20 fathoms (M.).

Thyone fusus (O. F. Müller).—On the west side of Upper Loch Fyne, in from 10 to 15 fathoms (M.).

Thyone raphanus, Dub. and Kor.—East Loch Tarbert (Mihi). East side of Upper Loch Fyne, in about 20 fathoms (M.).

Psolus phantapus (Strassenfeldt).—Upper Loch Fyne, at Minard, and on the east side of the loch, in 11 to 25 fathoms (M.).

Doubtful species.

Cribella aculeata.—Said to have been found on the shore of Upper Loch Fyne (M.).

Ophiocoma minuta.—Reported from Minard, Upper Loch Fyne (M.).

Ophiactis Ballii (Thomp.) ?—*Zostera* bed, East Loch Tarbert (B. & S.).

THE ACTINOZOA OF LOCH FYNE.

ALCYONARIA.

Alcyonium digitatum (Linné).—Upper Loch Fyne, at Minard, on both sides, in 10 to 25 fathoms, and also on the shore.

Sarcodictyon catenata, Forbes.—Upper Loch Fyne, in the centre near the head (M.).

Virgularia mirabilis (Linné).—Upper Loch Fyne, at Minard, and in the centre of the loch, in 11 to 70 fathoms.

ACTINIARIA.

Bolocera Tuediæ (Johnston).—Loch Fyne, off Tarbert; taken occasionally in the dredge (Mihi). In deep water between Penmore and Inveraray (G.). This is a large species with the tentacle scarcely retractile; it is of a red colour, and appears to be confined to moderately deep water. At Minard, and in the centre of the loch, in 15 to 70 fathoms (M.).

Anemonia sulcata, Pennant [*Anthea cereus* (Ellis and Solander)].— Frequent on the leaves of *Zostera* in East Loch Tarbert, at extreme low water (Mihi).

Adamsia palliata (Bohadsch.).—Minard, on both sides, and in the centre of the loch, in 10 to 70 fathoms (M.). Usually found adhering to the univalve shells inhabited by *Eupagurus Prideauxii.*

Actinia equina, Linné [*A. mesembryanthemum* (Ellis and Solander)]. —Common between tide marks.

Tealia crassicornis (Müller).—Frequent between tide marks among stones, and usually with small gravel adhering to the test.

Stomphia Churchiæ, Gosse.—Upper Loch Fyne, at Minard, on the east and west side, and in the centre, in 12 to 36 fathoms (M.).

Edwardsia callimorpha (Gosse).—Taken in Upper Loch Fyne, on the east side, in 10 to 12 fathoms (M.).

Edwardsia carnea (Gosse).—Port Loy, at low water, attached to stones (M.).

Caryophyllia Smithii (Stokes).—Loch Fyne ; rare (Mihi).

HYDROZOA OF LOCH FYNE.

The species recorded below were nearly all obtained by the steam-yacht 'Medusa'; their names are arranged in accordance with Hincks' *British Hydroid Zoophytes.*

Hydractinia echinata (Fleming).—Upper Loch Fyne, at Minard, on both sides, and in the centre, in 12 to 20 fathoms (M.).

Eudendrium rameum (Pallas).—Upper Loch Fyne, in the centre (M.).

Campanularia verticillata (Linné).—Upper Loch Fyne (M.). A moderately common species in deep water.

Lafoëa dumosa (Fleming).—At Minard and east side of Upper Loch Fyne, in 12 to 20 fathoms (M.).

Lafoëa fruticosa (M. Sars). At Minard, Upper Loch Fyne, in 12 to 20 fathoms (M.).

Halecium muricatum (Ellis and Solander).—At Minard, Upper Loch Fyne, in 15 to 20 fathoms (M.).

Sertularella rugosa (Linné).—At Minard, Upper Loch Fyne, in 12 to 20 fathoms (M.). The stem of *S. rugosa*, which may be found adhering to the fronds of *Laminaria* or *Flustra*, sends out more or less numerous and crowded shoots, scarcely an inch in height.

Sertularella fusiformis (Hincks).—On the east side of Upper Loch Fyne, in 15 fathoms (M.). This is a small species, measuring from a quarter of an inch to an inch in height.

Diphasia fallax (Johnston).—Loch Fyne (A. M. Norman, see *British Hydroid Zoophytes*, p. 251).

Sertularia pumila, Linné.—Upper Loch Fyne, between tide-marks (M.).

Sertularia filicula, Ellis and Solander.—Upper Loch Fyne, at Minard, in 17 to 25 fathoms (M.). This is one of the less common of the *Sertulariœ*.

Sertularia abietina, Linné.—At Minard, Upper Loch Fyne, in 12 to 20 fathoms (M.). A moderately common species.

Sertularia argentea, Ellis and Solander.—Upper Loch Fyne, in moderately deep water (M.). In this species the shoots extend to a foot or more in height.

Aglaophenia myriophyllum (Linné).—Loch Fyne (A. M. Norman,—see *British Hydroid Zoophytes*, p. 292).

SPONGOZOA.

Grantia compressa, Fleming.—Between tide-marks, Upper Loch Fyne (M.). East Loch Tarbert (Mihi).

Grantia ciliata, Fleming.—Upper Loch Fyne, east side, in 12 to 20 fathoms (M.).

Halichondria albescens, Johnston.—East and west sides of Upper Loch Fyne, and also between tide-marks.

Halichondria panicea (Pallas), variety *papillaris.* — East side of Upper Loch Fyne, and also between tide-marks (M.).

Suberites domuncula, Olivi.—Loch Fyne, in 48 fathoms ; found by Mr Pearcy (M.). (?) (*Hymeniacidon suberea,* of Bowerbank, is a synonym of this).

Suberites ficus (Johnston).—At Minard Narrows, Upper Loch Fyne, in 15 to 20 fathoms.

Halisarca Dujardinii, Johnston.—Upper Loch Fyne, on *Inachus dorsettensis,* in 55 fathoms ; found by Mr Pearcy (M.).

FORAMINIFERA OF LOCH FYNE.

MILIOLIDÆ.

Biloculina ringens (Lamarck).—Loch Fyne, in dredged materials (B. & S.).

Biloculina depressa, D'Orbigny.—East Loch Tarbert (B. & S.).

Biloculina elongata, D'Orbigny.—Loch Fyne (Mihi).

Spiroloculina limbata, D'Orbigny.
Spiroloculina canaliculata, D'Orbigny. } Loch Fyne, in dredged material (B. & S.).

Miliolina trigonula, Lamarck.
Miliolina tricarinata, D'Orbigny. } Loch Fyne, in dredged material (B. & S.).

Miliolina seminulum, Linné.

Miliolina subrotunda, Montague.

Miliolina secans (D'Orbigny).

Miliolina Ferussacii (D'Orbigny).

Miliolina agglutinans (D'Orbigny). } Dredged in East Loch Tarbert (B. & S.). *M. seminulum* and *M. subrotunda,* also off Inveraray (G.).

Miliolina contorta (D'Orbigny).—Dredged, Tarbert Bank, Loch Fyne, in 20 to 25 fathoms (Mihi).

Cornuspira foliacea (Philippi).—Dredged in East Loch Tarbert (B. & S.).

ASTRORHIZIDÆ.

Astrorhiza limicola, Sand.—Tarbert Bank, Loch Fyne, in 20 to 25 fathoms ; dredged (Mihi).

Litꭓolidæ.

Reophax scorpiurus, Montfort.

Haplophragmium canariense (D'Orbigny).

Haplophragmium pseudospirale (Williamson).

Ammodiscus gordialis, Parker and Jones.

Trochamina squamata (Parker and Jones).

> Dredged at Tarbert Bank, Loch Fyne, in 20 to 30 fathoms (Mihi).

Textulariidæ.

Textularia sagittula, Defrance.—East Loch Tarbert (B. & S.).

Textularia pygmœa, D'Orbigny.—Dredged in Loch Fyne, at Tarbert Bank, in 20 fathoms (Mihi).

Bulimina marginata, D'Orbigny.—Dredged in East Loch Tarbert (B. & S.).

Lagenidæ.

Lagena sulcata (Walker and Jacobs).—Dredged in East Loch Tarbert (B. & S.).

Lagena lævis (Montague).

Lagenal gracillima, Sagz.

Lagena globosa (Montague).

Lagena striata (D'Orbigny).

> Dredged in Loch Fyne, at Tarbert Bank, in 20 to 25 fathoms (B. & S.).

Lagena marginata (Walker and Jacobs).—Dredged in East Loch Tarbert (B. & S.).

Lagena melo (D'Orbigny).—Dredged in Loch Fyne (B. & S.).

Lagena squamosa (Montague).—Dredged in East Loch Tarbert (B. & S.).

Lagena hexagona (Williamson).—Loch Fyne, at Tarbert Bank (B. & S.).

Lagena Jeffreysii, Brady.—Dredged in Loch Fyne and East Loch Tarbert (B. & S.).

Nodosaria scalaris (Lamarck).—Not uncommon in Loch Fyne (B. & S.).

Nodosaria (Dentalina) communis (D'Orbigny).—Loch Fyne (B. & S.).

Cristellaria rotulata (Lamarck).

Cristellaria crepidula (Fichtel and Moll).

> Dredged at Tarbert Bank, Loch Fyne (Mihi).

Polymorphina lactea (Walker and Jacobs).—East Loch Tarbert ; a moderately common species (B. & S.)

Polymorphina tubulosa (D'Orbigny). ⎫
Polymorphina gibba, D'Orbigny. ⎬ These three species were obtained with the dredge in Loch Fyne, at Tarbert Bank, in 20 to 25 fathoms (Mihi).
Polymorphina rotundata, Born. ⎭

GLOLIGERINIDÆ.

Orbulina unicersa, D'Orbigny.—East Loch Tarbert (B. & S.).

ROTALIIDÆ.

Patellina corrugata, Williamson.—Loch Fyne (B. & S.).

Discorbina rosacea (D'Orbigny).—East Loch Tarbert (B. & S.).

Discorbina globularis (D'Orbigny).—Loch Fyne, at Tarbert Bank, in 20 to 25 fathoms (Mihi).

Planorbulina mediterranensis, D'Orbigny.—East Loch Tarbert (B. & S.).

Truncatulina lobatula (Walker and Jacobs). ⎫
Rotalia Beccarii (Linné). ⎬ Dredged in East Loch Tarbert (B. & S.).
Rotalia nitida (Williamson). ⎭

Gypsina inhærens, Schultze.— Loch Fyne ; rare (Mihi).

NUMMULINIDÆ.

Nonionina asterizans. ⎫
⎬ East Loch Tarbert, usually in dredged material ; common (B. & S.).
Nonionina depressula. ⎭

Operculina ammoides, Gronovius.—Tarbert Bank, Loch Fyne, dredged in 20 to 25 fathoms (Mihi).

Polystomella crispa (Linné). ⎫
⎬ East Loch Tarbert ; common (B. & S.).
Polystomella striato-punctata) Fichtel and Moll). ⎭

APPENDIX TO THE FAUNA OF LOCH FYNE.

Several more or less interesting marine organisms have recently been obtained within the Clyde area, which could not, for obvious reasons, be included in the preceding catalogue.   As it is desirable, however, that some, at least, of these should be noticed, it is proposed to do so here by way of an appendix.   The seaward boundary of what is here called the "Clyde Area" is a line extending from the Mull of Cantyre to Corsewall Point at the mouth of Loch Ryan.

FISHES.

Among the fishes obtained, the following may be mentioned :—

Lumpenas lampretiformis.—A specimen of this species was captured in the deep water between Arran and the Heads of Ayr, and another midway between the Island of Sanda and Bennan Head.

Triglops Murrayi, Günther.—A young specimen of this somewhat rare species was taken in the shrimp-trawl, a short distance to the eastward of Sanda.

The 'Black mouthed dog-fish' (*Pristiurus melanostomus*) and the 'Lesser spotted dog-fish' (*Syllium canicula*) were captured in the 'Garland' trawl; the former was taken off Arran, and the latter in the vicinity of Ailsa Craig.

A few specimens of the 'Flapper skate' (*Raia macrorhynchus*) were obtained by the 'Garland' in different parts of the seaward area of the Clyde; a specimen, captured in the deep water to the east of Arran, measured 40 inches across the pectoral fins.

MOLLUSCA.

Trochus granulatus, Born.—A single living specimen of this mollusc was taken off the Island of Sanda. This locality, though near the mouth of the estuary, is still within the Clyde area; and as this species does not appear to have been hitherto included among the recent mollusca of the Clyde, its occurrence here is of interest. Dr Jeffreys, in *British Conchology*, states that the species has been obtained off the Mull of Galloway in 50 fathoms, and also refers to it having been found in Belfast Bay; but adds that the two broken specimens found there had probably been accidentally introduced. These are the nearest to the Clyde, of the localities from which the species has been recorded.

CRUSTACEA.

Gonoplax angulata (Fabricius).—A male specimen of this crab was captured on the 3rd of November 1896, in the vicinity of Ailsa Craig, and on the following day a female was obtained a few miles southeastward of the Mull of Cantyre. This species has for many years past been included among the Clyde crustacea; but in recent years there seems to be a tendency to doubt the accuracy of the Clyde records. The capture of these two specimens, even though the localities be near the seaward boundary of the estuary, will ~~therefore~~ help to some extent to remove the uncertainty as to the correctness of these earlier records.

Nika edulis and *Pasaphœa sivado*, etc.—Both of these crustaceans are sometimes of more or less frequent occurrence near the mouth of the Clyde estuary.

Hippolyte prideauxiana, Leach.—A single specimen of this little shrimp (a female with ova) was obtained near the outer boundary of the Clyde in November last; it is an addition to the Clyde fauna. *H. prideauxiana* is sometimes captured off the coast of Devon. I am indebted to the Rev. T. R. R. Stebbing for the name of the species.

Siriella norvegica, G. O. Sars.—One or two specimens were recently identified among some surface tow-net material from the Clyde, collected a few miles to the east of the Island of Pladda. This is also an addition to the Clyde fauna.

Erythrops serrata, G. O. Sars.—During the past year this schizopod has been obtained in several parts of the Clyde area, sometimes alone, sometimes in company with another and smaller species,—*Erythrops elegans.* Neither of these are recorded in Prof. Henderson's *Decapod and Schizopod Crustacea of the Clyde.*

Monoculodes tuberculatus, G. O. Sars.—A single specimen of this amphipod occurred in a tow-net gathering, collected a few miles east of the Island of Arran. There does not appear to be any previous record of it in the British seas.

Epimeria tuberculata, G. O. Sars.—A single specimen was obtained in the vicinity of Ailsa Craig. There appears to be no previous British record of this species. I am indebted to the Rev. T. R. R. Stebbing for the names of both these amphipods.

Ephimedia Eblanæ, Spence Bate.—Two small specimens of an amphipod, which Mr Stebbing thinks may probably belong to this species, were obtained in the seaward part of the Clyde estuary. They resemble somewhat closely the figure of the species in Bate and Westwood's *British Sessile-Eyed Crustacea,* but are smaller than the size stated by these authors.

Philomedes (Cypridina) brenda (Baird).—Two specimens, both males, of this rare ostracod were obtained in the Clyde, in the deep water to the east of Arran. Dr Brady, who kindly examined one of the specimens, says : 'Your specimen is, I have no doubt, *P. brenda.*' The only previous British records for this ostracod appear to be the following :—Off the coast of Durham, near the Dogger Bank (Rev. A. M. Norman). Off Noss, in Shetland, 80–90 fathoms (M'Andrew). Ultra-British distribution, Greenland (Holsteinbourg Harbour) ; various parts of Norway and Sweden, from E. Finmark and Trondhjem to Drobak and Kullaberg.*

Aspidophryxus peltatus, G. O. Sars (Pl. III. fig. 19). Several specimens of this curious crustacean were obtained as parasites on *Erythrops serrata* and *Erythrops elegans* (already recorded), but chiefly on the former. Males appeared to be scarce, and any that were noticed were not adhering to the females, but were each some distance apart, and connected to the female by a slender filament, as shown by the drawing (fig. 19). Rev. Mr Stebbing also identified this species for me.

COPEPODA.

A considerable number of copepods have, in addition to those recorded from Loch Fyne, been discovered in various parts of the Clyde area. It is not my intention, however, to record all these at present ; that may probably be done more fully later on. Meantime, I only give detailed descriptions of three species that appear to be new to science.

Stenhelia intermedia, sp. n. (Pl. II. figs. 10–21).
Description of the Female.—Length about ·7 mm. ($\frac{1}{28}$th of an inch). Body robust. Antennules eight-jointed ; the first four moderately stout and sub-equal ; the fifth, sixth, and seventh small, but of nearly equal length, while the end joint is about as long as the fourth joint ; the first four joints are together fully twice the entire length of the last four

---

* *Mon. of the Marine and Fresh-Water Ostracoda of the N. Atlantic and North-Western Europe,* by Drs Brady and Norman (Second Part), p. 656 (1896).

(fig. 11). Antennæ, with secondary branches elongate and slender, three-jointed ; the middle joint very small, the other two sub-equal (fig. 13). Mandibles and other mouth organs somewhat similar to those of *Stenhelia hispida*, Brady (figs. 14, 15). The first pair of swimming feet are moderately stout ; the inner branches, which are rather longer than the outer, have the first joint equal to nearly twice the entire length of the second and third, which are short and sub-equal ; the joints of the outer branches are nearly of equal length, and armed with strong marginal spines ; a stout spine springs from both the exterior and interior angles of the second basal joint (fig. 16). The inner branches of the next three pairs are rather shorter than the outer branches ; in the fourth pair the outer branches are nearly one and a half times longer than the inner branches (fig. 18). The basal joints of the fifth pair are broadly ovate, and produced interiorly so as to extend to near the extremity of the secondary joints ; a short stout seta springs from the distal half of the inner margin of the basal joint, and four setæ from the broad and somewhat truncate apex, the two middle setæ being considerably longer than the others ; secondary joints sub-quadrangular, and furnished with five setæ of variable lengths round the distal end (fig. 19). Caudal stylets short, not half the length of the last abdominal segment ; the principal seta of each stylet very stout, and somewhat fusiform (fig. 21).

Description of the Male.—The male differs from the female in having the antennules modified to form powerful grasping organs (fig. 12). The inner branches of the second pair of feet are, like those of the males of *Stenhelia ima*, two-jointed, and rather shorter than the outer branches, and are each provided with two stout spine-like terminal appendages (fig. 17). The fifth pair, which are smaller than those of the female, have the basal joint armed interiorly with two stout apical spines ; the secondary joint is provided with two stout spines on the inner margin,—the posterior spine being longer than the other ; two sub-apical spines exteriorly, one short and one of moderate length, and a moderately long apical seta ; three setæ also spring from a small foliaceous appendage on the first abdominal segment, and immediately behind the fifth pair of feet (fig. 20).

Habitat.—Kilbrennan Sound ; not very common.

Remarks.—This species is somewhat intermediate between *Stenhelia hispida* and *Stenhelia ima*. It has the stout build of the first, while the structure of the second feet in the male somewhat resembles those of the male of *Stenhelia ima ;* but one of the characters which, at a glance, distinguishes this from the other species, is the remarkably stout seta on each of the caudal stylets.

Cletodes tenuipes, sp. n. (Pl. I. figs. 19–27).
Description of the Female.—Length about ·55 mm. ($\frac{1}{45}$th of an inch). Somewhat like *Cletodes propinqua* in general appearance (fig. 19). Antennule small, five-jointed ; the second and last joints are longer than the others, but the fourth is very small ; the last three joints bear moderately stout setiferous spines (fig. 20). The secondary branches and the antennæ are rudimentary ; they are each reduced to a single seta (fig. 21). Mandibles well developed ; palp small, one-jointed (fig. 22). Posterior foot-jaws moderately stout (fig. 23). All the swimming feet are slender, especially the inner branches,—those of the fourth pair being almost rudimentary. In the first pair the length of the inner branches is equal to about two-thirds the length of the outer branches ; the first joint is very short, but the second is elongate (fig. 24). In the next three pairs the inner branches, which, like those of the first pair, are all two-jointed, are much shorter than the outer branches ; those of the fourth pair are very slender,

and scarcely half the length of the outer branches ; the first joint is very minute, the second is elongate and sctiform, and bears a single terminal hair ; the outer branches are slender and elongate, and the marginal spines are also slender (fig. 25). In the fifth pair the basal joint is small, interiorly subquadrangular, but produced exteriorly to form the base of a moderately stout seta ; an elongate plumose seta springs from the inner angle. Secondary joint narrow suborate, the length being equal to fully three times the width at the broadest part ; it is furnished with six setæ, —three on the outer margin, one on the inner margin, and two at the apex (fig. 26). Caudal stylets elongate and narrow, rather longer than the last abdominal segment, and somewhat attenuated towards the distal end. Two small setæ spring from the outer edge of each stylet, and rather nearer the base than the apex. Terminal setæ, two—one small and one elongate (fig. 27).

Habitat.—Near Carradale, Kilbrennau Sound.

Remarks.—The species now described somewhat resembles *Cletodes propinqua*, Brady and Robertson ; but the structure and slender form of the swimming feet readily distinguish it from that form, and also from the next. No males were observed.

Cletodes hirsutipes, sp. n. (Pl. I. figs. 11–18).

Description of the Female.—Length about ·6 mm. ($\frac{1}{40}$th of an inch). Somewhat like *Cletodes propinqua* in general appearance (fig. 10). An tennules five-jointed, short ; the first three and the last joints sub-equal ; the fourth is very small ; the second, third, and fourth joints bear several strong spines (fig. 11). The antennæ are provided with short one-jointed secondary branches, each with two setæ (fig. 12). Mandibles rather feeble, armed with about four elongate slender teeth (fig. 13). Both branches of the first four pairs of swimming feet short and stout ; the outer branches, three, the inner two, jointed ; the inner branches of the first pair are equal to about three-fourths the length of the outer branches, and the end joint is scarcely twice the length of the first joint ; both the inner and the outer branches are densely fringed with short hairs (fig. 15). The next three pairs have also both branches fringed with short setæ ; in the fourth pair the outer branches are about one and a half times longer than the inner, and the apical setæ are very long and plumose (fig. 16). In the fifth pair the basal joint is produced interiorly, so as to form a narrow and sub-quadrangular lamina, the apex of which reaches beyond the middle of the secondary joint ; it is armed with three stout spiniform setæ,—one being at the apex and two near the middle of the inner margin ; the secondary joint is narrow oblong, the length being equal to nearly three times the width ; the outer margin is densely fringed with hairs, and four spiniform setæ spring from the broadly rounded apex (fig. 17). Caudal stylets, about as long as the last abdominal segment, foliaceus, and somewhat distorted ; the inner margin being broadly and obliquely rounded, while there is a slight concavity near the middle of the outer edge (fig. 18).

Habitat.—Kilbrennan Sound, near Carradale ; rather scarce.

Remarks.—The structure of the mandibles and the stout and hirsute thoracic feet are characters sufficiently distinct to distinguish this from other British species ; the fringe of hairs on the margin of the secondary joints of the fifth pair is frequently so coated with mud as to have the appearance of a continuous brownish-coloured border. No males were observed

SOME PARASITES OF *CALANUS FINMARCHICUS.*

FIRST.

Calanus with (?) *Microniscus calani,* G. O. Sars (Pl. III. fig. 20). The drawing shows a *Microniscus* in situ on a *Calanus,* and this is the position in which I usually find the parasite when it happens to be attached to a *Calanus ;* but the parasite is more frequently obtained free than in the position shown by the drawing, through its having, in one way or other, become detached from the copepod.

SECOND.

Calanus infested by a *Nematode* parasite (Pl. III. fig. 21). This drawing is that of a *Calanus,* bearing internally a *Nematode* parasite much longer than its host. It sometimes happens that a considerable proportion of the *Calani* contained in a tow-net gathering will be found infested by these parasites, while at other times the parasites are rarely observed.

THIRD.

Calanus with (?) *Infusorian* parasite (Pl. III. fig. 22). This drawing is that of a *Calanus* with an *Infusorian*-like parasite adhering to it. Specimens of *Calanus* are sometimes found with several of these organisms attached to them ; the parasites are found adhering to the body of the copepods, to the antennules, to the antennæ, and to other appendages, but usually about the head ; sometimes large numbers of *Calanus* will be found infested by these parasites.

All these three parasitic forms have been observed in the Clyde, up as far as the head of Loch Fyne, during the past year.

DESCRIPTION OF THE PLATES.

PLATE I.

Delavalia mimica, n. sp.

Fig.	1.	Female, side view,	× 50	diameters
Fig.	2.	Antennule, female,	× 507	,,
Fig.	3.	Mandible and Palp,	× 507	,,
Fig.	4.	Posterior foot-jaw,	× 760	,,
Fig.	5.	Foot of first pair of swimming feet,	× 380	,,
Fig.	6.	Foot of fourth pair of swimming feet,	× 337	,,
Fig.	7.	Foot of fifth pair of swimming feet, female,	× 380	,,
Fig.	8.	Foot of fifth pair of swimming feet, male, (A) Appendage of first abdominal segment, male,	× 380	,,
Fig.	9.	Last abdominal segments and caudal stylets,	× 253	,,

Cletodes hirsutipes, n. sp.

Fig. 10. Female, side view, . . . × 80 diameters.
Fig. 11. Antennule, ... × 380 ,,
Fig. 12. Antenna, ... × 380 ,,
Fig. 13. Mandible and Palp . . . × 760 ,,
Fig. 14. Posterior foot-jaw, . . . × 760 ,,
Fig. 15. Foot of first pair of swimming feet, . × 380 ,,
Fig. 16. Foot of fourth pair of swimming feet, × 507 ,,
Fig. 17. Foot of fifth pair, female, . . × 380 ,,
Fig. 18. Last abdominal segments and caudal
stylets, ... × 190 ,,

Cletodes tenuipes, n. sp.

Fig. 19. Female, side view, . . . × 160 ,,
Fig. 20. Antennule, ... × 760 ,,
Fig. 21. Antenna, ... × 760 ,,
Fig. 22. Mandible and Palp, . . . × 760 ,,
Fig. 23. Posterior foot-jaw, . . . × 760 ,,
Fig. 24. Foot of first pair of swimming feet, . × 760 ,,
Fig. 25. Foot of fourth pair of swimming feet, × 760 ,,
Fig. 26. Foot of fifth pair of swimming feet, . × 760 ,,
Fig. 27. Last abdominal segments and caudal
stylets, ... × 507 ,,

PLATE II.

Bradypontius Normani (B. and R.).

Fig. 1. Foot of first pair of swimming feet, . × 190 ,,
Fig. 2. Foot of fourth pair of swimming feet, × 126 ,,

Paramisophria cluthæ, n. g. and n. sp. (female).

Fig. 3. One of the antenna, . . . × 126 ,,
Fig. 4. Anterior foot-jaw, . . . × 190 ,,
Fig. 5. Posterior foot-jaw, . . . × 190 ,,
Fig. 6. Foot of first pair of swimming feet, . × 190 ,,
Fig. 7. Foot of fourth pair of swimming feet, × 126 ,,
Fig. 8. Foot of fifth pair of swimming feet, . × 190 ,,

Stephos gyrans (Giesbrecht).

Fig. 9. Antennule, . . . × 126 ,,

Stenhelia intermedia, n. sp.

Fig. 10. Female, lateral view, . . × 80 ,,
Fig. 11. Antennule, female, . . . × 380 ,,
Fig. 12. Antennule, male, . . . × 380 ,,
Fig. 13. Antenna, ... × 380 ,,
Fig. 14. Mandible and Palp, . . . × 380 ,,
Fig. 15. Posterior foot-jaw, . . . × 380 ,,
Fig. 16. Foot of first pair of swimming feet, . × 380 ,,
Fig. 17. Foot of second pair of swimming feet,
male, ... × 337 ,,

Fig. 18. Foot of fourth pair of swimming feet,   × 253 diameters.
Fig. 19. Foot of fifth pair of swimming feet,
       female, ...   × 380   ,,
Fig. 20. Foot of fifth pair of swimming feet,
       male   .   .   .   .   × 380   ,,
Fig. 21. Last abdominal segments and caudal
       stylets, ...   × 253   ,,

Plate III.

Bradypontius Normani (B. and R.).

Fig. 1. Female, dorsal view, .   .   .   ×   40   ,,
Fig. 2. Male, dorsal view, .   .   .   ×   52   ,,
Fig. 3. Antennule, female, .   .   .   × 168   ,,
Fig. 4. Antennule, male, ˙.   .   .   × 126   ,,
Fig. 5. Antenna,   .   .   ˌ   .   × 253   ,,
Fig. 6. Mandible, ...   × 190   ,,
Fig. 7. Maxilla, ...   × 190   ,,
Fig. 8. Anterior foot-jaw,   .   .   .   × 190   ,,
Fig. 9. Posterior foot-jaw, .   .   .   × 190   ,,
Fig. 10. Foot of fifth pair, female,   .   × 380   ,,
Fig. 11. Appendage of the first abdominal
       segment (male),   .   .   × 380   ,,

Hersiliodes littoralis (T. Scott).

Fig. 12. Female, dorsal view,   ×   67

Paramisophria cluthæ, n. g. and n. sp.

Fig. 13. Female, side view, .   .   .   ×   70   ,,
Fig. 14. Antennule, ...   × 127   ,,
Fig. 15. Maxilla, ...   × 135   ,,
Fig. 16. Last abdominal segments and caudal
       stylets,   .   .   .   × 104   ,,

Stephos gyrans (Giesbrecht).

Fig. 17. Female, side view, .   .   .   ×   52   ,,
Fig. 18. Foot of fifth pair, .   .   .   × 380   ,,
Fig. 19. *Aspidophryxus peltatus*, G. O. Sars,
       male and female,   .   ×   70   ,,
Fig. 20. *Calanus*, with *Microniscus calani* at-
       tached,   .   .   .   ×   35   ,,
Fig. 21. *Calanus*, with parasitic nematode, .   ×   35   ,,
Fig. 22. *Calanus*, with (?) Infusorian parasite,   ×   35   ,,

PLATE I.

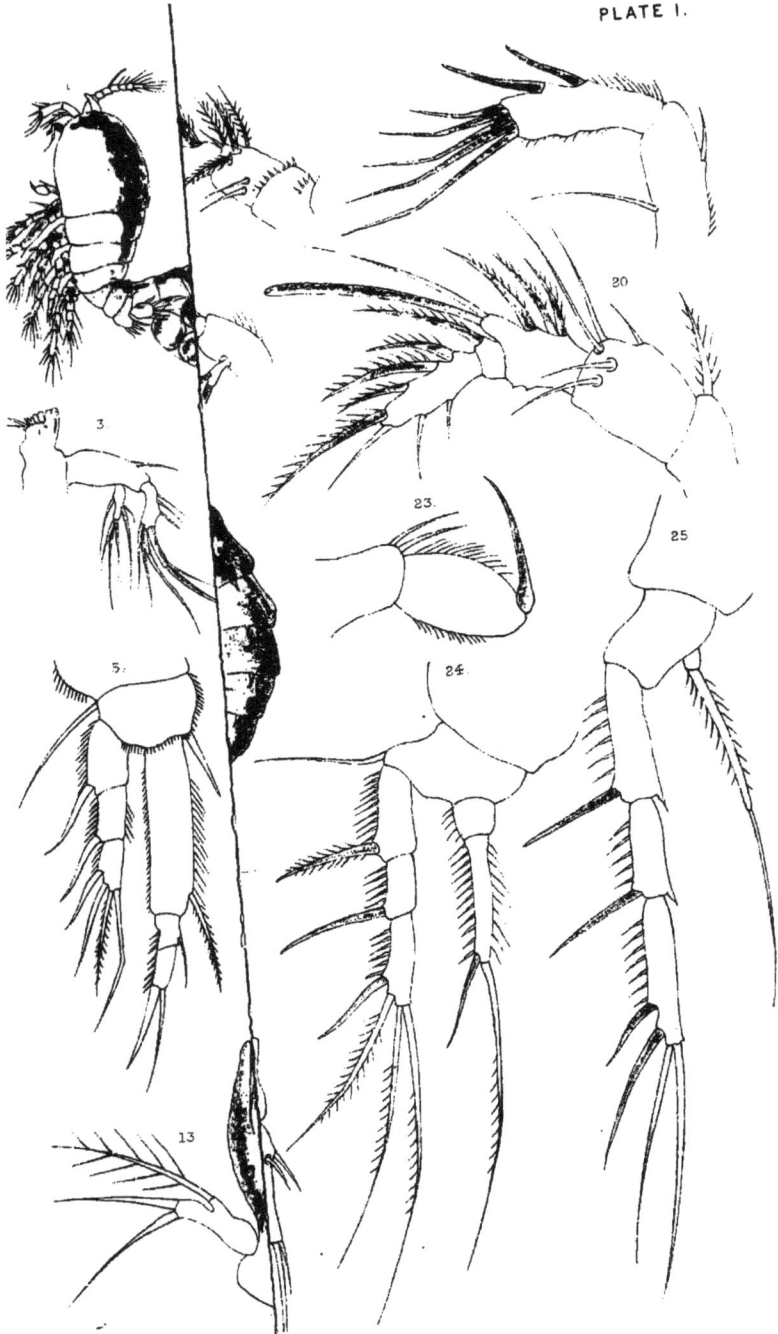

...uipes, sp.n.

A. Scott del. ad nat.　　　Figs. 1-9.—*Dalenella monica*, sp.n.　　　Figs. 10-18.—*Citrella bicaudipes*, sp.n.　　　Figs. 19-27.—*Scionless brevipes*, sp.n.

PLATE II.

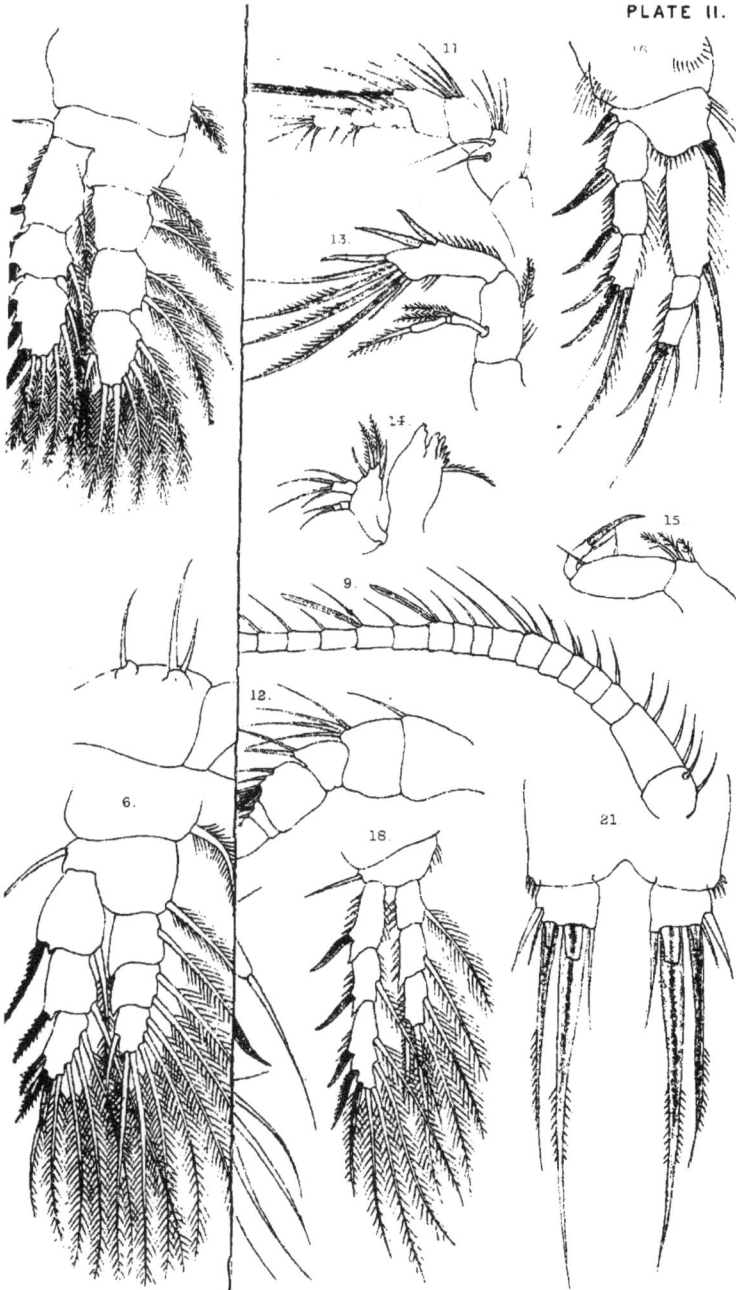

Scott *del. ad nat.*   F1os| F1os. 10-21.—*Stenhelia intermedia*, sp. n.

PLATE II.

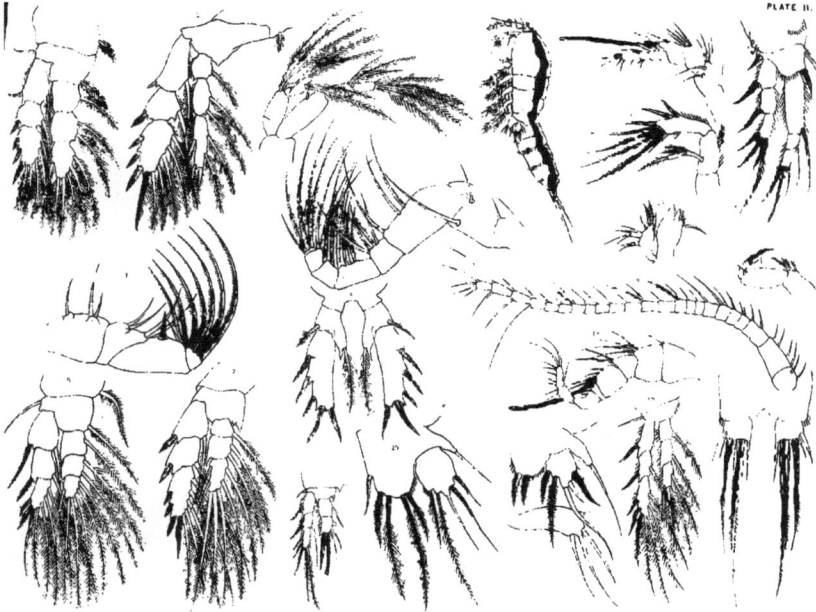

A. Scott del. ad nat.   Figs. 1 and 2.—*Bradyanus Normani* (B. & R.).   Figs. 3-8.—*Parumisophria clarkei*, G. et sp. n.   Fig. 9.—*Misophria gracilis* (Claus).   Figs. 10-12.—*Misophria var. rapidis*, sp. n.

PLATE III.

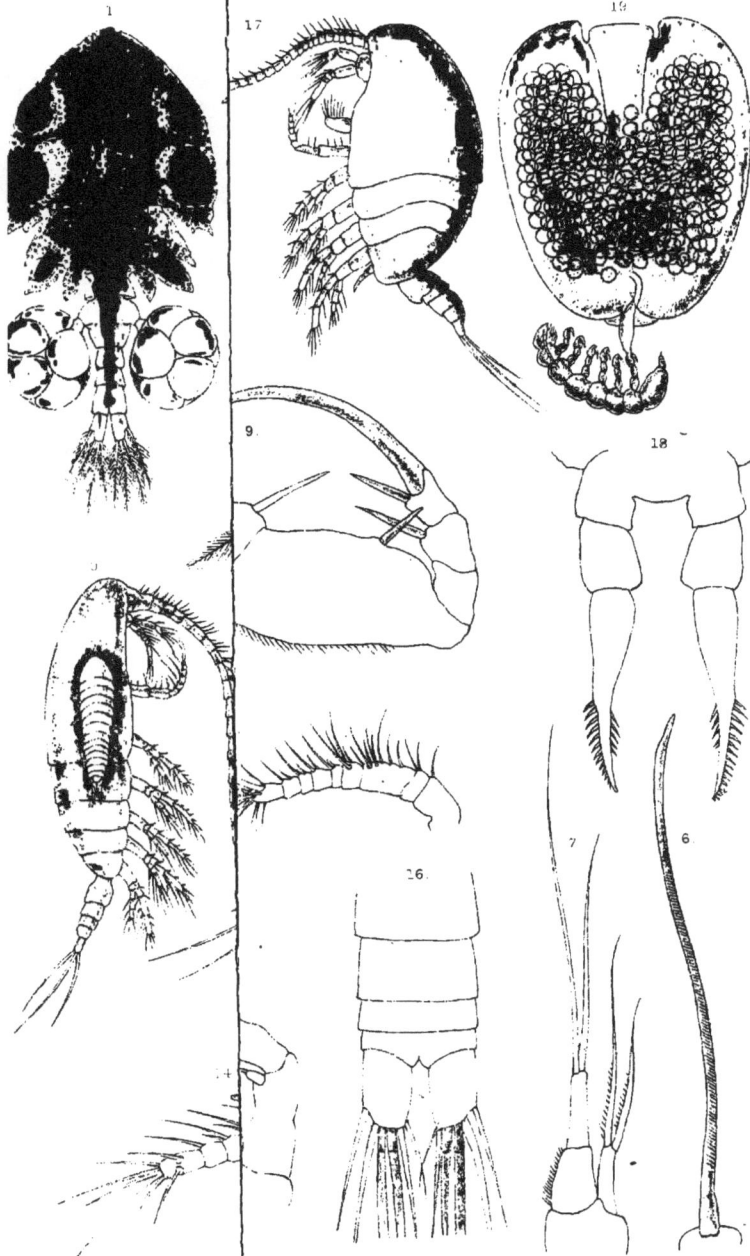

A. Scott del. et pinx. ad nat. ...hria cluthæ, g. et. sp.n.
...marchicus with parasites

PLATE III.

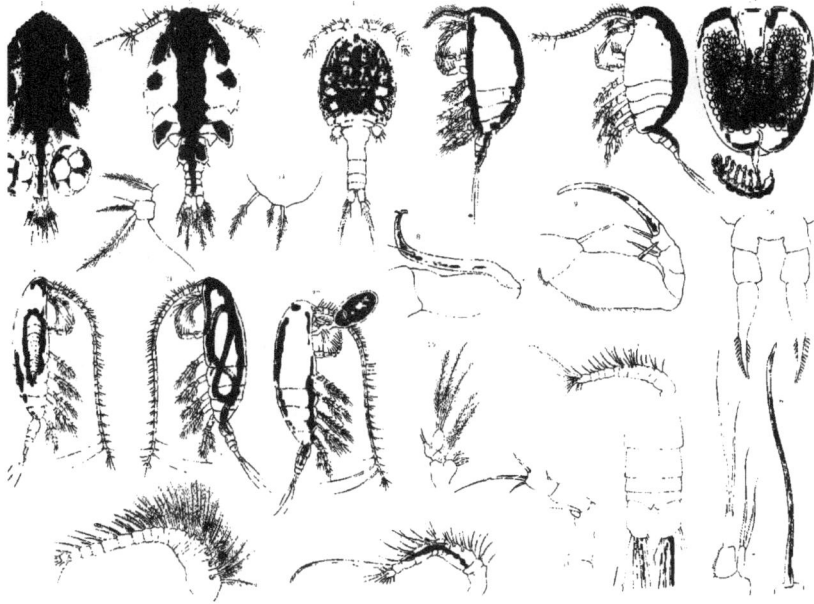

XI.—NOTES ON THE ANIMAL PLANKTON FROM H.M.S. 'RESEARCH.' By Thomas Scott, F.L.S.

This collection, which consisted of twenty-four tow-net gatherings, was contained in twenty-two small bottles : all the gatherings (with one exception) were collected at the surface of the water, in the Shetland-Faröe Channel, at the end of July and the beginning of August last year, by H.M.S. 'Research.' The collections were made by a tow-net made of fine silk bolting-cloth.

Detailed Description of the Collection.

1. This gathering consisted of the contents of one haul of the tow-net, collected at the surface at ' Jackal Station No. II.,' lat. 61°45′N. long. 0°59′W., on July 30th, 1896, at 10 a.m. (net down 15 minutes). The total contents were equal to about 9 c.c., of which 0·75 c.c. consisted of flocculent matter (Diatoms, Infusoria, *Ceratium*, etc.), the remainder being composed of larger organisms, chiefly Entomostraca. The plant plankton has been described by Professor Cleve (p. 297). The following organisms were observed by me :—

Calanus finmarchicus (Gunner).	*Acartia Clausii*, Giesbrecht.
Pseudocalanus elongatus, Boeck.	*Oithona* (?) *similis*, Claus.
Temora longicornis (Müller).	*Podon intermedius*, Lilljeborg.

The number of Entomostraca in this gathering was, approximately, about 3500, and for every 26 *Calani* there were about 6 *Pseudocalani*, 32 *Temoræ*, and 200 *Acartiæ*, while *Oithona* and *Podon* were very rare.

2. This gathering consisted of the contents of one haul of the tow-net, collected at the surface, at the same station as above, and on the same date, at 1 p.m. (net down 15 minutes). The total contents measured 8·75 c.c , of which about ·75 c.c. consisted of flocculent matter. The following organisms were observed :—

Calanus finmarchicus (Gunner).	*Acartia Clausii*, Giesbrecht.
Pseudocalanus elongatus, Boeck.	*Oithona* (?) *similis*, Claus.
Centropages typicus, Kröyer.	*Podon intermedius*, Lilljeborg.
Temora longicornis (Müller).	*Parathemisto* (?) *gracilipes*, Norman.
Eucalanus elongatus (Dana).	*Limacina retroversa* (Fleming).

The number of Entomostraca in this gathering was about the same as the last, and the proportion of the species was also similar. There were only one or two specimens of *Parathemisto* and of *Centropages*, and one specimen of *Eucalanus*. *Limacina* was very rare.

3. This gathering consisted of the contents of one haul of the tow-net, collected at the surface, at the same station as the other two, and on the same date, at 3 p.m. (net down 15 minutes).

The total contents of this haul measured 10 c.c., about 1 c.c. consisting of flocculent matter. The following organisms were observed in this gathering :—

Calanus finmarchicus (Gunner).	*Acartia Clausii*, Giesbrecht.
Pseudocalanus elongatus, Boeck.	*Oithona* (?) *similis*, Claus.
Euchæta norvegica, Boeck.	*Podon intermedius*, Lilljeborg.
Temora longicornis (Müller).	*Parathemisto* (?) *gracilipes*, Norman.

U

The number of Entomostraca was, approximately, about 4000, and the proportions of the species, the one to the other, nearly as in the formula—

Calanus,	Pseudocalanus,	Temora,	Acartia,	Oithona,	Podon.
50	7	24	180	12	2

There were two specimens of *Euchæta* and one of *Parathemisto*. Besides the usual Diatomacea, Infusoria, and Radiolaria, there were a few specimens of *Globigerina* in this gathering.

4. This gathering consisted of one haul of the tow-net, collected at the surface, at 'Jackal Station No. XIII.,' lat. 61°1′ N. long. 3°12′ W., on July 31st, 1896, at 10 a.m. (net down 15 minutes).

The total contents of this haul measured about 10·5 c.c., including about ·5 c.c. of flocculent matter. The following organisms were observed in this gathering :—

Calanus finmarchicus (Gunner).   *Oithona* (?) *similis*, Claus.
Pseudocalanus elongatus, Boeck.   *Podon intermedius*, Lilljeborg.
Temora longicornis (Müller).   *Parathemisto gracilipes*, Norman.
Acartia Clausii, Giesbrecht.   *Limacina retroversa* (Fleming).

The number of the Entomostraca was about 4500, and the proportional numbers of the species nearly as in the last gathering. *Limacina* was very rare.

5. This gathering consisted of one haul of the tow-net, collected at the same station, and on the same date as the last, at 11.30 a.m. (net down 15 minutes).

The total contents of this haul measured about 11·5 c.c., including about 0·5 c.c. of flocculent matter. The following were the organisms observed :—

Calanus finmarchicus (Gunner).   *Acartia Clausii*, Giesbrecht.
Pseudocalanus elongatus, Boeck.   *Oithona* (?) *similis*, Claus.
Temora longicornis (Müller).   *Podon intermedius*, Lilljeborg.
Centropages typicus, Kröyer.   *Parathemisto* (?) *gracilipes*, Norman.

The total number of Entomostraca was about 5000, the proportional numbers of the species being somewhat similar to those of previous gatherings.

6. This gathering consisted of one haul of the tow-net, collected at the surface, at the same station as Nos. 4 and 5, and at the same date, at 2 p.m. (net down 15 minutes).

The total contents of this haul measured about 12·5 c.c., 1·5 c.c. of which consisted of flocculent matter. The following are the organisms observed in this gathering :—

Calanus finmarchicus (Gunner).   *Podon intermedius*, Lilljeborg.
Pseudocalanus elongatus, Boeck.   *Evadne Nordmanni*, Lovén.
Temora longicornis (Müller).   *Parathemisto* (?) *gracilipes*, Norman.
Acartia Clausii, Giesbrecht.   *Limacina retroversa* (Fleming).
Oithona (?) *similis*, Claus.   *Sagitta.*
                              Globigerina.

The total number of Entomostraca and the proportional numbers of the different species were similar to the last.

7, 8. These two small gatherings were put together in the same bottle : they consisted of the contents of two hauls of the tow-net, collected at the surface, at ' Knight Errant Station No. 33,' lat. 60°3′ N. long. 5°51′ W. at 10 a.m. and noon on August 1st, 1896 (net down 15 minutes each time).

The total contents of the two hauls measured about 13 c.c., but about half of this quantity consisted of flocculent matter, the other half being made up of Entomostraca and *Salpæ.* The following are the organisms observed :—

Calanus finmarchicus (Gunner).	*Acartia Clausii*, Giesbrecht.
Eucalanus crassus, Giesbrecht.	*Oithona* (?) *similis*, Claus.
Pseudocalanus elongatus, Boeck.	*Ectinosoma atlanticum* (Brady and
Temora longicornis (Müller).	Robertson).
Metridia hibernica (Brady and	*Podon intermedius*, Lilljeborg.
Robertson).	*Salpa* (?) *vulgaris.*
Paracalanus parvus, Claus.	

Calanus and *Pseudocalanus* were scarce, *Acartia* was frequent, *Salpa* was also frequent, but all the others were more or less rare.

9, 10. These two gatherings were put together in the same bottle : they consisted of the contents of two hauls with the tow-net, at the surface of the water, at the same station as the last, and collected on the same date, at 2 p.m. and 4 p.m. (the net being down for 15 minutes each haul).

The total contents of the two hauls measured only about 5 c.c., and consisted chiefly of *Salpa* and some flocculent matter. Entomostraca were scarce, only five species being observed, of which the following are the names :—

Calanus finmarchicus (Gunner), rare.	*Podon intermedius*, Lilljeborg, few.
Pseudocalanus elongatus, Boeck, few.	*Salpa* (?) *vulgaris*, frequent.
Acartia Clausii, Giesbrecht, few.	(?) *Arachnactis*, sp., rare.
Oithona (?) *similis*, Claus, few.	Ctenophora, rare.

11, 12. These two gatherings were put together in the same bottle : they consisted of the contents of two hauls with the tow-net, at the surface of the water, at the same station as the last ; they were collected at 2 p.m. and 3.30 p.m. on August 2nd, 1896 (the net being down 15 minutes each haul).

The total contents of these two hauls measured about 11 c.c., including about 1 c.c. of flocculent matter. The Entomostraca numbered, approximately, about 4000. The following are the names of the organisms observed :—

Calanus finmarchicus (Gunner).	*Podon intermedius*, Lilljeborg.
Eucalanus elongatus (Dana).	*Limacina retroversa* (Fleming).
Aëtidius armatus, Brady.	*Sagitta.*
Euchæta norvegica, Boeck.	*Tomopteris.*
Metridia hibernica (Brady and	Ctenophora.
Robertson).	

The formula shows the proportional numbers nearly—

Calanus, Pseudocalanus, Metridia, Podon.
90   6   50   40

The others were more or less rare.

13. This gathering consisted of one haul of the tow-net, at the same ' Knight Errant Station ' as the last, collected at the surface at 10 a.m. on August 3rd, 1896 (the net being towed for 15 minutes).

The total contents of the haul measured about 9 c.c., and consisted mostly of flocculent matter : several species of Entomostraca were observed, but individuals were few or rare. *Acartia Clausii* was of more frequent occurrence than any of the others.

The following are the species that have been observed :—

Calanus finmarchicus (Gunner), rare.	*Acartia Clausii*, Giesbrecht, frequent.
Pseudocalanus elongatus, Boeck, one.	*Evadne Nordmanni*, Lovén, few.
Paracalanus parvus (Claus), two.	Ctenophora, frequent.
Centropages typicus, Kröyer, one.	(?) *Arachnactis*.

14. This gathering consisted of one haul of the tow-net, at the same station, and on the same date, as No. 13 : it was collected at the surface at 2 p.m. (the net being towed for 15 minutes).

This gathering only measured about 5 c.c., including 1 c.c. of flocculent matter.

Calanus finmarchicus (Gunner), few.	*Acartia Clausii*, Giesbrecht, frequent.
Pseudocalanus elongatus, Boeck, rare.	*Oithona* (?) *similis*, Claus, rare.
Centropages typicus, Kröyer, one.	*Podon intermedius*, Lilljeborg, rare.
Temora longicornis (Müller), rare.	*Evadne Nordmanni*, Lovén, rare.

Globigerina.

* (?) *Arachnactis*.

There were also a few Ctenophora and larval Crustacea, as well as one or two young Gasteropod molluscs.

15. This gathering consisted of one haul with the tow-net at ' Jackal Station No. XIV.,' lat. 61°20'N. long. 4°22'W., taken at 11 a.m. on the 4th of August 1896, at the surface of the water (the net being towed for 15 minutes).

The gathering measured about 16·5 c.c., 1·5 c.c. of which consisted of flocculent matter, the remaining 15 c.c. being mostly Copepoda. The number of Copepods in the gathering would be approximately about 5000, the greater part consisting of *Calanus finmarchicus*. Other species (the names of which are given below) were rare :—

Calanus finmarchicus (Gunner).	*Euchœta norvegicus*, Boeck.
Paracalanus parvus (Claus).	*Acartia Clausii*, Giesbrecht.
Temora longicornis (Müller).	*Oithona* (?) *similis*, Claus.

Globigerina, rare.

16. This gathering consisted of one haul with the tow-net at the same station as No. 15 : it was collected at the surface at 2 p.m. on the 4th August (the net being towed for 15 minutes).

This was comparatively a small gathering : it measured about 5 c.c., including 2·5 of flocculent matter. The number of Copepoda in the gathering was about 1500. *Calanus finmarchicus* was more common than any of the others, as shown by the formula—

Calanus, Pseudocalanus, Acartia, Oithona, Podon.
　160　　　　4　　　　60　　　9　　　17

Names of the species :—

Calanus finmarchicus (Gunner).	*Acartia Clausii*, Giesbrecht.
Pseudocalanus elongatus, Boeck.	*Oithona* (?) *similis*, Claus.
Metridia hibernica (Brady and Robertson), two.	*Podon intermedius*, Lilljeborg.
	Evadne Nordmanni, Lovén, one.

17. This gathering, which was contained in two bottles, consisted of one haul with the tow-net at the same station, and on the same date, as Nos. 15 and 16 : it was collected at 8 to 12 feet below the surface of the water, at 4 p.m. (the net, in this experiment, was only towed for 10 minutes).

This gathering, which was considerably larger than any of the others, measured about 96 c.c., nearly 89 c.c. of this consisting entirely of *Calanus finmarchicus,* which was the only Entomostracan observed in this gathering. The number of *Calanus* would be, approximately, about 30,000. The remaining portion of the gathering consisted of *Ctenophora* (rare), Diatomacea, Infusoria, and Radiolaria.

18. This gathering consisted of one haul of the tow-net at the same station as No. 17, and on the same date : it was collected at 9 p.m., at the surface of the water (the net being towed for 15 minutes).

The total contents of this gathering measured 65 c.c., and consisted chiefly of *Calanus.* A few other species were observed, but these were of more or less rare occurrence. The number of Copepoda in this gathering would be, approximately, about 22,000. The following are the names of all the species observed :—

Calanus finmarchicus (Gunner).	*Acartia Clausii,* Giesbrecht.
Pseudocalanus elongatus, Boeck.	*Parathemisto* (?) *gracilipes,* Norman.
Temora longicornis (Müller).	*Sagitta.*
Euchæta norvegica, Boeck.	Ctenophora.

Larval decapod Crustacea, rare.

19. This gathering consisted of one haul of the tow-net at 'Jackal Station No. XVII.' : it was collected at the surface of the water at 10 a.m. the 5th of August 1896 (the net being towed for 15 minutes).

This gathering measured about 10·5 c.c., the larger portion of which consisted of *Calanus.* The only other species observed were *Pseudocalanus elongatus* and *Acartia Clausii,* both of which were rare. The number of Copepoda in this gathering was, approximately, 3500.

20. This gathering consisted of one haul of the tow-net at the same station as No. 19, and at the same date : it was collected at the surface at 2 p.m. (the net being towed for 15 minutes).

The contents of the tow-net measured about 14·5 c.c., about a fourth part consisting of Radiolaria, Infusoria, and Diatomacea. The Entomostraca, which numbered about 3600, consisted chiefly of *Calanus ;* the other species that were observed were all of more or less rare occurrence. Their names are as follows :—

Calanus finmarchicus (Gunner).	*Metridia hibernica* (Brady and Robert-
Pseudocalanus elongatus, Boeck.	son).
Temora longicornis (Müller).	*Acartia Clausii,* Giesbrecht.

Parathemisto (?) *gracilipes,* Norman.

also a few Ctenophora and larval Crustacea.

21. This gathering consisted of one haul of the tow-net at the same station, and on the same date, as the last : it was collected at 3.30 p.m., and was a surface gathering (the net was towed 15 minutes).

The contents of the tow-net measured 23 c.c., including about 1·5 c.c. of diatomaceous matter. The remaining 21·5 c.c. consisted almost entirely of Entomostraca, chiefly *Calanus.* The following are the names of the species observed, all of which, with the exception of *Calanus,* were more or less rare :—

Calanus finmarchicus (Gunner).	*Acartia Clausii,* Giesbrecht, rare.
Eucalanus elongatus (Dana), one.	*Oithona* (?) *similis,* Claus, very rare.
Pseudocalanus elongatus, Boeck,	*Evadne Nordmanni,* Lovén, rare.
very rare.	*Microniscus calani,* G.O. Sars, three.
Temora longicornis (Müller), very	(?) *Clavelina,* one small group.
rare.	

22. This gathering consisted of one haul of the tow-net at 'Knight Errant Station No. 28,' lat. 60°2′ N. long. 7°11′ W., taken on August 6th, at 11.30 a.m. It was a surface gathering, and the time the net was towed was 15 minutes.

The gathering measured about 10·5 c.c., 3 c.c. of which consisted of flocculent matter : the remainder consisted chiefly of Entomostraca, and numbered, approximately, about 3000. The names of the species observed are :—

Calanus finmarchicus (Gunner).   *Oithona* (?) *similis*, Claus.
Paracalanus parvus (Claus).   *Podon intermedius*, Lilljeborg.
Temora longicornis (Müller).   *Parathemisto* (?) *gracilipes*, Norman.
Acartia Clausii, Giesbrecht.   *Limacina retroversa* (Fleming).
        Globigerina, Ctenophora, *Evadne Nordmanni*, Lovén.

In 626 specimens, the following proportions were found :—

Calanus, Paracalanus, Temora, Acartia, Oithona, Podon, Evadne.
   310   1   46   146   7   15   1

23. This gathering consisted of one haul of the tow-net at the same station, and on the same date, as the last : it was a surface gathering, and was collected at 6 p.m. (the net being towed for 15 minutes).

The contents of this gathering measured about 7·5 c.c., and contained a considerable number of *Salpæ*. Though several species of Entomostraca were observed, individuals were few in number. *Acartia Clausii* was more frequent than any of the others. The following are the names of the species observed :—

Eucalanus elongatus (Dana) one.   *Euchæta norvegica*, Boeck, five or six.
Paracalanus parvus (Claus), one.   *Acartia Clausii*, Giesbrecht, frequent.
Pseudocalanus elongatus, Boeck, three. *Podon intermedius*, Lilljeborg, frequent.
Metridia hibernica (Brady and   *Salpa* (?) *vulgaris*, frequent.
   Robertson), several.   *Diphys* (?) *campanulifera*, rare.
Scolocithrix abyssalis, Giesbrecht, Ctenophora, rare.
   one or two.

24. This gathering consisted of one haul of the tow-net at the same station, and on the same date, as Nos. 22 and 23 : like the others it was a surface gathering, and was collected at 8 p.m. (the net being towed 15 minutes).

This gathering, which consisted chiefly of Ctenophora and *Salpa*, measured about 7·5 c.c. Entomostraca were comparatively few in number. The following are the names of the species observed :—

Calanus finmarchicus (Gunner),   *Metridia hibernica* (Brady and Robert-
   frequent.   son), two.
Eucalanus elongatus (Dana), one.   *Acartia Clausii*, Giesbrecht, frequent.
Paracalanus parvus (Claus), one or   *Oithona* (?) *similis*, Claus, few.
   two.   *Podon intermedius*, Lilljeborg, frequent.
Pseudocalanus elongatus, Boeck, rare. *Evadne Nordmanni*, Lovén, rare.
Temora longicornis (Müller), one.   *Sergestes* (?) *atlanticus*, M. Edw., one
                                          (jun.).

Salpa, several. ½ (?) *Arachnactis*. (?) *Diphys.*

LIST OF THE SPECIES NAMED IN THE PRECEDING DESCRIPTION.

MACRURA.

Sergestes (?) *atlanticus,* M. Edwards.

COPEPODA.

Calanus finmarchicus (Gunner).
Eucalanus elongatus (Dana).
Paracalanus parvus (Claus).
Pseudocalanus elongatus, Boeck.
Aëtidius armatus, Brady.
Scolocithrix abyssalis, Giesbrecht.
Temora longicornis (Müller).

Mitridia hibernica (Brady and Robertson).
Euchæta norvegica, Boeck.
Centropages typicus, Kröyer.
Acartia Clausii, Giesbrecht.
Oithona (?) *similis.* Claus.
Ectinosoma atlanticum (Brady and Robertson).

CLADOCERA.

Podon intermedius, Lilljeborg.   *Evadne Nordmanni,* Lovén.

AMPHIPODA.

Parathemisto (?) *gracilipes,* Norman.

EPICARIDÆ.

Microniscus calani, G.O. Sars.

PTEROPODA.

Limacina retroversa (Fleming)

TUNICATA.

Salpa (?) *vulgaris.*
(?) *Clavelina.*

VERMES.

Sagitta bipunctata, G. and G.   *Tomopterus onisciformis,* Esch.

CŒLENTERATA.

Diphys, sp., and one or two others not identified.

NOTES ON SOME OF THE SPECIES.*

Sergestes (?) *atlanticus,* M. Edwards.—A single immature specimen of *Sergestes,* which the Rev. Mr Stebbing (who examined the specimen for me) thinks may probably belong to *S. atlanticus,* was obtained in the last of the tow-net gatherings collected by H.M.S. 'Research'; the specimen measured about 18 mm. (¾ inch) in length. The distribution of *Sergestes atlanticus,* according to Spence Bate—see his work on the *Macrura* of the 'Challenger' Expedition—is almost world-wide; the 'Challenger' specimens were obtained in such widely divergent locali-

---

* I am largely indebted to Dr Giesbrecht's valuable work, *Der Pelagischen Copepoden des Golfes von Neapal,* for information concerning the distribution of the Copepoda. Dr Brady's *Monograph of British Copepoda,* and various other works, have also been consulted.

ties as off Japan ; off Matuka, Fiji Islands ; south of Australia ; off Monte Video ; near Teneriffe ; and at 300 miles off the Chesapeake. Moreover, it was taken at the surface and also at the depth of 2425 fathoms. *Sergestes arcticus*, Kröyer, which Spence Bate considers to be identical with *S. atlanticus*, was obtained off the coast of Greenland. The length of the 'Challenger' specimens varied from 20 to 50 mm. ; while, according to S. Smith, a specimen of *Sergestes arcticus*, obtained off the east coast of the United States of America, measured 90 mm. in length.

Calanus finmarchicus (Gunner).—The distribution of this species is also nearly world-wide, as the following other records will show :—Arctic Ocean, North Atlantic, and European Seas (Brady) ; Mediterranean, West Coast of South America, Hongkong (Giesbrecht) ; Australasia and South Pacific (Brady) ; Sulu Sea (Dana).

Eucalanus elongatus (Dana).—This is clearly Dana's species ; it agrees perfectly with the description and figures of it in Dr Giesbrecht's monograph on the *Copepoda of the Gulf of Naples*. This also is an apparently widely-distributed species : Dana records it from the Sulu Sea, and T. Street from north of the Celebes ; Dr Claus obtained it in the Mediterranean, and Dr Giesbrecht notes its occurrence westward of Gibraltar ; on the West Coast of South America, from Valparaiso northwards ; and in the Pacific between lat. 3° S. and 14° N., westward to long. 132° W. But, though the species is of wide distribution, its occurrence in the Shetland-Faröe Channel is of interest, as there seems to be no previous record of its being found so far north in the Atlantic (unless perhaps, some of the North Atlantic records of *Eucalanus attenuatus* : Dana may, owing to an oversight, really refer not to that species, but to *Eucalanus elongatus*). *Eucalanus elongatus* does not yet appear to have been recorded from the North Sea, at least within the British area.

Eucalanus crassus, Giesbrecht.—Only a single specimen (a female) of this species was obtained, and it was captured at 'Knight-Errant Station No. 33.' *Eucalanus crassus* is a more robust species than either *Eu. elongatus* or *Eu. attenuatus*. The abdomen has the same number of segments as in that of the last-named species, but the genital segment is considerably shorter, and is laterally more dilated and broadly rounded : the second abdominal segment is very short. The occurrence of *Eucalanus crassus* in the Shetland-Faröe Channel is somewhat interesting, as bearing on the distribution of the species. Dr Giesbrecht records the occurrence of the species off Rio de Janeiro and north-east thereof ; to the west of South America between lat. 14° and 26° S., and further between long. 175° W. and 138° E., and lat. 19° to 20 N. Moreover, while re-examining some specimens of *Eucalanus* from the Gulf of Guinea, for the purpose of comparing them with that from the Shetland-Faröe Channel, I found one of this species : this specimen was taken at a depth of 50 fathoms, lat. 70° 54′ N., long. 17° 25′ W., and the temperature of the water at the 50 fathoms was 56°·58 Fahr. The above is all that I know as regards the distribution of this *Eucalanus*. It is just possible, however, that some of the records of *Eucalanus attenuatus* may, as with *Eu. elongatus*, really refer to *Eucalanus crassus*.

Paracalanus parvus (Claus).—This is a small and widely distributed species, and readily distinguished by the structure of the fifth pair of thoracic feet in ♂ and ♀. The following is a brief summary of what is known as to its distribution :—Heligoland (Claus), North Sea (Mobins), Trieste (Claus ; Car.), Teneriffe (I. C. Thompson), West Coast of South America, Hongkong (Giesbrecht), Plymouth (Bourne), Irish Sea (I. C. Thompson), Firth of Forth (Mihi).

Pseudocalanus elongatus, Boeck.—This Copepod is usually more or less common all round the British coasts. According to Dr Giesbrecht, the known distribution of this species extends approximately from lat. 50° to 60° N., and from the Baltic to about long. 10° W.

Aëtidius armatus, Brady.—Several specimens (all female) of this species were obtained on the same date and also at the same station ('Knight-Errant Station No. 33') as *Eucalanus crassus*, but at a somewhat later hour of the day. The occurrence of *Aëtidius armatus* in the Shetland-Faröe Channel is, like that of *Eucalanus crassus*, of considerable interest as bearing on the distribution of the species. *Aëtidius armatus* was first described by Prof. G. S. Brady in his work on the 'Challenger' Copepoda, where he records its occurrence in gatherings from the following places :—Indian Ocean (lat. 46° 46′ S., long. 45° 31′ E.), Torres Straits, off Port Jackson, Australia ; Chinese Sea (lat. 17° 54′ N., long. 1·7° 14′ E.) ; in lat. 32° 24′ S., long. 13° 5′ W. ; and in lat. 3° 10′ N., long. 14° 51′ W. Dr Giesbrecht, in his work on the *Copepoda of the Gulf of Naples*, records it as follows :—Gibraltar, 99°–124° W., 3° S.–11° N. Mr I. C. Thompson records it from Malta, and in my report on 'Some Entomostraca from the Gulf of Guinea, West Coast of Africa,' there are a few records of the species for that district. It may also be stated that the specimens recorded by Dr Giesbrecht were obtained at a depth of 2300 metres ; those from the Gulf of Guinea were obtained at depths varying from 5 to 460 fathoms (fully 1400 metres). The capture of *Aëtidius armatus* at this northern station is therefore of interest, as it extends greatly the limits of its distribution, and indicates that its distribution is almost world-wide. The presence of the *Aëtidius armatus* at this 'Knight-Errant' station might of course be due to the action of oceanic currents transporting the specimen beyond the normal limits of the distribution of the species.

Scolocithrix abyssalis, Giesbrecht.—The occurrence of this species in the 'Research' collection is of interest. In my report on some Entomostraca from the Gulf of Guinea (pub. 1894), a Copepod is described as *Scolocithrix tumida*, which appears to be identical with *Scolocithrix abyssalis*, Giesbrecht. The 'Research' specimen agrees perfectly with the form described as *S. tumida*. The specimens from the North Pacific and the Gulf of Guinea were obtained in moderately deep water, the former from 1000 to 4000 metres, the latter from 85 to 460 fathoms ; but the 'Research' specimen is from a surface gathering. The known distribution of this *Scolocithrix* appears to be very limited ; the only record given by Dr Giesbrecht, and referred to above, is as follows :— '124°–132° W., 11°–14° N. ; in 1000–4000 metres, Tiefe,' and it was obtained at only two stations in the Gulf of Guinea, viz. : at about 100 miles to the west of Lonago (lat. 4° 27′ 7″ S., long. 10° 1′ 8″ E.), in 85 and 235 fathoms ; and at about the same distance west of Princes Island (1° 55′ 5″ N., 5° 55′ 5″ E.), in 460 fathoms. Its occurrence at the surface of the water at the Shetland-Faröe Channel is therefore interesting from its bearing on the distribution of the species, both horizontally and bathymetrically.

Temora longicornis (Müller).—The known distribution of this species appears to be confined to a comparatively limited area of the North Atlantic—viz., from lat. 50° to 60° N., westward to about 10° W. long. It has no doubt been recorded from the Adriatic Sea by Dr Claus, and from Marseilles by Dr Gouret ; but Dr Giesbrecht, the eminent authority on the Copepoda, seems to be of the opinion that these records may not refer to this, but to some closely allied form. In my report on ' Some Entomostraca from the Gulf of Guinea, West Coast of Africa,' *Temora longicornis* is included in the list of Copepoda obtained in the collections from that district, but attention is directed to the fact that the fifth pair of thoracic feet in the male differ somewhat from the same appendages of the males of that species from the British Seas, and a figure is given on Plate IX. (fig. 13) showing the nature of the difference. I am now inclined to consider the *Temora* recorded under this name from the Gulf of Guinea to belong, not to *T. longicornis*, but to *T. (Calanus) turbinatus* (Dana)—a form closely related to the other.

Metridia hibernica (Brady and Robertson).—This species was described in 1873, in the *Annals and Magazine of Natural History*, as *Paracalanus hibernicus* by Dr Brady and Mr (afterwards Dr) Robertson ; it was subsequently ascribed to *Metridia armata* of Boeck, and also to the genus *Pleuromma* of Claus. Dr Giesbrecht, in his valuable monograph on the *Copepoda of the Gulf of Naples*, has shown that, while the form discovered by Brady and Robertson is a true Metridia, it belongs to no previously described species ; he therefore restores the specific name first given to it by Drs Brady and Robertson. The following is a brief summary of what is known regarding the distribution of this species, viz. : North Atlantic, from lat. 51° 22′ N., and long. 12° 25′ W. to Rockall Bank, Scilly Isles (Dr Brady). Arctic examples, apparently belonging to this species, but twice the size of British specimens, were obtained by the ' Alert ' and ' Discovery ' Expeditions (Dr Brady). Firth of Forth, Firth of Clyde, and Moray Firth (Mihi). Irish Sea (I. C. Thompson).

Euchæta norvegica, Boeck.—It will be noted that this Copepod has been obtained in a few of the surface tow-net gatherings ; it appears, however, to be found in the greatest abundance in deep water. The distribution of *Euchæta norvegica* appears to be restricted to a somewhat limited area. Dr Mobius has obtained it off the south-west coast of Norway. Professor Sars reports it as occurring generally in the North Atlantic and Arctic Sea, between Norway, Greenland, and Iceland. It occurs abundantly in Loch Fyne, but sparingly in other parts of the Clyde estuary. The Clyde appears to be near the southern limit of its distribution. It is only within comparatively recent years that *Euchæta norvegica* has been observed in the British seas.

Centropages typicus, Kröyer.—This species is of more or less general occurrence all round the British Islands, but is somewhat restricted in its distribution. It appears to be confined to the North Atlantic, the North Sea, and the Mediterranean. The following are some of the localities from which *Centropages typicus* has been recorded, viz. : Canary Islands (I. C. Thompson) ; this seems to be near the southern limit of the species. Mediterranean—[Malta (I. C. Thompson) ; Trieste (Car), Gulf of Naples (Giesbrecht), off Nice (Claus)]. Cape Finisterre (Kröyer), West of Ireland (Brady and Robertson), Shetland Islands (Norman), Faröe Channel (Brady), south-west and south of Norway (Boeck), Heligoland (Claus), Wimereux (Canu).

Acartia Clausii, Giesbrecht.—All the specimens of *Acartia* observed belonged to the one species, viz., *A. Clausii.* This is quite a distinct form, especially as regards the structure and armature of the fifth feet, and in this respect it differs very markedly from *Acartia longiremis,* Lilljeborg. So far as known, the distribution of *Acartia Clausii* appears to be even more extensive than, though probably not so general as, *Centropages typicus.* The area of its known distribution includes the Mediterranean (Giesbrecht, Claus, Gourret, I. C. Thompson), Libreville, Gaboon River, West Africa (Mihi), Canary Islands (I. C. Thompson, Wimereux (Canu), Plymouth (Bourne), Loch Fyne, Scotland (Norman), Firth of Forth, Scotland (Mihi). The present seems to be one of the most northerly records hitherto reported for this species.

Oithona (?) *similis,* Claus.—All the 'Research' specimens of *Oithona* appear to belong to the species described by Dr Giesbrecht as *Oithona similis,* Claus. Dr Giesbrecht seems to consider that most of the British records of *Oithona spinifrons* are referable to this species, and he is satisfied that *Oithona spinirostris,* Giesbrecht, is also identical with it. The authentic distribution of *Oithona similis* is limited to a comparatively few places ; as, for example, Nice, Trieste, and Gulf of Naples in the Mediterranean (Claus, Car, Giesbrecht), Kieler Fohrde (Giesbrecht), Bay of Wismar, Baltic (Braun), West Baltic (Hensen), the Indian and Pacific Oceans (Giesbrecht).

Ectinosoma atlantica (Brady and Robertson).—This minute species has to all appearance a world-wide distribution. Besides the various records of its occurrence around the British Islands, it has also been reported from the North and South Atlantic and the Pacific Oceans (the Gulf of Guinea, West Coast of Africa, near Ascension Island ; and near the Galapagos Island). *Ectinosoma atlantica* is so minute that it has doubtless often escaped observation : it is sometimes a moderately common species in Loch Fyne.

Podon intermedius, Lilljeborg.—All the specimens of *Podon* observed belonged to this species, which is readily distinguished from *P. polyphemoides* by the number of hairs of the second antenna—one branch having seven and the other six hairs.

Parathemisto (?) *gracilipes,* Norman.—None of the specimens of *Parathemisto* obtained appeared to be mature, but, as far as could be made out, they all belonged to *P. gracilipes,* which is a smaller species than *P. oblivia.*

V.—ON THE DISTRIBUTION OF PELAGIC INVERTEBRATE FAUNA OF THE FIRTH OF FORTH AND ITS VICINITY during the Seven Years from 1889 to 1895, both inclusive. By THOMAS SCOTT, F.L.S., Mem. Soc. Zool. de France. (Plates IV., V., VI., VII.)

CONTENTS.

INTRODUCTORY.

The scientific investigations that have until recently been carried on in the Firth of Forth, and also outside the limits of the estuary, included among other things the examination of the various groups of pelagic invertebrates that form so important a part of the food of fishes. The examination was carried on chiefly in connection with the trawling operations of the *Garland,* the steam cruiser set apart specially for scientific work. The trawling experiments of the *Garland* were made over certain fixed portions of the estuary and neighbourhood which were denominated "stations." The method adopted for the examination of the pelagic invertebrates was somewhat as follows :—During the time that each station was being trawled two tow-nets were kept constantly at work. One of these was towed at about half a fathom below the surface of the water, and was distinguished as the surface tow-net ; the other was fastened to the end of the beam of the trawl in such a way as just to clear the bottom, and was distinguished as the bottom tow-net. Sometimes other nets were used at intermediate depths, but not with the same regularity as the first two. The gatherings collected with these mid-water nets are therefore not specially referred to in this paper. The tow-nets at first in use were comparatively of small size, but subsequently a larger kind, somewhat similar to those that had been used on board the *Challenger,* were adopted, and are still used in connection with the Board's scientific work.

At first nine experimental stations were selected for the carrying out of special scientific work, but this number was afterwards increased to ten. It has not, however, been considered necessary for the purposes of this paper to take note of the published statistics for all these stations, and therefore six of them have been fixed upon as representative of the others—namely, Stations I., III., IV., V., VIII., and IX.   Stations I. and III. extend from the east side of Inchkeith eastward, and are comparatively near to each other.   Station IV. is an inshore station, and extends for a considerable distance round the "South Bay."   Station V. extends from May Island westward along the centre of the Firth till nearly abreast of Elie. Stations VIII. and IX. are beyond May Island, and outside the limits of the estuary.

The results obtained by the use of the tow-nets were at first recorded in a somewhat general form, but as the work proceeded the differentiation of the various organisms became less difficult, and made it possible for a more detailed description of the results being placed on record.   It will be obvious, therefore, that in the study of the distribution of the pelagic invertebrates the earlier records will not be so useful as those of subsequent date.   Moreover, the investigations were not carried on during the first year or two so continuously as they were afterwards, and for these reasons I propose to limit the consideration of the pelagic invertebrates to those collected (by means of the surface and bottom tow-nets already described) during the seven years from 1889 to 1895 inclusive, and recorded in the several Annual Reports for these years.

But though the examination of the pelagic invertebrates within the area under consideration was carried on more or less continuously and regularly during the years referred to, there are times when, for various reasons, no tow-net gatherings were collected, and for which there are no records, so that the records for some years are less complete than for others.   In 1889, for example, there are no records of tow-net gatherings for any of the stations in March, April, and December; in 1890, in January and June; in 1892, in August; in 1894, in March, May, July, September, and October; and in 1895, in March, June, and November. Moreover, there may be records for a particular month for some of these stations and not for others.   Thus in 1889 though there are records for October for Station I. there are none for any of the other stations; but while that is so there are still a large number of statistics available for at least a partial study of the distribution of the pelagic invertebrates of the Firth of Forth.

The time available for the preparation of this paper did not permit of the detailed consideration of every species or kind of invertebrate recorded in the lists of tow-net gatherings, so that only those that are of numerical importance, or that are otherwise of importance from a fisheries' point of view, are so dealt with, while an endeavour is also made to invite attention to any interesting point observed in connection with the distribution of the other species.   Tables illustrating in time and space the distribution of these pelagic forms are introduced wherever they are considered necessary.

Moreover, four charts are added at the end of the paper for the purpose of supplementing and extending the information given in the Tables concerning the distribution of a few of the more, numerically or otherwise, important species or groups of invertebrates comprised in the tow-net lists.   These charts not only show when and where the species was abundant, or scarce, or absent, but they also allow of a comparison being made between the surface and bottom distribution of the species.   The importance of this will be evident when it is remembered that though a species may be rare at the surface it may at the same time be common or even abundant at the bottom, as well as *vice versa.*

It may further be stated that in proceeding to consider the distribution of the invertebrates mentioned in the various lists of tow-net gatherings collected at the six selected stations during the seven years, those that are higher in the scale of classification are referred to first, and afterwards those that are lower.

It may also be as well to state that there has been a considerable advance in the knowledge of marine zoology since the *Garland* commenced work in the Firth of Forth twelve years ago,[*] and one of the results of this advance has been the introduction of many changes in the nomenclature of marine organisms. Owing to these changes several of the names mentioned in the *Garland's* tow-net lists are replaced by those that are now considered to be the rightful designations of the species to which they refer; and one advantage of this alteration will be that the results of the tow-net work of the *Garland* will be brought more into line with present-day knowledge.

INVERTEBRATES CAPTURED BY THE TOW-NETS.

The invertebrates captured by means of the surface and bottom tow-nets during the seven years from 1889 to 1895, at the representative stations already referred to, comprised a large number of forms of different kinds, but many of these were only of rare occurrence. The greater proportion by far of the contents of all the tow-net gatherings consisted of only a few species, which may be summed up as follows :—Schizopoda (*Thysanoessa* chiefly, *Erythrops*) ; Amphipoda (*Hyperia, Hyperoche,* and *Parathemisto*) ; Copepoda (*Calanus, Temora, Acartia*) ; *Sagitta ;* Cœlenterata (*Pleurobrachia, Beroë,* and others) ; and young Crustacea (Decapoda, Schizopoda, and Cirrepedia). Before proceeding to discuss the distribution of these six groups, as brought out by an examination of the tow-net records, I propose to refer briefly to several of the more uncommon forms which have been captured from time to time at the selected stations and during the years already specified.

(1.) THE MOLLUSCA.

There are very few records of Mollusca which have been captured by the tow-nets, and only two really pelagic forms have occurred. The Mollusca recorded comprise two small *Eolis,* the species of which was not determined. One was obtained in the bottom tow-net at Station V. in 1891, and the other at Station I. in May 1895 ; also *Doris (?) repanda,* which was obtained in the bottom net in November 1890 at Station III., and *Doto coronata,* obtained in December 1890 at Station IV., also in the bottom net. A few species of *Cuspidaria cuspidata* are recorded under Station IX., having been brought up in the bottom tow-net at that station in April 1893, but the presence of this species was owing, no doubt, to the net having accidentally dipped into the mud at the bottom. The two pelagic species are the Pteropods *Limacina retroversa* and *Clione borealis.* The first is recorded on two different occasions in 1890, having been obtained in April in the bottom tow-net at Station V., and in May in a surface-net gathering at Station I. It was also obtained in the bottom tow-net at Station III. in July 1891, and in the bottom tow-net at Station IX. in December 1892. The other (*Clione borealis*) was captured

[*] It is pleasing to know that the little steamer belonging to the Fishery Board for Scotland has had some share in promoting this advance, though perhaps the time has not yet come for the full appreciation of the *Garland's* work.

in the surface tow-net at Station III. in January 1889, in the bottom tow-net at Station V. in November 1893, and in the surface net at Station VIII. in December of the same year.   *Limacina retroversa* is a very generally distributed species, but seems to be more common on the West than it is on the East Coast.   On the West Coast it sometimes occurs in immense shoals, and at times forms a considerable part of the food of the herring.   I have found the stomachs of herrings sent to me from the West Coast for examination filled with little else than these Pteropods, numbers of which appear to have been swallowed wholesale, as some of the shells were practically uninjured.   *Clione*, which has no shell, is a northern species, and is a rare visitant to the Firth of Forth.   All the three recorded instances of its occurrence happened during the winter months.   This was probably owing to the temperature of the water at that season being more favourable to its wandering habits, or the specimens may have been carried south by currents coming down from the North Atlantic.   Another point connected with the capture of the *Clione* is, that on the two occasions in which it occurred in the surface tow-net the surface of the sea was smooth or but slightly disturbed, but on the occasion when it was found in the bottom tow-net the weather was stormy and the sea rough.

(2.) THE CRUSTACEA.

(*a*) DECAPODA.

The Decapoda captured by the tow-nets comprised the following species :—

Macropodia rostrata (Linn.),

Crangon allmanni, Kinahan,	*Hippolyte fascigera*, Gosse,
Egeon fasciatus, Risso,	*Spirontocaris pusiola* (Kroyer),
Cheraphilus nanus (Kroyer),	*Pandalus montagui*, Leach,
Hippolyte varians, Leach,	*Pandalus brevirostris*, Rathke,

all of which, with the exception of the first, belong to the Carida.   They were all captured in the bottom net at various times and at one or more of all the selected stations.

Macropodia is only recorded twice during the seven years—once in October 1890, from Station IV., and again from the same station in August 1891.

Crangon allmanni occurred more frequently than any of the Decapods mentioned, and there are records of its capture with the bottom tow-nets in all but one of the seven years, 1889 being the only year in which no distinct record of the species occurs.   In 1890 it is recorded twice, once in February and once in November, both records being for Station V. It is also recorded twice in 1891, once in March for Station V. and in September for Station III.   There are seven records for 1892—viz., in January, March, and November for Station I., in February for Station III., in February for Station IV., and in January and February for Station V.   There is only one record for 1893—viz., one for Station V. in

November. In 1894 there are three records, one in February for Station IX., and in August one each for Stations I. and III. In 1895 there are six records, one in February for Stations III., IV., V., and VIII., and one in September for Stations VIII. and IX., or twenty-one records in all for this species of Crangon. Nine (or nearly 43 per cent.) of the records are for February alone, and if those for January and November are included they amount to two-thirds of the whole number, the other records being divided almost equally among the three months March, August, and September.

Egeon (Crangon) fasciatus is only recorded once during the seven years—viz., for Station III. in September 1891.

Cheraphilus (Crangon) nanus. There are several records of this species during the prescribed period. Three of these occur in 1891, one in March for Station I. and one in the same month for Station V. The third record occurs in September for Station III. In 1892 there are four records, two for Station I. in January and March, one for Station III. in February, and one for Station V. in January. There is one in February 1894 for Station III., and one in the same month of 1895 for Station VIII., or nine records in all for *Cheraphilus nanus*, and, with the exception of the one in September for Station III., they all occur in the colder months.

Hippolyte varians. There is but one record of this species—viz., in December 1890 for Station V.

Hippolyte fascigera. This species is recorded for Station VIII. in January 1894, and in February 1895 for the same station.

Spirontocaris (Hippolyte) pusiola is recorded only once—viz., in February 1895 for Station IV.

Pandalus montagui (= *P. annulicornis* of Bell's *Stalk-eyed Crustacea*). The number of records for this species is seven, one for Station VIII. in February 1890 ; one in September 1891 for Station III. In 1892 there is one record in January for Station I. and also for Station V., and one in February for Station III. In 1893 there is one for Station V. in November, and one in February 1895 for Station VIII.

Pandalus brevirostris. For this species there are two records, and both are for 1891 ; one is for Station V. in March, and the other for Station III. in September.

It thus appears that nine species of named Decapod Crustacea have been captured during the six years from 1890 to 1895 (one or two Decapods are recorded for 1889, but only by their generic names, and are therefore not included here). These nine species were all captured with the bottom tow-net—not a single instance is recorded of the occurrence of any of them in the surface nets. The records for the whole nine species for the six years number 46. They are tabulated under eight out of the twelve months, and are characterised by the great majority of them being included in the statistics for the five months that are usually the colder months of the year. The subjoined formulæ will show some of these details more clearly :—

(1) The months in which the capture of Decapoda are recorded	Jan.	Feb.	Mar.	Aug.	Sept.	Oct.	Nov.	Dec.
Number of records for each of the months for the seven years	7	17	6	3	7	1	4	1
(2) The Stations where the Decapoda were captured	I.	III.	IV.	V.	VIII.	IX.		
Number of records for each station	7	11	5	13	7	3		

The number of records for February alone is 17, or 27 per cent. of the whole; and the number for the five colder months, January, February, March, November, and December, is 35, or fully 76 per cent. The reason that these Decapods are found usually in the bottom tow-nets, and not in those worked at the surface, is that they are veritable bottom feeders ; and were it not for their peculiar habit of springing upward as well as backward when disturbed, their capture would be much less frequent. When any suspicious object is approaching these Crustaceans (I refer to the Carida or "shrimp" group) they face round towards it and intently watch its movements, and at the same time swim gently backwards, or they may at once bury themselves in the sand or mud. If the danger, however, should approach them suddenly, they spring quickly backwards and also obliquely upwards, and it is then that they come within the sweep of the bottom tow-net. But though the occurrence of these organisms in the bottom tow-net may be thus more or less satisfactorily explained, the reason for their being captured chiefly during the colder months seems to be less easy of explanation. It may be observed, however, that only five of the records are for the inshore station, and that about half of the total number are for Stations III. and V., the two stations in the middle of the estuary, the one extending east from Inchkeith the other west from May Island.

(b) SCHIZOPODA.

The Schizopoda are well represented as regards the number of species, and a few of them, especially those of the Euphausiidæ, appear to have been at times pretty numerous at the station under consideration. The Schizopods referred to in the records and distinguished by their names are the following :—*

EUPHAUSIIDÆ.

Boreophausia raschii (M. Sars).   *Thysanoessa neglecta* (Kroyer).

MYSIDÆ.

Siriella jaltensis, Czern. (= Siriella   *Leptomysis gracilis*, G. O. Sars.
          crassipes, G. O. Sars).   ,,   *lingoura*, G. O. Sars.
     ,,   *armata* (M. Edw.).   *Hemimysis lamornæ* (Couch).
Gastrosaccus spinifer (Goës).   *Macropsis slabberi* (Van Ben.).
Heteromysis formosa, S. I. Smith.   *Praunus flexuosus* (Müller). (=
Erythrops goësii, G. O. Sars.   *Mysis chamæleon*, Varg.).
Mysidopsis didelphys (Norman).   ,,   *inermis* (Rathke).
     ,,   *gibbosa*, G. O. Sars.   *Schistomysis spiritus*, Norman.
     ,,   *angusta*, G. O. Sars.   ,,   *ornata* (G. O. Sars).
          Neomysis vulgaris (J. V. Thompson).

---

* The names and their arrangement are in accordance with Rev. A. M. Norman's revision of the British Schizopoda, published in the 'Annals and Magazine of Natural History' for June, August, and September 1892.

Owing to the difficulty of distinguishing the species of Schizopoda off-hand, their occurrence is frequently recorded in the lists of tow-net fauna simply under the family names, "Euphausiidæ" or "Mysidæ," or the generic names "Boreophausia sp.," "Mysis sp.," as the specimens observed happened to belong to either group ; and before taking up the consideration of the detailed species, it is proposed to consider briefly the distribution of the Schizopoda under these names, as follows :—

THE EUPHAUSIIDÆ.—The records of the Schizopoda under the name of Euphausiidæ or *Boreophausia* sp. amount to one hundred and twelve for the seven years, and taking the one year with the other, they are distributed over all the twelve months ; but the largest number occur in the first four and the last three months, as indicated by the Table annexed, which shows the number of the records of the Euphausiidæ that are referred to simply under their family or generic names for each station, and for each month of the year for all the seven years.

TABLE I., showing the Distribution of the Euphausiidæ, as indicated by the Records for the Years, Months, and Stations :—

Years.	Jan.	Feb.	March.	April.	May.	June.	July.	August.	Sept.	Oct.	Nov.	Dec.	Totals for the Years.	Years.	I.	III.	IV.	V.	VIII.	IX.	Totals for the Stations.
1889,	1	-	-	-	-	1	1	1	-	1	-	-	5	1889,	2	-	-	2	-	1	5
1890,	-	3	6	3	-	-	3	-	-	3	2	1	21	1890,	2	5	2	6	3	3	21
1891,	4	3	2	1	7	4	-	-	-	-	-	1	22	1891,	4	7	5	2	4	-	22
1892,	3	4	5	2	4	-	1	-	1	-	-	-	20	1892,	2	5	1	3	3	6	20
1893,	7	4	-	1	-	-	-	-	1	4	4	6	27	1893,	5	7	3	6	3	3	27
1894,	2	6	-	1	-	-	-	-	-	-	-	1	10	1894,	3	1	1	2	2	1	10
1895,	-	2	-	-	-	-	-	-	-	1	-	5	8	1895,	1	3	1	-	1	2	8
Totals for the Months,	17	22	13	8	11	5	5	1	2	9	6	14	113	Totals for the Stations,	19	28	13	21	16	16	113

Of these records thirty-one are for the Euphausiidæ observed in the surface tow-nets, while the other eighty-one are bottom tow-net records. In few of the surface gatherings were Euphausiidæ frequent or common, but in a number of the gatherings collected with the bottom tow-nets they were fairly numerous. The results brought out by these Tables regarding their distribution are referred to further on in connection with the detailed account of the different species.

Meanwhile I proceed to notice the records of the second group of the Schizopoda as they appear under the general name of *Mysis* sp. or Mysidæ. I find that there are only thirty-three separate notices of Mysidæ, and three of these are entered under the name *Leptomysis* sp. The reason for this is that most of the Mysidæ were usually not so plentiful, and were identified at the time or shortly after they were collected. The records of Mysidæ, though few in number, show a distribution very similar to the Euphausiidæ—that is, they occur chiefly among the statistics for the colder months. There are no records of Mysidæ for. the three summer months, while those for February, November, and December are together equal to 71 per cent. of the whole number. I shall now leave these more general groups for the present, and proceed to consider the distribution of those forms recorded under distinctive specific names.

L

(i.) The Euphausiidæ.

Boreophausia raschii. The records of this species are comparatively few, and are all included in the statistics for 1889 and 1890. In 1889 there are eleven records of this species, one each for Stations I., III., and IV. in January, one each for Stations I. and V. in February, and one each for all the six stations in November. There are three records for 1890, one each for Stations I. and VIII. in February, and one for Station IV. in March.

Thysanoessa neglecta. This is the Schizopod that is usually referred to in the records as "Thysanoessa sp." It was observed in 1891 at Station III. in October, and again at the same station in November; it was also observed in November at Stations V. and VIII. In 1892 the records for Thysanoessa are as follows:—For Station I. in September, and again in November and December; for Station III. in December, and in the same month at Station IV.; at Station V. in October, and again in November and December; once at Station VIII. in October, and on two different occasions in December (on one of these occasions the species was observed in both the surface and bottom tow-nets); at Station IX. it was also taken twice in December, one on the 2nd and again on the 22nd. There are only two records for Thysanoessa in 1893—viz., in January for Station III. and in September for Station I. The only records in 1894 are two in December, when the species was taken in both the surface and bottom tow-nets at Station III. In 1895 there is no distinct record for Thysanoessa, and it probably was not observed during that year.

(ii.) The Mysidæ.

Siriella jaltensis. (This is equal to S. crassipes, G. O. Sars, of earlier papers.) There is but one record for this among the faunal lists for the seven years; it was obtained at Station V. in March 1893.

Siriella armata is another species that is seldom met with in the Firth of Forth, and only one record of it is published in the lists of tow-net fauna of the selected stations; it occurred at Station III. in March 1892.

Gastrosaccus spinifer. There are records of this species from one or other of the selected stations for all the seven years except 1891. The first record is for Station VIII. in November 1889. In 1890 the species is recorded for Station VIII. in December, and for Station IX. in August. In 1892 for Station I. in January, and Station V. in March. There is only one record in 1893—viz., for Station I. in January, and one in 1894 for Station III. in February. In 1895 there are two records, one for Station III. in February, and one for Station VIII. in December.

Heteromysis formosa is one of the rare species recorded for the Firth of Forth, and has only been observed twice during seven years—viz., at Station I. in August 1894, and at Station IV. in February 1895.

Erythrops goësii. This Mysid, though not known to occur in Britain previous to the institution of the Board's scientific investigations, is the most common of the Mysidæ in the seaward part of the Forth estuary, and there are several records of it for all the selected stations. In 1889 it is recorded in January for Station III., in February for Station I., and in November for Station VIII. In 1890 there are records of it for Station I. in October and November; for Station V. in February, September, November, and December; for Stations VIII. and IX. in February, and again in April; for Station IX. in August; and for Station VIII. in December. In 1891 the records include one for Station I. in March, and for Station III. in September and November, and for Station V. in March and Station VIII. in June. In 1892 *Erythrops* is recorded for Station I. in January, March, and November, and for Station III. in February. It is also recorded for Station V. in January, February, March, May, September, and December; for Station VIII. once in March and September, and twice in December; and for Station IX. in May, July, and December. The records for 1893 are comparatively few in number, and include one for Station III. in December, one for Station V. in March and November, for Station VIII. in December, and Station IX. in February and October. In 1894 there are only three records, one for Station III. in February, and two for Station V., one each in February and August. There are several records for 1895, and they are distributed as follows :— One for Station I. in April, one for Station III. in February, and again in October, one for Station V. in February, one for Station VIII. in February, May, and December, one for Station IX. in April, and one each for Stations VIII. and IX. in September. The annexed Tables show (1) the number of records for the months and years and (2) the number for the years and stations.

TABLE II., showing the Distribution of *Erythrops goësii*, as indicated by the Records for the Years, Months, and Stations :—

Years.	Jan.	Feb.	March.	April.	May.	June.	July.	August.	Sept.	Oct.	Nov.	Dec.	Total for each Year.	Years.	I.	III.	IV.	V.	VIII.	IX.	Total for each Year.
1889, .	1	1	-	-	-	-	-	-	-	-	1	-	3	1889, .	1	1	-	-	1	-	3
1890, .	-	3	-	2	-	-	-	1	1	1	2	2	12	1890, .	2	-	-	4	3	3	12
1891, .	-	-	2	-	-	1	-	-	1	-	1	-	5	1891, .	1	2	-	1	1	-	5
1892, .	2	2	3	-	2	-	1	-	2	-	1	4	17	1892, .	3	1	-	6	4	3	17
1893, .	-	1	1	-	-	-	-	-	-	1	1	2	6	1893, .	-	1	-	2	1	2	6
1894, .	-	2	-	-	-	-	-	1	-	-	-	-	3	1894, .	-	1	-	2	-	-	3
1895, .	-	3	-	2	1	-	-	-	2	1	-	1	10	1895, .	1	2	-	1	4	2	10
Total for each Month,	3	12	6	4	3	1	1	2	6	3	6	9	56	Totals for the Six Stations,	8	8	-	16	14	10	56

Mysidopsis didelphys. There are only three records for this species, one for Station VIII. in December 1890, and two in March 1891—one for Station I. and another for Station V.

Mysidopsis gibbosa is recorded for Station III. in January 1889 ; for Station IX. in August 1890 ; for Station I. in March 1891, and for Station III. in January and September of the same year. In 1892 there is a record for Station I. in January and Station IV. in March, while the only record in 1894 is one for Station III. in February.

Mysidopsis angusta. The only records of this species are all for 1890 except one, and comprise one for Stations I. and III. in October, and one for Station V. in October, November, and December in 1890, and one for Station III. in 1891.

Leptomysis lingoura is recorded once during the seven years—viz., for Station III. in March 1892.

Hemimysis lamornæ. This Schizopod is recorded three times in 1892— viz., twice for Station I., once in January and once in March, and for Station IX. in December. The species is recorded again in 1893 for Station V. in January, and in 1895 for Station III. in February and Station VIII. in December.

Macropsis slabberi, remarkable among British Schizopods for the length of the eye-stalks, is of frequent occurrence over a considerable portion of the Forth estuary, and specially in the upper reaches. While several of the species mentioned here may be regarded as only visitors, this appears to be a resident, at least in that part of the Firth which extends westward from Queensferry. In this part of the Forth, which may be considered the headquarters of the species on the East Coast of Scotland, it may sometimes be found in large numbers, and its presence seaward may be due to the overcrowding in these waters causing a greater amount of migration to take place, though it is also possible that the species may be resident in the seaward as well as the landward part of the Firth. The records of the species that have to be noted in connection with the present inquiry are the following :—The first reference to *Macropsis* occurs in February 1890 for Station IV., and again for the same station in March. There is no further record of it till October, when it is again recorded for that station and also for Stations I. and III. It is again recorded for the same three stations in November, and the November record for Station IV. is for both surface and bottom tow-nets, while the others are for the bottom tow-nets only. In 1891 there are records of this species for Stations III. and IV. in January, for Stations I. and III. in October, and in the same months for Station IV. The October record for Station IV. is for the surface tow-net. In July of the same year *Macropsis* is recorded as frequent in a surface tow-net gathering at Station VIII. in July, which seems to indicate a considerable migration of the species seaward, either from the upper parts of the estuary or elsewhere. In 1892 it is recorded for Station I. in January and for Station IV. in March. In 1893 the first record is for Station IV. in January, while the next records are for Stations I. and III., the first being for November, the other for October. There are no records of *Macropsis* for 1894. The records for 1895 include one for Station III. in February and another for the same station in October. There is one for Station IV. in February, and two for the same station in October, one for the surface tow-net and one for the bottom tow-net ; and another for the same station in December.

Praunus flexuosus. This species, which is more familiar by its older name of *Mysis chamæleon* or *Mysis flexuosa,* is, though a common one, infrequent among the tow-net invertebrates from the selected stations. In 1892 it is recorded for Station IV. in January and March; in 1893 for Station III. in December ; in 1894 for Station III. in February ; and in 1895 for Station IV. in January, and for the same station in February. *Praunus flexuosus* was observed only among bottom tow-net gatherings,

Praunus (Mysis) inermis. The only records for this species occur in 1895, in February, and comprise one each for Stations V., VIII., and X., all being bottom tow-net gatherings.

Schistomysis spiritus. This species is recorded for Station IX. in November 1889; for Station I. in October 1890; and in the same year for Station V. in November and December. In 1891 it is recorded for Station I. in March, for Station IV. in January, and for Station V. in November. There are no records for 1892. In 1893 there is a record for Station III. in January, and one in February 1894 for the same station. There is no record of *S. spiritus* for 1895.

Schistomysis ornata. This, which is one of the most richly coloured of the British Mysidæ, is also at times moderately frequent in the Firth of Forth, and there are several records of its occurrence at most of the selected stations. The first record is for Station I. in August 1889. In 1890 there are records of it for Station I. in October, for Station V. in February, November, and December, for Station VIII. in February, April, and December, and for Station IX. in February and October. In 1891 it is recorded for Station I. in March, and for Station III. in September. It is also recorded for Station IV. in March and November. In 1892 it is recorded in January and March for the two Stations I. and V. There are no records for 1893. In 1894 it is recorded for Station III. in February. In February 1895 it is again recorded for Station III., and for Station IV. in January. All the records are only for bottom tow-nets.

Neomysis vulgaris. The only record for this species is one for Station IV. in December 1892, from the surface tow-net.

Table III. shows the distribution of the whole of the species of Mysidæ referred to in the preceding notes for the years and months and also for the stations; while Table IV. shows, in the same way, the distribution of the whole of the Schizopoda.

TABLE III., showing the Distribution of all the species of Mysidæ for the Years, Months, and Stations.

Years.	Jan.	Feb.	March.	April.	May.	June.	July.	August.	Sept.	Oct.	Nov.	Dec.	Totals for the Years.	Years.	I.	III.	IV.	V.	VIII.	IX.	Totals for the Years.
1889,	2	1	–	–	–	–	–	1	–	–	3	–	7	1889,	2	2	–	–	2	1	7
1890,	–	7	1	3	–	–	–	3	1	10	9	8	42	1890,	7	3	5	12	8	7	42
1891,	4	–	9	–	–	1	1	–	3	4	3	–	25	1891,	7	8	5	3	2	–	25
1892,	8	3	12	–	2	–	1	–	2	–	1	6	35	1892,	10	3	5	9	4	4	35
1893,	4	1	2	–	–	–	–	–	–	2	2	3	14	1893,	2	4	2	3	1	2	14
1894,	–	7	–	–	–	–	–	2	–	–	–	–	9	1894,	1	6	–	2	–	–	9
1895,	2	13	–	1	1	–	–	–	2	4	–	4	27	1895,	1	7	8	2	7	2	27
Totals for the Months,	20	32	24	4	3	1	2	6	8	20	18	21	159	Totals for the Stations,	30	33	25	31	24	16	159

TABLE IV., showing the Distribution of the Schizopoda, as indicated by the Tow-net Records for the Years, Months, and Stations.

Three Inner Stations.	Years.	Jan.	Feb.	March.	April.	May.	June.	July.	Aug.	Sept.	Oct.	Nov.	Dec.	Number of Records for years and stations.
	1889, .	1	1	–	–	–	–	–	1	–	–	1	–	4
	1890, .	–	1	1	–	–	–	–	–	–	1	–	1	4
	1891, .	1	1	1	1	–	–	–	–	–	1	–	–	5
I.	1892, .	1	1	1	–	–	–	–	–	1	–	1	1	6
	1893, .	2	1	–	1	–	–	–	–	2	1	1	–	8
	1894, .	1	1	–	–	–	–	–	1	–	–	–	1	4
	1895, .	–	–	–	1	–	–	–	–	–	–	–	1	2
Number of Records for all the Months,		6	6	3	3	–	–	–	2	3	3	3	4	33
	1889, .	1	–	–	–	–	–	–	–	–	–	1	–	2
	1890, .	–	1	1	1	–	–	1	–	–	1	–	1	6
	1891, .	1	1	–	2	–	2	–	–	1	1	1	1	10
III.	1892, .	–	1	1	2	–	–	–	–	–	–	–	1	5
	1893, .	1	2	–	–	–	–	–	–	–	1	2	2	8
	1894, .	–	1	–	–	–	–	–	1	–	–	–	3	5
	1895, .	–	1	–	–	–	–	–	–	–	1	–	1	3
Number of Records for all the Months,		3	7	2	5	–	2	1	1	1	4	4	9	39
	1889, .	1	–	–	–	–	–	–	–	–	–	1	–	2
	1890, .	–	1	2	–	–	–	1	–	–	1	–	2	7
	1891, .	1	1	1	1	–	–	–	–	–	1	1	–	6
IV.	1892, .	–	1	1	–	–	–	–	–	–	–	–	1	3
	1893, .	2	–	–	–	–	–	–	–	–	1	1	–	4
	1894, .	–	1	–	–	–	–	–	–	–	–	–	–	1
	1895, .	1	1	–	–	–	–	–	–	–	2	–	2	6
Number of Records for all the Months,		5	5	4	1	–	–	1	–	–	5	3	5	29

TABLE IV.—*continued.*

Three Outer Stations.	Years.	Number of Records for each Month of each Year.												Number of Records for years and stations.
		Jan.	Feb.	March.	April.	May.	June.	July.	August.	Sept.	Oct.	Nov.	Dec.	
	1889, .	–	1	–	–	–	1	–	1	–	1	1	–	5
	1890, .	–	1	–	–	–	–	–	–	1	1	1	1	5
	1891, .	–	–	1	1	1	–	–	–	–	–	1	–	4
V.	1892, .	1	1	1	1	–	–	–	–	1	1	1		8
	1893, .	2	–	1	–	–	–	–	–	–	1	1	1	6
	1894, .	–	1	–	1	–	–	–	.1	–	–	–	–	3
	1895, .	–	1	–	–	–	–	–	–	–	–	–	–	1
Number of Records for all the Months,		3	5	3	3	1	1	–	2	2	4	5	3	32
	1889, .	–	–	–	–	–	–	–	–	–	–	1	–	1
	1890, .	–	1	1	1	–	–	–	–	–	–	–	1	4
	1891, .	–	–	–	1	1	2	1	–	1	–	1	–	7
VIII.	1892, .	–	1	–	1	1	–	–	–	1	1	–	3	8
	1893, .	1	–	–	–	–	–	–	–	–	–	–	2	3
	1894, .	1	1	–	–	–	–	–	–	–	–	–	–	2
	1895, .	–	1	–	–	–	–	–	–	1	–	–	1	3
Number of Records for all the Months,		2	4	1	3	2	2	1	–	3	1	2	7	28
	1889, .	1	–	–	–	–	1	–	–	–	–	1	–	3
	1890, .	–	1	1	1	–	–	–	1	–	1	–	–	5
	1891, .	–	–	–	–	–	–	–	–	–	–	–	–	–
IX.	1892, .	1	1	–	1	2	–	1	–	–	–	–	2	8
	1893, .	1	–	1	–	–	–	–	–	–	1	–	1	4
	1894, .	–	1	–	–	–	–	–	–	–	–	–	–	1
	1895, .	–	1	–	1	1	–	–	–	1	–	–	1	5
Number of Records for all the Months,		3	4	2	3	3	1	1	1	1	2	1	4	26

It will be observed from the various Tables, and also from what has been already stated concerning the distribution of these crustaceans in the Firth of Forth district, that, generally speaking, they are much more numerous in the winter and spring months than in summer; and not only is it the case that a greater variety of species appear to be present in the estuary during those colder months, but it is evident that some of them at least are also individually much more numerous. The recorded results of the *Garland* researches, even during that part of the year which seems to be most favourable to the presence of these organisms, shows that their movements are more or less erratic. On February 14, 1890, Euphausiidæ are recorded as "common" in a bottom tow-net gathering at Station V., but at Station VII., a little to the south of Station V., trawled the previous day, they are described as "very rare," while, on the other hand, the Mysidæ weré more common, and in much greater variety, at this station. Moreover, at Station I., which is in line with Station V., but a little farther west, and also at Stations II., III., and IV., which were all tow-netted within a few days of each other, Euphausiidæ are either altogether absent from the tow-net lists or they are described as "rare." These Schizopods were again present in the Forth in considerable numbers in the following March; they were common at Stations I., III., IV., VIII., and IX., but few or rare at Stations II., V., VI., and VII. During the following summer Euphausiidæ were entirely absent from, or were very scarce in, the tow-net gatherings which were then collected; but the records for the ensuing month of October describe these organisms as frequent at some of the stations during that month, and they continue to be represented in the tow-net gatherings in varying numbers, from one or other of the stations, all through the winter and on to the spring of 1891. During the summer and winter months of that year comparatively few records of the Euphausiidæ occur in the lists of tow-net gatherings, and not till the beginning of 1892 is there much increase in the number of the records; but, though in February and March there is a decided increase, the captures of them even then are usually described as "few" or "rare." With one or two exceptions, Schizopods continued to be scarce in the tow-net gatherings till the winter of 1893-94, when they appear to have become more than usually numerous, and, in a considerable proportion of the gatherings collected during December and January of that winter, they are referred to as "common" or "frequent." With the advent of summer, however, and from that time on into 1895, the numbers of these crustaceans are again considerably reduced. From a study of all the facts at our disposal in the published records in the Fishery Board's Annual Reports for 1889 to 1896, there seems to be a fair amount of evidence tending to show the existence of a more or less regular yearly increase and decrease in the numbers of Schizopoda present in the estuary of the Forth, but the reason of this increase and decrease is not sufficiently clear to permit at present of a satisfactory explanation.

Besides this apparent yearly variation referred to, there also appears to be a fluctuation, which may or may not be regular, that extends over a longer time. I refer to the presence of Schizopoda in considerable numbers in the winter of 1889-90, and again in the winter of 1893-94; but to ascertain whether this variation is merely accidental or not would have required the investigations to be carried on over a much longer period than they have been. In considering this question of the seasonal variations of the Schizopoda, I have dealt mainly with the Euphausiidæ, because this family is the one most numerously represented in the Firth of Forth, and the one which has shown the greatest tendency to a regular periodic increase and decrease. The Mysidæ, also, to some extent show a similar tendency to variation, but the variation is scarcely so well marked as in the case of Euphausiidæ.

(c) CUMACEA.

The species of the Cumacea which have been identified are :—

Iphinoë trispinosa (Goodsir).   *Diastylus rugosa*, G. O. Sars.
Cumopsis goodsiri (V. Beneden).   *Pseudocuma cercaria* (V. Beneden).
Leucon nasicus, Kroyer.   *Campylaspis rubicunda*, Lilljeborg.
Eudorella truncatula (Spence Bate).   *Petalosarsia declivis*, G. O. Sars.

Others are recorded simply under the family or generic names. As these crustaceans usually live on or close to the bottom of the water, the records of them in the tow-net lists are, as might be expected, comparatively few, and may be briefly stated as under :—

Iphinoë trispinosa is only recorded for Station I. in July 1889, and for the same station in February and July in 1890.

Cumopsis goodsiri. The only record for the species is for Station III. in September 1891.

Leucon nasicus. In 1890 this species is recorded for Station I. in April, for Station V. in March, and for Station VIII. in December; in 1892, for Station V. in April, and for Station VIII. in May 1895.

Eudorella truncatula. The records for this species are four in number, one for Station V. in September and one for Station VIII. in December 1890; one for Station I. in February 1892, and one for the same station in June 1893.

Diastylus rugosa is recorded under its full name only for Station I. in April 1890, and for Station V. in March 1891 ; but there are several records of *Diastylus* sp. included in the additional references to the group under the more general name of Cumacea, etc.

Pseudocuma cercaria. This is a moderately common Cumacean, though, for the reason stated about the habits of these creatures, its name does not occur very frequently in the lists of tow-net fauna. The records are as follows :—In 1890 there are four in April—viz., one for Stations I., V., VIII., and IX., and it is recorded again for Station IX. in August and October. In 1891 there is a record for Station IX. in February, and for the same station again in September, and there is also one for Station VIII. in September. In 1892 there is a record for Station I. in January, and again for the same station in September. There is only one record for 1893—viz., for Station VIII. in April. In 1894 there are records of *Pseudocuma* for Station V. in August, and for Station IX. in February and April; and in 1895 for Station I. in April and October, for Station III. in October, for Station V. in April and August, and in the same two months for Station IX.

Campylaspis rubicunda. This species, which at that time had not been recorded for Britain, was observed at Station IX. in April 1893, and at Station V. in 1894.

Petalosarsia declivis, another rare Cumacean, was observed at the same time and place as *Campylaspis.* This concludes the records of the named species, but there are several other records of Cumaceans under the more general names of Cumacea, *Cuma*, etc., which I shall now refer to.

The records of Cumaceans under these general names are as follows :— In 1890 the first is for Station V. in March ; in April there are records for Stations III., VIII., and IX., and for Station VIII. in December. In 1891 the records are—one for Station VIII. in February, for Station I. in March, for Station VIII. in April, and for Station III. in November. In 1892 there is a record for Station VIII. in March, for Station IX. in May, for Stations V. and IX. in July, for Station VIII. in September, and for Station IX. in December. In 1893 there are four records of Cumacea—viz., two in June for Stations I. and V., one in August for Station V., and one in October for Station IX. The only records in 1894 are—one for Station V. in February, and another for Station IX. in the same month. There are five records in 1895, one for Station IV. in February, one for Stations III., VIII., and IX. in February, and one for Station VIII. in December.

There does not appear to be any feature of special interest in the distribution of the Cumacea as brought out by the published records, except that a few of the species are rare or new in the British seas.

(d) Isopoda.

The next group of organisms that fall to be noticed is the Isopoda. This group is represented in the tow-net lists by six species, the names of which are as follow :—

Gnathia (?) *maxillaris* (M. Edw.).   *Idothea emarginata* (Fabr.).
Eurydice pulchra, Leach.   „   *linearis* (Linn.).
Idothea baltica (Pallas).   *Janira maculosa*, Leach.

With the exception of *Idothea baltica*, these Isopods were all extremely rare in the tow-net gatherings under consideration, and the records for them are as under :—

Gnathia (?) *maxillaris* is recorded once in 1890, for Station VIII. in March.

Eurydice pulchra is recorded in the same year for Station III. in November.

Idothea baltica is recorded in 1889 for Station VIII. in February, and for Station IX. in January. In 1890 the same species is recorded in April for Stations IV. and V., and again for Station V. in October. In 1891 it is recorded for Station I. in January and April, and for Station IV. in March. It is recorded again for Station IV. in May 1892. In 1893 it is recorded in February for Station III., and in November for Stations I. and IV. In 1894 there is one record—viz., for Station IV. in April. In 1895 this Isopod is recorded in April for Stations I., III., IV., V., and VIII., and for Station V. in September. It will be seen that nine out of the nineteen records of *Idothea baltica*, or fully 47 per cent., are for the month of April; and it has also further to be noted that, curiously enough, the whole number of the records, with only a single exception, are from surface tow-net gatherings.

Idothea linearis. There is only one record for this species—viz., for Station IV. in January 1893.

Idothea emarginata. There is also only one record for this species —for Station I. in May 1891.

·· *Janira maculosa* is also only recorded once, for Station IX. in April 1890.

TABLE V., showing the Distribution of the Isopoda :—

Stations.	Years.	Jan.	Feb.	March.	April.	May.	June.	July.	August.	Sept.	Oct.	Nov.	Dec.	Total for Years.	Total for Stations	
I.	1891	1	–	–	1	1	–	–	–	–	–	–	–	3	5	
	1893	–	–	–	–	–	–	–	–	–	–	1	–	1		
	1895	–	–	–	1	–	–	–	–	–	–	–	–	1		
III.	1890	–	–	–	–	–	–	–	–	–	–	1	–	1	3	Three Inner Stations.
	1893	–	1	–	–	–	–	–	–	–	–	–	–	1		
	1895	–	–	–	1	–	–	–	–	–	–	–	–	1		
IV.	1890	–	–	–	1	–	–	–	–	–	–	–	–	1	7	
	1891	–	–	1	–	–	–	–	–	–	–	–	–	1		
	1892	–	–	–	–	1	–	–	–	–	–	–	–	1		
	1893	1	–	–	–	–	–	–	–	–	–	1	–	2		
	1894	–	–	–	1	–	–	–	–	–	–	–	–	1		
	1895	–	.–	–	1	–	–	–	–	–	–	–	–	1		
V.	1890	–	–	–	1	–	–	–	–	–	1	–	–	2	4	
	1895	–	–	–	1	–	–	–	–	1	–	–	–	2		
VIII.	1889	–	1	–	–	–	–	–	–	–	–	–	–	1	3	Three Outer Stations.
	1890	–	–	1	–	–	–	–	–	–	–	–	–	1		
	1895	–	–	–	1	–	–	–	–	–	–	–	–	1		
IX.	1889	1	–	–	–	–	–	–	–	–	–	–	–	1	2	
	1890	–	–	–	1	–	–	–	–	–	–	–	–	1		
Total for Months,		3	2	2	10	2	–	–	–	1	1	3	–	24	24	

(e) Amphipoda.

We shall now proceed to consider the Amphipoda recorded in the lists of tow-net fauna, and at the very outset we meet with a marked differentiation in regard to the distribution and numbers of certain of the species. Twenty-two species of the Amphipods mentioned in the lists of tow-net gatherings are more or less fully defined by name. The distribution, as regards the numbers and the frequency of three of the species, is so different from the others as to indicate a more or less decided difference in their habits and mode of life. The three species I refer to are *Hyperia galba*, *Hyperoche tauriformis* (Spence Bate), and *Parathemisto oblivia*. Moreover, while these three differ more or less distinctly from the other Amphipods mentioned in the tow-net lists, they are also in their own habits somewhat dissimilar to each other. *Parathemisto*, for example (and perhaps also *Hyperoche*), is to a large extent a free-swimming species; on the other hand, the *Hyperiæ* are also characterised as messmates of large Medusæ, and live a considerable part of their lives under the shelter afforded them by these curious organisms. Hence the distribution of *Hyperia* in the Firth of Forth depends to some extent on the presence or absence of Medusæ, whereas *Parathemisto* being a "free swimmer," its movements are not so limited as those of the other.

The following is a list of the Amphipods referred to by name in the lists of tow-net fauna :—

Hyperia galba (Montagu).
Hyperoche tauriformis (Bate).
Parathemisto oblivia (Kroyer).
Euthemisto (?) *compressa* (Goës).
Callisoma crenata (Bate).
Hippomedon denticulatus (Bate).
Bathyporeia sp.
Ampelisca sp.
Stenothoë marina (Bate).
Argissa hamatipes (Norman).
Metopa alderi, Bate.
Perioculodes longimanus (Bate).

Iphimedia obesa, Rathke.
Apherusa bispinosa (Bate).
   ,,   *borealis* (Boeck).
Paratylus swammerdami (M. Edw.).
Dexamine sp.
Melphidippella macera (Norman).
Amathilla homari (Fabr.).
Gammarus sp.
Gammaropsis erythrophthalma, Lillj.
Pariambus typicus (Kr.).

As it is not my intention to describe in detail the distribution of all the Amphipods named in the list, I shall confine my remarks chiefly to the Hyperiidæ, because of their great importance both numerically and as fish food.

Hyperia galba. This Amphipod is occasionally present in the Firth of Forth, but usually in limited numbers, and the records of its occurrence seem to indicate that it is as frequently captured by the bottom tow-net as it is by the surface net. In 1889 the records of it for the selected stations are very few. It is recorded in the bottom tow-net lists for July, and again in November. The November record describes it as common. It is recorded for Station V. in August and November, and described as frequent in the bottom tow-net gatherings and few in the November surface-net gatherings; and it occurred sparingly in the bottom tow-net gatherings from Stations VIII. and IX. in August. But the infrequency of the records of this and other species in 1889 is no doubt partly due to there being no lists of tow-net fauna for March, April, and May (with one exception—viz., for Station V. for surface net in May), and for nearly all the months of October and December. In 1890, though there are frequent records of the occurrence of Amphipods under the name of *Hyperia* sp., *H. galba* is only two or three times noted, once in February for Station I., in July for Station III., and in December for Station VIII. The species referred to under the name of *Hyperia* sp. is probably *Hyperoche tauriformis*, recorded in 1891 *et seq. ann.* In 1891 no direct reference to *Hyperia galba* occurs among the tow-net faunal lists. There are five records in 1892, but one or two of them are doubtful. In 1893 there are two records for Station III., one in August and one in November, that appear to refer to *Hyperia galba*. There are no records of it in 1894, but there are eight in 1895, one for Station I. in October and December, two (one surface and one bottom) for Station III. in October, one for Station IV. in August and October, one for Station V. in September, and one for Station VIII. in October.

Hyperoche tauriformis. This species, which is the *Hyperia tauriformis* of Spence Bate, was recognised as a member of the Forth fauna in 1891, and has been observed more or less frequently every year since. In previous tow-net tests it was recorded simply as *Hyperia* sp., and the records of Amphipoda under that name in Part III. of the Annual Reports probably refer chiefly to *Hyperoche*, though they may also occasionally include immature forms of *Hyperia galba* which could not at the time be otherwise satisfactorily disposed of. In the Report for 1891 there are about fourteen to sixteen records of this Amphipod, chiefly from Stations V., VIII., and IX. In 1892 there are eighteen records. They extend over nearly all the stations, and the majority of them are for the autumn and winter months. The following are the records for 1892 :— For Station I. in November and December; for Station III. in July, October, and November; for Station IV. in October, November, and December; for Station V. in February and March, and again in October and November; for Station VIII. in July, October, and November; and for Station IX. in October and December. In 1893 the number of records for *Hyperoche tauriformis* is fourteen. It was recorded for Station I. in January, June, August (surface and bottom tow-nets), and September (surface and bottom tow-nets); for Station III. in June (surface and bottom tow-nets); for Station V. in August and September; for Station VIII. in July and September; and for Station IX. in January and August. There are ten records in 1894—viz., for Station I. in August and November; for Station III. in June, August (surface and bottom tow-nets), and in November (surface and bottom tow-nets); for Station IV. in February, and for Station V. in June and August. In 1895 there are eighteen records, divided as follows :—One for Station I. in October and December (for surface and bottom tow-nets); for Station III. in February, August, October (for surface and bottom tow-nets), and December (also for surface and bottom tow-nets); for Station IV. in October (for surface and bottom tow-nets) and December (also for surface and bottom tow-nets); for Station V. in February; for Station VIII. in February, October, and December; and for Station IX. in December.

Parathemisto oblivia. This Amphipod was more common and more generally distributed in the estuary than any of the other species, and it is one of the species that enter largely into the food of certain fishes, as the herring. At times the stomachs of herring contain large numbers of this crustacean.

In 1889—the first of the series of seven years—the records of *Parathemisto* are few, there being only seven for the whole of the year; but there are, as already explained, several of the months of 1889 for which there are no lists of tow-net fauna, and this no doubt accounts to a considerable extent for the records for that year being so few in number. Among the records for 1889 there is one for Station I. in January and another for Station IV.; for Station V. the records are, one in January and two in February, and one also for Station VIII. in January, and for Station IX. in January and November.

In 1890 there are comparatively few records. *Parathemisto* was scarce in the surface and bottom tow-nets. For Station III. in April (in the surface tow-net), and at Station I. in May, it was rare ; in the surface net for Station IV. in December; at Station V. it was obtained in April, November, and December. The December record includes both the surface and bottom tow-net gatherings. There are also the following records for Station VIII.—viz., two in April and December (surface and bottom tow-nets), and one in May and in August (bottom tow-net only). There are two records for Station IX., one in April and one in August.

In 1891 the records of *Parathemisto* are much more frequent than in the two preceding years. There are twelve records for Station I., one in January, two in February, March, and April, one in May, June, and October, and two in November. There are also twelve records for Station III., one in January, February, and April, two in May, one in June, August, and October, and two in November and December. There are ten records for Station IV., two in January, February, and March, one in April, two in May, and one in June. There are ten records for Station V., two in February and March, one in April and May, and two in November and December. The records for Station VIII. are two in February, one in June, and two in November ; while those for Station IX. are two in February, and one in April, May, and November—the total number of records for 1891 being fifty-four.

The records of *Parathemisto* for 1892 include the following, viz. :—Two for Station I. (surface and bottom tow-nets) in January, February, and March, and one in April for the bottom tow-net, and in November and December for the surface tow-net. The records for Station III. are two in January, February, and March, and one in October, November, and December. For Station IV. there are two records for January, February, and March, and one for April, November, and December. For Station V. there is one in January and February, two in March, April, and October, and one in November and December. The records for Station VIII. are two in January, February, and March, one in May, June, July, and October, and two in December ; and for Station IX. there are two records in July, February, and March, one in April, May, July, and October, and two in December. Altogether there are sixty records of *Parathemisto* for 1892, and over fifty of them occur in the first three and last three months of the year. It may also be noted that the records for Stations I., III., and IV. describe the species as common in the surface tow-nets in January or February, while the records for Stations VIII. and IX. describe it as common in the bottom tow-nets in January.

In 1893 the records of *Parathemisto* are much fewer in number than for the previous year. Those for Station I. are two in January and one in April, October, and November. For Station III. there is one in January, February, and October, and two in December. The December records describe *Parathemisto* as frequent in the surface and common in the bottom tow-net. For Station IV. there is one record in January, February, April, and May, two in October, and one in November. For Station V. there is one in February and June, and two in October and December. For Station VIII. there is one in January, April, May, and October, and two in December ; while for Station IX. there is one in

April, two in October, and two in December. The October records for Station IX. describe the species as common in the surface tow-net and frequent in the bottom net, and in December as frequent in both nets There are in all thirty-four records of *Parathemisto* for 1893, twenty seven of them being for the first three or last three months.

In 1894 the number of records are as follow :—For Station I. there is one in February, two in April and August, and one in November. For Station III. there is one in February and April, two in August, one in November, and two in December. For Station IV. there is only one record—viz., for surface tow-net in February, in which *Parathemisto* is described as common. There are only two records for Station V.—viz., for surface and bottom tow-nets in February. There are three records for Station VIII., one in January and February, and one in April. The only record for Station IX. is one in February. The number of records of *Parathemisto* for 1894 is twenty, twelve of which are for the first two and last two months, the other eight being divided equally between April and August.

In 1895 the records for Station I. are one in February and two in December, and in the surface tow-net record for December *Parathemisto* is described as common. For Station III. there is one in February and April, and two in December. For Station IV. there is one record in February and one in December. There is one record in April for Station V. For Station VIII. there is one in February and April, and two in December; while for Station IX. the records are two in February and one in December—or seventeen records altogether for 1895, thirteen, or fully 70 per cent., of them being for the two months February and December.

The following Table shows the distribution of the records of *Parathemisto* as described in the preceding notes :—

TABLE VI., showing the Distribution of *Parathemisto*, as indicated by the Tow-net Records for the Years, Months, and Stations.

Three Inner Stations.	Years.	Jan.	Feb.	March.	April.	May.	June.	July.	Aug.	Sept.	Oct.	Nov.	Dec.	Total for the Years and Stations.
	1889, .	1	–	–	–	–	–	–	–	–	–	–	–	1
	1890, .	–	–	–	–	1	–	–	–	–	–	–	–	1
	1891, .	1	2	2	2	1	1	–	–		1	2	–	12
I.	1892, .	2	2	2	1	–	–	–	–	–	–	1	1	9
	1893, .	2	–	–	1	–	–	–	–	–	1	1	–	5
	1894, .	–	1	–	2	–	–	–	2	–	–	1	–	6
	1895, .	–	1	–	–	–	–	–	–	–	–	–	2	3
Totals for the Months, ...		6	6	4	6	2	1	–	2	–	2	5	3	37

Table VI.—*continued*.

Inner Stations— continued.	Years.	Jan.	Feb.	March.	April.	May.	June.	July.	Aug.	Sept.	Oct.	Nov.	Dec.	Total for the Years and Stations.
	1889,	-	-	-	-	-	-	-	-	-	-	-	-	-
	1890,	-	-	-	2	-	-	-	-	-	-	-	-	2
	1891,	1	1	-	1	2	1	-	1	-	1	2	2	12
III.	1892,	2	2	2	-	-	-	-	-	-	1	1	1	9
	1893,	1	1	-	-	-	-	-	-	-	1	-	2	5
	1894,	-	1	-	1	-	-	-	2	-	-	1	2	7
	1895,	-	1	-	1	-	-	-	-	-	-	-	2	4
Totals for the Months,		4	6	2	5	2	1	-	3	-	3	4	9	39
	1889,	1	-	-	-	-	-	-	-	-	-	-	-	1
	1890,	-	-	-	-	-	-	-	-	-	-	-	1	1
	1891,	2	2	2	1	2	1	-	-	-	-	-	-	10
IV.	1892,	2	2	2	1	-	-	-	-	-	-	1	1	9
	1893,	1	1	-	1	1	-	-	-	-	2	1	-	7
	1894,	-	1	-	-	-	-	-	-	-	-	-	-	1
	1895,	-	-	-	1	-	-	-	-	-	-	-	1	2
Totals for the Months,		6	6	4	4	3	1	-	-	-	2	2	3	31

Three Outer Stations.	Years.	Jan.	Feb.	March.	April.	May.	June.	July.	Aug.	Sept.	Oct.	Nov.	Dec.	Total for the Years and Stations.
	1889,	1	2	-	-	-	-	-	-	-	-	-	-	3
	1890,	-	-	-	1	-	-	-	-	-	-	1	2	4
	1891,	-	2	2	1	1	-	-	-	-	-	2	2	10
V.	1892,	1	1	2	2	-	-	-	-	-	2	1	1	10
	1893,	-	1	-	-	-	1	-	-	-	2	-	2	6
	1894,	-	2	-	-	-	-	-	-	-	-	-	-	2
	1895,	-	-	-	1	-	-	-	-	-	-	-	-	1
Totals for the Months,		2	8	4	5	1	1	-	-	-	4	4	7	36

TABLE VI.—*continued.*

Outer Stations-continued.	Years.	Jan.	Feb.	March.	April.	May.	June.	July.	Aug.	Sept.	Oct.	Nov.	Dec.	Total for the Years and Stations.
	1889,	1	-	-	-	-	-	-	-	-	-	-	-	1
	1890,	-	-	-	2	1	-	-	1	-	-	-	2	6
	1891,	-	2	-	-	-	1	-	-	-	-	2	-	5
VIII.	1892,	2	2	2	-	1	1	-	-	-	1	-	2	11
	1893,	1	-	-	1	1	-	-	-	-	1	-	2	6
	1894,	1	1	-	1	-	-	-	-	-	-	-	-	3
	1895,	-	1	-	1	-	-	-	-	-	-	-	2	4
Totals for the Months, ...		5	6	2	5	3	2	-	1	-	2	2	8	36
	1889,	1	-	-	-	-	-	-	-	-	-	1	-	2
	1890,	-	-	-	1	-	-	-	1	-	-	-	-	2
	1891,	-	2	-	1	1	-	-	-	-	-	1	-	5
IX.	1892,	2	2	2	1	1	-	1	-	-	·1	-	2	12
	1893,	-	-	-	1	-	-	-	-	-	2	-	2	5
	1894,	-	1	-	-	-	-	-	-	-	-	-	-	1
	1895,	-	1	-	-	-	-	-	-	-	-	-	1	2
Totals for the Months, . . .		3	6	2	4	2	-	1	1	-	3	2	5	29

The Table shows that the total number of records for Stations I., III., V., and VIII. is nearly the same ; and it is of interest to note that Stations IV. and IX., which are also nearly equal but show a smaller number of records than the others, might have been expected to have yielded very different results, as the first is an inshore station, while the other is the farthest seaward of all the Forth stations, being situated considerably east of the May Island. Another point of interest is, that besides the comparative absence of these Amphipods during the summer months there is evidence of a somewhat remarkable increase of those Crustaceans in the Forth in 1891 and 1892 compared with other years. With reference to the remarkable increase in the number of *Parathemisto* in the Forth during these two years, it is specially worthy of note that in February 1892 an immense shoal of *Euthemisto compressa*—a species closely allied to *Parathemisto* in structure and habits—was observed off the Yorkshire coast. So great in number were they that the sea was described as "literally alive with them," and it is stated that "heaps" of them "were afterwards washed ashore by sea-winds, and afforded a feast for starlings and other frequenters of the tidal line." [*] Some of these *Euthemisto* found their way into the Forth estuary, and are mentioned among the tow-net records for that year. They were first observed in the Forth in February,

[*] T. H. Nelson in 'Naturalist' for May 1892.

M

and they either remained in the estuary all the summer or returned in the autumn, for they were observed again in November.

Whether this irruption of the Hyperiidæ was merely accidental and temporary, or due to a periodic migration at more or less distant intervals of time, the records are not sufficiently extensive to show. Whatever be the reason, there seems to be no doubt that some time during 1891 and 1892 there was a greater number than usual of these Crustaceans off our East Coast, part of which found their way inshore, with the result that the number of records of *Parathemisto* was increased beyond those of the previous or following years.

In the following analysis of the Table, some of the results referred to are more clearly brought out—the three formulæ subjoined show the total numbers of the records (1) for the stations, (2) the months, and (3) the years, thus :—

Formula (1) shows the total number of the records of *Parathemisto* for each of the stations for all the months and years :—

Numbers of the Stations,	I.	III.	IV.	V.	VIII.	IX.
Total number of Records for each Station,	37	39	31	36	36	29

Formula (2) shows the total number of the records of *Parathemisto* for each of the months for all the years, (*a*) for Stations I., III., and IV.; (*b*) for Stations V., VIII., and IX.; and (*c*) for all the six stations : -

Names of the Months,	Jan.	Feb.	Mar.	April.	May.	June.	July.	Aug.	Sept.	Oct.	Nov.	Dec.
Total number of Records for each Month, for (*a*) the Inner Stations,	16	18	10	15	7	3	0	5	0	7	11	15
(*b*) the Outer Stations,	10	20	8	14	6	3	1	2	0	9	8	20
(*c*) all the Six Stations,	26	38	18	29	13	6	1	7	0	16	19	35

Formula (3) shows the total number of the records of *Parathemisto* for each year, for (*a*) the inner stations, I., III., and IV.; (*b*) the outer stations, V., VIII., and IX.; and (*c*) for all the six stations :—

Numbers of Years,	1889.	1890.	1891.	1892.	1893.	1894.	1895.
Total number of Records for each Year, for (*a*) the Inner Stations,	2	4	34	27	17	14	9
(*b*) the Outer Stations,	6	12	20	33	17	6	7
(*c*) all the Six Stations,	8	16	54	60	34	20	16

Euthemisto compressa. This species (referred to above) was recorded for the first time in the Firth of Forth in the "Tenth Annual Report of the Fishery Board for Scotland," from specimens obtained at Station V. in February 1892. It was collected again in the same year in November; and the next record of it is in November 1893, when it was obtained at Station V.

The other Amphipods mentioned in the list are, with the exception of *Paratylus* and *Apherusa*, only occasionally referred to among the records of tow-net invertebrates, and are therefore, except from a natural history point of view, of less importance than those already mentioned. Consequently I do not propose to enter very minutely into the consideration of their distribution.*

Callisoma crenata. There are few records of this species, and they extend from 1890 to 1895.

Hippomedon denticulatus is only recorded once among the tow-net lists for the selected stations—viz., for Station IX. in February 1894.

---

* A tabulated list of species at the end of this paper shows some further details of the distribution of these Amphipods.

Bathyporeia and *Ampelisca* have been recorded, but the species were apparently not identified.

Stenothoë marina has one or two records, but it appears to have been comparatively rare in the tow-net gatherings under consideration.

Argissa hamatipes is occasionally recorded in the lists of tow-net invertebrates collected at the three outer stations, but less frequent in those collected at the inner stations.

Metopa alderi appears to have been comparatively frequent at nearly all the stations.

Perioculodes longimanus is recorded chiefly from the outer stations.

Iphimedia obesa is only recorded once—in 1892 for Station V. in January.

Apherusa (two species) and *Paratylus* are referred to several times in the lists of tow-net invertebrates; neither of them appears to be rare, though less common than some of the others.

Melphidippella macera is only once recorded—viz., at Station VIII. in March 1892.

Amathilla homari is also recorded once—viz., for Station IV. in December 1892.

Gammaropsis erythrophthalmus is recorded once—for Station V. in December 1890.

Dulichia sp. and *Pariambus typicus* are both rarely mentioned among the lists of tow-net gatherings.

(*f*) CLADOCERA.

Evadne nordmanni is recorded from nearly all the six stations, and was occasionally moderately frequent.

Podon (?) *polyphemoides*, though not so common as *Evadne*, is recorded several times, and for at least four different stations.

(*g*) COPEPODA.

The Copepoda, though usually present in nearly all the tow-net gatherings, being sometimes common or even abundant, do not usually show much variety of species. The following is a list of the Copepods recorded in the various tow-net gatherings collected during the seven years at the six selected stations :—

Calanus finmarchicus (Gunner).
Pseudocalanus elongatus, Boeck.
Temora longicornis (Müller).
Centropages hamata, Lillj.
Centropages typicus, Kr.
Metridia hibernica (B. & R.).
Candace pectinata, Brady.
Acartia sp. (? *Clausii* and *longiremis*).

Anomalocera patersoni, Templ.
Parapontella brevicornis (Lub.).
Oithona (?) *similis*, Claus.
Thalestris longimanus, Claus.
Thalestris serrulatus, Brady.
Alteutha depressa (Baird).
Dermatomyzon nigripes (B. and R.).
Monstrilla sp.
Caligus rapax, Müller

A few of the species named above are found occurring with more or less frequency in nearly all the lists of tow-net invertebrates, but the larger number of them were infrequent or rare, and it is those only that are of numerical importance that need be specially referred to here. The first of them that I propose to notice is *Calanus finmarchicus*. *Calanus finmarchicus* is, because of its numbers and its general distribution, one of the most important of the group to which it belongs, at least in the British seas. Being usually so numerous, it no doubt furnishes a rich supply of food to the young of all kinds of fishes, and, as a matter of fact, it frequently forms the principal food of the herring. This species is, as a rule, usually more numerous than the other pelagic Copepods, and it occurs at times in considerable abundance in the estuary. Its distribution, as indicated by the tow-net records, is as follows :—In 1889 *Calanus* is recorded twice for Station I. in January ; once in July, August, October, and November for Station III. ; once in January, July, and August for Station IV. ; once in January, May, June, August, and September for Station V., while for the same station in November *Calanus* is recorded as abundant in the surface and frequent in the bottom tow-nets ; for Station VIII. it is recorded once in January, twice in June and July, once in August, and once in November ; for Station IX. there are two records in June, one in August, and two in November. In the surface tow-net record for November *Calanus* is described as abundant, and as common in the bottom tow-net.

In 1890 the records of *Calanus* for Station I. are two in February, March, April, May, July, October, and November, and one in December. It is described as abundant in one and common in four of these gatherings. The records for Station III. are two in February, March, April, May, July, October, and November, and one in August ; for Station IV. there are two records in March, May, November, and December, and one in February, April, July, August, and October. The records for Station V. are two in February, March, April, May, September, October, November, and December, and one in July. For Station VIII. there are two records in March, April, May, September, and December, and one in February, July, and October ; while for Station IX. there are two records in March, April, May, October, and December, and one in February, July, and September. The records for *Calanus* in 1891 are for Station I. two in February, March, April, May, June, and August, and one in January, July, October, and November. For Station III. there are two records in February, April, May, June, July, and December, and one in January, August, September, October, and November. For Station IV. there are two records in January, February, March, April, May, June, and October, and one in August and November. The records for Station V. are two in February, March, April, May, July, October, November, and December, and one in June and August. Station V. having been trawled twice in October adds a third record for that month. There are two records for Station VIII. in February, April, May, June, and November, and one in July, August, September, and October ; while for Station IX. there are two records in February, May, June, July, August, and November, and one in April, September, and October—making ninety-seven records in all for 1891. The number of the records of *Calanus* for 1892 is about equal to that of the previous year. Those for Station I. are two in April, May and December, and one in January, February, March, July, September, and November. For Station III. there are two records in February, March, April, May, November, and December, and one in January, July,

September, and October. The records for Station IV. are two in March, April, May, June, October, and December, and one in January, February, July, September, and November. For Station V. there are two records in January, April, May, October, November, and December, and one in March, July, and September. The records for Station VIII. are two in March, April, and September, and one in January, February, May, June, July, and November. In December this station was trawled twice, and furnished four records of *Calanus* for that month. The records for Station IX. are two in January, February, April, May, September, and October, and one in March, June, and July. There are also three records in December, the station having been trawled twice that month.

The records for 1893 are as follows :— For Station I. there are two in January, April, May, September, and November, and one in February, June, July, August, October, and December. For Station III. there are two in January and December, and one in February, May, June, July, August, September, October, and November. For Station IV. there are two in January, February, April, June, October, and November, and one in May and December. There are for Station V. two records in February, May, August, October, and December, and one in March, June, July, September, and November. There are two records for Station VIII. in January, April, July, and December, and one in February, May, August, September, and October, while for Station IX. there are two records in January, February, May, June, July, October, and December, and one in April and August.

The records of *Calanus* for 1894, which are fewer than for the three preceding years, include two for Station I. in April and December, and one in January and February. There are two for Station III. in August, and one in February, April, and June, while in December there are four records owing to the station having been twice trawled that month. For Station IV. there are two records in April and one in January and February. The records for Station V. are two in February and April and one in August. For Station VIII. there are two records in February and one in January, April, and August, and those for Station IX. are two in February and April. The records for this year number thirty-four.

The records of *Calanus* for 1895, which are also few in number, include two for Station I. in April and December, and one in February, May, and October. There are two for Station III. in February, April, October, and December, and one in May. There are two for Station IV. in February, October, and December, and one in January, April, and May. The records for Station V. are two in February, April, May, and August, two for Station VIII. in February, May, October, and December, and one in April and September, while for Station IX. there are two records in February, April, and December, and one in May, August, and September. The number of records for 1895 is fifty, while the total number for the seven years is four hundred and eighty-four.

The distribution of *Calanus finmarchicus*, as described in the preceding notes and shown by the annexed tabular arrangement of the number of records, is more uniform throughout the six stations than that of *Parathemisto*. It may be observed that there is a tendency towards an increase in the number of records during the earlier months, and also to a small extent during the later months of the year, but this tendency shows itself principally in the greater number of double records for these months : that is, records which include both the bottom and surface tow-net gatherings. This is shown more clearly by the Table.

TABLE VII., giving the Distribution of *Calanus finmarchicus*, as shown by the Tow-net Records for the Years, Months, and Stations.

Three Inner Stations.	Years.	Jan.	Feb.	March.	April.	May.	June.	July.	Aug.	Sept.	Oct.	Nov.	Dec.	Totals for Years and Stations.
	1889,	2	–	–	–	–	–	1	1	–	1	1	–	6
	1890,	–	2	2	2	2	–	2	–	–	2	2	1	15
	1891,	1	2	2	2	2	2	1	2	–	1	1	–	16
I.	1892,	1	1	1	2	2	–	1	–	1	–	1	2	12
	1893,	2	1	–	2	2	1	1	1	2	1	2	1	16
	1894,	1	1	–	2	–	–	–	–	–	–	–	2	6
	1895,	–	1	–	2	1	–	–	–	–	1	–	2	7
Total Records for the Months,		7	8	5	12	9	3	6	4	3	6	7	8	78
	1889,	1	–	–	–	–	1	1	1	–	–	1	–	5
	1890,	–	2	2	2	2	–	2	1	–	2	2	–	15
	1891,	1	2	–	2	2	2	2	1	1	1	1	2	17
III.	1892,	1	2	2	2	2	–	1	–	1	1	2	2	16
	1893,	2	1	–	–	1	1	1	1	1	1	1	2	12
	1894,	–	1	–	2	–	1	–	2	–	–	–	4	10
	1895,	–	2	–	2	1	–	–	–	–	2	–	2	9
Total Records for the Months,		5	10	4	10	8	5	7	6	3	7	7	12	84
	1380,	1	–	–	–	–	–	1	1	–	–	1	–	4
	1890,	–	1	2	1	2	–	1	1	–	1	2	2	13
	1891,	2	2	2	2	2	2	–	1	–	2	1	–	16
IV.	1892,	1	1	2	2	2	2	1	–	1	2	1	2	17
	1893,	2	2	–	2	1	2	–	–	–	2	2	1	14
	1894,	1	1	–	2	–	–	–	–	–	–	–	–	4
	1895,	1	2	–	1	1	–	–	1	–	2	–	2	10
Total Records for the Months,		8	9	6	10	8	6	3	4	1	9	7	7	78

[TABLE.

TABLE VII.—*continued.*

Three Outer Stations.	Years.	Number of Records for each Month of each Year.												Totals for Years and Stations.
		Jan.	Feb.	March.	April.	May.	June.	July.	Aug.	Sept.	Oct.	Nov.	Dec.	
	1889,	1	–	–	–	1	1	–	1	1	–	2	–	7
	1890,	–	2	2	2	2	–	1	–	2	2	2	2	17
	1891,	–	2	2	2	2	1	2	1	–	3	2	2	19
V.	1892,	2	–	1	2	2	–	1	–	1	2	2	2	15
	1893,	–	2	1	–	2	1	1	2	1	2	1	2	15
	1894,	–	2	–	2	–	–	–	1	–	–	–	–	5
	1895,	–	2	–	2	2	–	–	2	–	–	–	–	8
Totals for the Months, . . .		3	10	6	10	11	3	5	7	5	9	9	8	86
	1889,	1	–	–	–	–	2	2	1	–	–	1	–	7
	1890,	–	1	2	2	2	–	1	–	2	1	–	2	13
	1891,	–	2	–	2	2	2	1	1	1	1	2	–	14
VIII.	1892,	1	1	2	2	1	1	1	–	2	1	–	4	16
	1893,	2	1	–	2	1	–	2	1	1	1	–	2	13
	1894,	1	2	–	1	–	–	–	1	–	–	–	–	5
	1895,	–	2	–	1	2	–	–	–	1	2	–	2	10
Totals for the Months, . . .		5	9	4	10	8	5	7	4	7	6	3	10	78
	1889,	–	–	–	–	–	2	–	1	–	–	2	–	5
	1890,	–	1	2	2	2	–	1	–	1	2	–	2	13
	1891,	–	2	–	1	2	2	2	2	1	1	2	–	15
IX.	1892,	2	2	1	2	2	1	1	–	2	2	–	3	18
	1893,	2	2	–	1	2	2	2	1	–	2	–	2	16
	1894,	–	2	–	2	–	–	–	–	–	–	–	–	4
	1895,	–	2	–	2	1	–	–	1	1	–	–	2	9
Totals for the Months, . . .		4	11	3	10	9	7	6	5	5	7	4	9	80

This Table also shows that the difference in the total number of records for each of the stations for the seven years is comparatively small. The highest number (86) is for Station V., while the lowest numbers occur at Stations I., IV., and VIII.—each of these stations having a total of 78 records for the seven years. The number of records for each of the stations may be stated as a formula, thus :—

Formula (1) shows the total number of records of *Calanus* for each station for all the seven years :—

	Inner Stations.			Outer Stations.		
Numbers of the Stations, ...	I.	III.	IV.	V.	VIII.	IX.
Total number of Records for each Station, -	78	84	78	86	78	80

The total numbers of the monthly records exhibit a greater amount of divergence than those for the stations as shown by the next formula.

Formula (2) shows the total number of records of *Calanus* for each of the stations for the seven years,.(*a*) for the inner stations, I., III., and IV.; (*b*) for the outer stations, V., VIII., and IX.; and (*c*) for all the six stations :—

Names of the Months, -	Jan.	Feb.	Mar.	April.	May.	June.	July.	Aug.	Sept.	Oct.	Nov.	Dec.
Total number of Records for each Month, for {(*a*) the Inner Stations, -	20	27	15	32	25	14	16	14	7	22	21	27
(*b*) the Outer Stations, -	12	30	13	30	28	15	18	16	17	22	16	27
(*c*) all the Six Stations, -	32	57	28	62	53	29	34	30	24	44	37	54

Yet although there is a certain amount of variation, with the exception of one or two months—as March and September—this variation is comparatively trifling, showing that *Calanus* was not only common, but also, on an average, very equally distributed throughout the estuary. The total number of records of this Copepod for the three inner and three outer stations, for the twelve months and for all the seven years, is 240 and 244 respectively, or practically the same for the two groups of stations. Another point which may be noticed here, and which is shown in the Table, is that the totals of the monthly records for the seven years and for each of the six stations for April are, with one exception, exactly alike—the number for April for each station being ten, except at Station I., where the number is twelve. It would also appear from the tabular enumeration of the records that during six out of the seven years (excluding 1889) *Calanus* was more common and generally diffused throughout the district referred to here, in the month of April, than in any of the other months. The reason for this uniformity may not be easily explained. It may be due to a periodic migration of *Calanus*, or to some influence inducing them to crowd more in towards the shore at this season—probably for spawning.

Formula (3) shows the total number of records of *Calanus* for each of the seven years, (*a*) for the inner stations, I., III., and IV.; (*b*) for the outer stations, V., VIII., and IX.; and (*c*) for all the six stations :—

Numbers of the Years, -	1889.	1890.	1891.	1892.	1893.	1894.	1895.
Total number of Records for each Year, for {(*a*) the Inner Stations,	15	43	49	45	42	20	26
(*b*) the Outer Stations,	19	43	48	49	44	14	27
(*c*) all the Six Stations,	34	86	97	94	86	34	53

The greatest difference between the yearly totals for the inner and outer stations is that for 1894, while the total number for the seven years for each group of stations shows only a difference of four—the numbers being for the inner stations 240 and for the outer stations 244.

But though *Calanus finmarchicus* is the most common Copepod in the Firth of Forth, there are one or two other species included in the lists of tow-net gatherings which are more or less generally distributed, and at times moderately frequent, and which no doubt compete closely with *Calanus* as important sources of fish-food, especially as the food of young fishes. *Temora longicornis* is one of these ; so also, though in a somewhat less degree, are one or two species of *Acartia*—viz., *A. longiremis* and *A. clausii* (both of which are found in the Forth estuary). I do not propose to refer at length to the distribution of these Copepods, as the Tables which follow show by comparison with that of *Calanus* its more important details, as well as some other of the more noticeable differences in this respect between them and that species.

TABLE VIII., showing the Distribution of *Temora longicornis*, as indicated by the Tow-net Records for the Years, Months, and Stations

Three Inner Stations.	Years.	Number of Records for each Month of each Year.												Totals for Years and Stations.
		Jan.	Feb.	March.	April.	May.	June.	July.	Aug.	Sept.	Oct.	Nov.	Dec.	
	1889,	–	–	–	–	–	–	–	1	–	1	1	–	3
	1890,	–	–	1	2	2	–	1	–	–	2	2	–	10
	1891,	–	1	1	2	2	2	1	2	–	–	–	–	11
I.	1892,	–	–	–	2	2	2	1	–	1	–	–	–	8
	1893,	–	–	–	2	2	1	–	–	–	–	–	–	5
	1894,	–	–	–	1	–	–	–	–	–	–	–	–	1
	1895,	–	–	–	1	2	–	–	–	1	1	–	–	5
Totals for the Months,		–	1	2	10	10	5	3	3	2	4	3	–	43
	1889,	–	–	–	–	–	1	1	1	–	–	–	–	3
	1890,	–	–	–	2	1	–	1	–	–	1	2	–	7
	1891,	–	1	–	2	2	2	1	–	–	–	–	–	8
III.	1892,	–	–	–	2	2	2	–	–	1	1	–	–	8
	1893,	–	1	–	–	1	2	1	1	–	1	–	1	8
	1894,	–	–	–	2	–	1	–	–	–	–	–	1	4
	1895,	–	1	–	2	2	–	–	–	–	1	–	–	6
Totals for the Months,		–	3	–	10	8	8	4	2	1	4	2	2	44
	1889,	–	–	–	–	–	–	–	–	–	–	1	–	1
	1890,	–	–	–	1	2	–	–	–	–	–	2	1	6
	1891,	–	1	2	2	2	–	–	2	–	–	–	–	9
IV.	1892,	–	–	–	2	2	1	–	–	1	2	1	–	9
	1893,	–	–	–	–	2	1	–	–	–	–	–	1	4
	1894,	–	–	–	1	–	–	–	–	–	–	–	–	1
	1895,	1	1	–	1	2	–	–	1	–	2	–	–	8
Totals for the Months,		1	2	2	7	10	2	–	3	1	4	4	2	38

[TABLE.

TABLE VIII.—continued.

Three Outer Stations.	Years.	Number of Records for each Month of each Year.												Totals for Years and Stations.
		Jan.	Feb.	March.	April.	May.	June.	July.	Aug.	Sept.	Oct.	Nov.	Dec.	
V.	1889,	-	-	-	-	1	1	-	2	-	-	1	-	5
	1890,	-	2	2	2	2	-	1	-	2	2	2	-	15
	1891,	-	-	2	2	2	-	1	2	-	-	-	-	9
	1892,	-	-	-	1	2	2	-	-	1	2	-	-	8
	1893,	-	-	-	-	2	1	1	2	1	-	1	-	8
	1894,	-	-	-	-	-	-	-	1	-	-	-	-	1
	1895,	-	-	-	1	2	-	-	-	-	-	-	-	3
Totals for the Months,		-	2	4	6	11	4	3	7	4	4	4	-	49
VIII.	1889,	-	-	-	-	-	1	-	1	-	-	1	-	3
	1890,	-	-	2	2	2	-	1	-	2	1	-	1	11
	1891,	-	1	-	2	2	1	1	2	2	-	-	-	11
	1892,	-	-	-	2	2	2	-	-	1	1	-	-	8
	1893,	-	-	-	1	1	1	1	1	1	-	-	-	6
	1894,	-	-	-	-	-	-	-	1	-	-	-	-	1
	1895,	-	-	-	1	1	-	-	-	1	-	-	-	3
Totals for the Months,		-	1	2	8	8	5	3	5	7	2	1	1	43
IX.	1889,	-	-	-	-	-	-	2	2	-	-	1	-	5
	1890,	-	-	2	2	1	-	2	-	1	2	-	-	10
	1891,	-	1	-	2	2	1	-	2	2	1	-	-	11
	1892,	-	-	-	1	2	2	-	-	1	1	-	1	8
	1893,	-	-	-	-	1	1	-	1	-	-	-	-	3
	1894,	-	-	-	1	-	-	-	-	2	-	-	-	3
	1895,	-	-	-	1	1	-	-	-	1	-	-	-	3
Totals for the Months,		-	1	2	7	7	4	4	7	5	4	1	1	43

This Table contains an enumeration of the number of records of *Temora* for all the stations and for the seven years, arranged in the same way as those of *Calanus*. An examination of the Table shows that the distribution of *Temora* differs little from that of *Calanus*, except that the number of records is rather fewer. Like *Calanus*, the highest number of yearly records of *Temora* occur in 1891, 1892, and 1893 ; and also, like that species, there is with one exception comparatively little difference in the total number of records for each of the stations : the principal difference is found at Station V., where the number is greater, by five records, than the highest number next to it, at Station III., while it is eleven records greater than the minimum number at Station IV. It may also be noted that the station with the highest number of records of *Temora* and *Calanus* is the same—viz., Station V. But the Table shows a slight diversity in the distribution of *Temora* throughout the different months of the year as compared with that of *Calanus*. Thus, the maxima of monthly records are found in April and May, that for May being distinctly higher than that for April, and the minima in January, February, March, and December; while in regard to *Calanus* the maximum number is that for April, while the number for February is smaller than the maximum by only five records. These points are more clearly indicated in the annexed formulæ.

Formula (1) shows the total number of records of *Temora* for each station for all the seven years :—

	(a) Inner Stations.			(b) Outer Stations.		
Numbers of the Stations,	I.	III.	IV.	V.	VIII.	IX.
Total number of Records for each Station,	43	44	38	49	43	43

Formula (2) shows the total number of records of *Temora* for each of the twelve months, (a) for the inner stations, (b) for the outer stations, and (c) for all the six stations :—

Names of the Months,	Jan.	Feb.	Mar.	April	May	June	July	Aug.	Sept.	Oct.	Nov.	Dec.
Total number of Records for each Month, for (a) the Inner Stations,	1	6	4	27	28	15	7	8	4	12	9	4
(b) the Outer Stations,	0	4	8	21	26	13	10	19	16	10	6	2
(c) all the Six Stations,	1	10	12	48	54	28	17	27	20	22	15	6

In the distribution of *Calanus* the month that shows the smallest number of records is September, but in the case of *Temora* the smallest number is that for January, and the proportional difference between the lowest and highest number of records is considerably greater for *Temora* than for *Calanus*.

Formula (3) shows the total number of records of *Temora* for each of the seven years, (a) for the inner stations, I., III., and IV. ; (b) for the outer stations, V., VIII., and IX. ; and (c) for all the six stations :—

Numbers of the Years,	1889.	1890.	1891.	1892.	1893.	1894.	1895.
Total number of Records for each Year, for (a) the Inner Stations,	7	23	28	25	17	6	19
(b) the Outer Stations,	13	36	31	24	17	5	9
(c) all the Six Stations,	20	59	59	49	34	11	28

There seems to be a greater variation in the total numbers for the years in the distribution of *Temora* than is the case with *Calanus*. The above formula shows that in the case of *Temora* the highest total number of yearly records is fully five times greater than the lowest, but in *Calanus* the difference is less than three times ; and further, the yearly maxima for *Temora* occur in 1890 and 1891, and in 1891 and 1892 for *Calanus*. The total number of records for *Temora* for the six stations during the seven years is two hundred and sixty.

ACARTIA SP.

The next Table shows the distribution of *Acartia* sp.,* in a manner similar to that of *Calanus* and *Temora*.

---

* Careful examination is necessary in order to distinguish the species of *Acartia*, and for various reasons it was not always possible to carry out a minute and sufficiently careful examination of doubtful species on board ship ; therefore the specimens of *Acartia* observed were frequently recorded merely under their generic name.

TABLE IX., showing the Distribution of *Acartia* sp., as indicated by the Tow-net Records for the Years, Months, and Stations.

Three Inner Stations.	Years.	Jan.	Feb.	Mar.	April.	May.	June.	July.	Aug.	Sept.	Oct.	Nov.	Dec.	Totals for the Years and Stations.
	1889,	-	-	-	-	-	-	-	-	-	-	1	-	1
	1890,	-	~	-	-	-	-	1	1	-	-	1	-	3
	1891,	-	2	1	1	2	1	-	1	-	-	-	-	8
I.	1892,	-	1	-	1	-	-	1	-	2	-	-	-	5
	1893,	-	1	-	1	1	1	-	-	-	-	-	-	4
	1894,	-	-	-	-	-	-	-	1	-	-	-	1	2
	1895,	-	-	-	-	-	-	-	-	-	-	-	-	-
Totals for the Months,		-	4	1	3	3	2	2	3	2	-	2	1	23
	1889,	-	-	-	-	-	1	-	1	-	-	-	-	2
	1890,	-	-	-	-	-	-	1	1	-	1	-	-	3
	1891,	-	2	-	1	1	2	-	1	-	-	1	-	8
III.	1892,	-	-	-	1	-	-	1	-	1	-	2	-	5
	1893,	-	2	-	-	-	2	-	-	-	-	-	-	4
	1894,	-	-	-	-	-	1	-	1	-	-	-	-	2
	1895,	-	-	-	1	1	-	-	-	-	-	-	-	2
Totals for the Months,		-	4	-	3	2	6	2	4	1	1	3	-	26
	1889,	-	-	-	-	-	-	-	-	-	-	1	-	1
	1890,	-	-	-	-	1	-	-	1	-	1	1	-	4
	1891,	-	1	1	2	-	1	-	1	-	2	-	-	8
IV.	1892,	-	-	-	-	2	-	1	-	1	2	-	-	6
	1893,	-	-	-	2	2	2	-	-	-	-	-	-	6
	1894,	-	-	-	-	-	-	-	1	-	-	-	-	1
	1895,	-	-	-	1	1	-	-	-	-	-	-	-	2
Totals for the Months,		-	1	1	5	6	3	1	3	1	5	2	-	28

[TABLE.

TABLE IX.—*continued.*

Three Outer Stations.	Years.	Jan.	Feb.	Mar.	April	May	June	July	Aug.	Sept.	Oct.	Nov.	Dec.	Totals for the Years and Stations.
	1889,	–	–	–	–	–	–	–	–	–	–	1	–	1
	1890,	–	–	–	1	–	–	1	–	1	1	1	–	5
	1891,	–	1	1	1	2	1	–	1	1	1	–	–	9
V.	1892,	–	–	–	–	1	1	1	–	–	1	–	–	4
	1893,	–	–	–	–	1	–	–	2	–	–	–	–	3
	1894,	–	–	–	–	–	–	–	1	–	–	–	–	1
	1895,	–	–	–	–	1	–	–	1	–	–	–	–	2
Totals for the Months,		–	1	1	2	5	2	2	5	2	3	2	–	25
	1889,	–	–	–	–	–	–	–	–	–	–	1	–	1
	1890,	–	–	1	–	1	–	–	–	1	1	–	–	4
	1891,	–	2	–	1	1	1	–	2	1	–	1	–	9
VIII.	1892,	–	–	–	–	1	1	1	–	–	1	–	–	4
	1893,	–	–	–	–	–	–	–	1	–	–	–	–	1
	1894,	–	–	–	–	–	–	–	1	–	–	–	–	1
	1895,	–	–	–	–	1	1	–	1	–	–	–	–	3
Totals for the Months,		–	2	1	2	4	2	1	5	2	2	2	–	23
	1889,	–	–	–	–	–	–	–	1	–	–	1	–	2
	1890,	–	–	–	1	–	–	1	–	1	–	–	1	4
	1891,	–	2	–	2	1	–	–	2	–	–	–	–	7
IX.	1892,	–	–	–	–	2	1	1	–	–	1	–	–	5
	1893,	–	–	–	–	1	–	–	–	–	–	–	–	1
	1894,	–	–	–	–	–	–	–	2	–	–	–	–	2
	1895,	–	–	–	–	–	–	–	–	–	–	–	–	–
Totals for the Months,		–	2	–	3	4	1	2	5	1	1	1	1	21

It will be observed from this Table that the distribution of *Acartia* varies to a greater extent from that of *Calanus* than that of *Temora* was found to do. The greatest number of yearly records of *Acartia* shown in the Table is that for 1891, and not only is the yearly total number for the six stations for 1891 the greatest, but the same thing is observed in the total number of each of the stations for that year. If, on the other hand, a comparison be made between the highest and lowest total numbers for the six stations, the difference is rather less than in *Temora*, and more similar to that observed in *Calanus*. It is also noticeable that *Acartiæ* are comparatively usually more frequent during the summer months. It will be observed, for example, that the highest of the total monthly records for each of the three outer stations are those for May and August; while at Station IV. the highest is that for May, and at Station III. for June; but at Station I. the highest is that for February. The next highest numbers are those for April, May, and August. Some of these points are more clearly indicated by the three formulæ.

Formula (1) shows, as before, the total number of records of *Acartia* for each of the stations :—

	(a) Inner Stations.			(b) Outer Stations.		
Numbers of the Stations,	I.	III.	IV.	V.	VIII.	IX.
Total number of Records for each Station,	23	26	28	25	23	21

The station with the lowest number is IX., while that with the highest number is IV., the difference between them being seven.

Formula (2) shows the total number of records of *Acartia* for each of the twelve months for the seven years, (a) for the inner stations, I., III., and IV. ; (b) for the outer stations, V., VIII., and IX. ; (c) for all the six stations :—

Names of the Months,	Jan.	Feb.	Mar.	April.	May.	June.	July.	Aug.	Sept.	Oct.	Nov.	Dec.
Total number of Records for each Month, for (a) the Inner Stations,	0	9	2	11	11	11	5	10	4	6	7	1
(b) the Outer Stations,	0	5	2	7	13	5	5	15	5	6	5	1
(c) all the Six Stations,	0	14	4	18	24	16	10	25	9	12	12	2

If these figures be closely examined and compared, it will be found that there is a more or less evident maximum in the total number of records about every three months. This is also found to be the case, though not so distinctly, when the numbers are divided so as to correspond with the inner and outer stations. But taking the total numbers first, it will be observed that, commencing with February, the total number for which is fourteen, there occurs in March a sudden drop, after which the numbers increase till May, when the second highest total for the whole year is reached. In June and July the numbers again fall, to be followed by a marked accession in August, so much so that the maximum for all the twelve months occurs then ; in September there is again a marked decrease, after which a slight recovery takes place in October and November, while in December the number is, next to January, the lowest for the year. This increase and decrease, though not so apparent where the numbers are divided to correspond with the inner and outer stations, can still be traced, especially in the series corresponding to the outer stations. The reason for this apparent rhythmical arrangement is not very clear, and it may be that the seemingly periodic recurrence of *Acartia* may be merely accidental.

Formula (3) shows the total number of records of *Acartia* for each year, for (a) the inner stations, I., III., and IV. ; (b) the outer stations, V., VIII., and IX. ; and (c) all the six stations :—

Numbers of the Years,	1889.	1890.	1891.	1892.	1893.	1894.	1895.
Total number of Records for each Year. (a) the Inner Stations,	4	10	24	16	14	5	4
(b) the Outer Stations,	4	13	25	13	5	4	5
(c) all the Six Stations,	8	23	49	29	19	9	9

It will be observed that the maximum number—that for 1891—is proportionally considerably greater in comparison with the others than is observed either in the case of *Calanus* or *Temora*.

A few of the other species of Copepoda referred to in the lists of tow-net invertebrates, such as *Pseudocalanus elongatus*, *Centropagus* sp., *Anomalocera patersoni*, and *Oithona* sp., are also at times more or less frequent in the Firth of Forth, but the published records of them show that they are usually less common than those already noticed.

The only points of interest in connection with the distribution of *Pseudocalanus elongatus,* so far as can be made out from the tow-net records of its occurrence, seem to be these—(1st) if an average of all the records of this Copepod for the seven years be taken, the numbers for April and May will be seen to be greater than those for all the other months added together ; and (2nd) that the maxima of the records for the seven years occur in 1891, 1892, and 1893.

Anomalocera patersoni, which from its very large size and beautiful colour is easily detected among the tow-net gatherings, is recorded a considerable number of times during the seven years. The greatest total number of records for any of the years is that for 1891, the number for that year being nearly double that for any other of the whole seven, as shown by the formula.

Formula (1) shows the number of records of *Anomalocera* for each of the seven years :—

Numbers of the Years,	1889.	1890.	1891.	1892.	1893.	1894.	1895.
Total number of Records for each Year,	7	13	17	6	9	4	6

This formula shows a gradual increase to the maximum, then a somewhat irregular decrease.

Formula (2) shows the total number of records of *Anomalocera* for each of the twelve months :—

Names of the Months,	Jan.	Feb.	Mar.	April.	May.	June.	July.	Aug.	Sept.	Oct.	Nov.	Dec.
Number of Records for each Month,	0	0	0	3	20	18	8	7	1	4	0	1

It will be observed that by far the largest number of records for this species occur in May and June ; they are much fewer in July, August, and October, and rarely, or none at all, in the other months. It would appear from this that *Anomalocera* is not resident in the estuary, but is only a summer visitor, and, as Formula (1) shows, its visits are fairly regular.

Formula (3) shows the total number of records of *Anomalocera* for each of the six stations :—

	Inner Stations.			Outer Stations.		
Numbers of the Stations,	I.	III.	IV.	V.	VIII.	IX.
Number of Records for each Station,	5	5	1	15	16	19

The difference between the inner and the outer stations is very marked here, but is only what might be looked for if *Anomalocera* be only a "visitor" to the estuary and its vicinity. This also explains why Station IV. appears to have been so rarely visited, for, being a migrant, *Anomalocera* is more likely to keep as much as possible to the open water, and avoid inshore localities.

The distribution of the other species mentioned in the lists of tow-net fauna does not call for special analysis ; but a few of them are of interest because of their beauty, their rarity, or because of their structural differences and habits.

Metridia hibernica, which is of an elongate form, and has a long and moderately slender abdomen, is recorded only once during the seven years—viz., at Station IX. in May 1892, when a few specimens were obtained. *Metridia* is a comparatively large Copepod.

Candace pectinata is also moderately large, and is readily noticed by its dark chocolate-coloured swimming feet. There are several records of this species in the lists of tow-net invertebrates.

Parapontella brevicornis appears to be only once recorded for the seven years, but the species is not very rare locally.

Dermatomyzon nigripes is also only once mentioned in the tow-net lists for the selected stations.

Monstrilla sp.   The most curious of the pelagic Copepods is the *Monstrilla*.   Two species—*M. helgolandica* and *M. rigida*—have been captured in the estuary.   In swimming it differs from the usual Copepod habit by keeping its antennules straight out in front, instead of holding them at right angles.

Oithona sp.   The *Oithona* that is sometimes not uncommon in the Firth of Forth is probably the *O. similis* Claus, as stated by Dr. Giesbrecht.

Caligus rapax is frequently found " free-swimming " in the estuary, and appears to be able to adopt that mode of life at will.

(3.) THE VERMES.

THE SAGITTA (*S. bipunctata*).

Leaving the Copepoda, I now pass on to consider the distribution of the *Sagitta*.   A glance through the lists of pelagic invertebrates which form part of the trawling records shows that the *Sagitta* forms an item of considerable importance amongst the various objects collected by the tow-nets.   Moreover, when the voracity of these curious organisms is taken into account, it is evident that the vast swarms of them that occasionally visit our shores must form a destructive agency of no mean order, and especially if in the vicinity of their swarms fish larvæ happen to be more or less common.   It has been clearly proved that *Sagitta* prey to a considerable extent on larval and post-larval fishes—chiefly round fishes.   This habit was observed by me several years ago, and before much attention had been given to it otherwise ; and it was found that the *Sagitta* not only preyed on the larvæ of round fishes, but they also attacked the smaller Crustacea, and were even observed to devour their own kind— the larger *Sagitta* devouring the smaller.

The distribution of the *Sagitta* in the Firth of Forth during the seven years will be best seen by a reference to the annexed Table.

TABLE X., showing the Distribution of *Sagitta*, as indicated by the Tow-net Records for the Years, Months, and Stations.

Three Inner Stations.	Years.	Number of Records for each Month of each Year.												Totals for the Years and Stations.
		Jan.	Feb.	March.	April.	May.	June.	July.	Aug.	Sept.	Oct.	Nov.	Dec.	
	1889.	1	2	–	–	–	1	1	1	–	1	1	–	8
	1890.	–	2	2	–	2	–	–	–	–	1	1	1	9
	1891.	1	1	1	2	2	2	1	1	–	1	1	–	13
I.	1892.	1	1	1	1	1	–	1	–	1	–	1	2	10
	1893.	2	2	–	1	–	1	1	1	1	1	2	1	13
	1894.	1	2	–	1	–	–	–	1	–	–	–	2	7
	1895.	–	1	–	1	1	–	–	–	1	–	–	2	6
Totals for the Months, . . .		6	11	4	6	6	4	4	4	3	4	6	8	66

TABLE X.—*continued.*

Inner Stations—continued.	Years.	Number of Records for each Month of each Year.												Totals for the Years and Stations.
		Jan.	Feb.	March.	April.	May.	June.	July.	Aug.	Sept.	Oct.	Nov.	Dec.	
	1889,	1	–	–	–	–	–	–	–	–	–	1	–	2
	1890,	–	2	–	1	1	–	1	–	–	1	1	–	9
	1891,	1	1	–	2	2	1	–	1	–	1	2	–	11
III.	1892,	1	2	1	1	1	–	1	–	1	–	1	2	11
	1893,	2	1	–	–	1	1	1	1	1	1	2	2	13
	1894,	–	1	–	1	–	–	–	1	–	–	–	4	7
	1895,	–	1	–	1	–	–	–	–	–	1	–	2	5
Totals for the Months,		5	8	3	6	5	2	3	3	2	4	7	10	58
	1889,	1	–	–	–	–	–	–	–	1	–	–	1	3
	1890,	–	2	–	–	1	–	1	–	–	1	2	1	8
	1891,	2	2	1	–	2	–	–	1	–	1	1	–	10
IV.	1892,	1	2	2	2	–	–	1	–	1	1	1	1	12
	1893,	2	2	–	1	–	1	–	–	–	1	1	2	10
	1894,	1	1	–	1	–	–	–	1	–	–	–	–	4
	1895,	1	2	–	–	1	–	–	1	–	2	–	2	9
Totals for the Months,		8	11	3	4	4	1	2	4	1	6	6	6	56

Three Outer Stations.	Years.	Number of Records for each Month of each Year.												Totals for the Years and Stations.
		Jan.	Feb.	March.	April.	May.	June.	July.	Aug.	Sept.	Oct.	Nov.	Dec.	
	1889,	2	2	–	–	1	1	–	2	–	–	1	–	9
	1890,	–	1	–	2	1	–	2	–	–	1	2	2	11
	1891,	–	–	1	1	1	1	1	1	2	2	2	2	14
V.	1892,	1	1	2	1	1	1	1	–	1	1	2	2	14
	1893,	2	1	–	–	1	1	2	2	1	2	1	2	15
	1894,	–	2	–	1	–	1	–	1	–	–	–	–	5
	1895,	–	2	–	1	2	–	–	1	–	–	–	–	6
Totals for the Months,		5	9	3	6	7	5	6	7	4	6	8	8	71

N

TABLE X.—continued.

Outer Stations—continued.	Years.	Number of Records for each Month of each Year.												Totals for the Years and Stations.
		Jan.	Feb.	March.	April.	May.	June.	July.	Aug.	Sept.	Oct.	Nov.	Dec.	
VIII.	1889,	2	-	-	-	-	2	2	1	-	-	1	-	8
	1890,	-	1	1	1	1	-	1	-	1	2	-	1	9
	1891,	-	2	-	2	1	-	1	1	1	2	2	-	12
	1892,	2	2	2	2	2	2	1	-	2	1	-	4	20
	1893,	2	2	-	-	2	1	1	1	1	1	-	2	13
	1894,	1	2	-	1	-	-	-	-	-	-	-	-	4
	1895,	-	2	-	1	2	-	-	1	1	2	-	2	11
Totals for the Months, . . .		7	11	3	7	8	5	6	4	6	8	3	9	77
IX.	1889,	2	-	-	-	-	2	1	1	-	-	1	-	7
	1890,	-	1	2	1	2	-	2	-	2	1	-	2	13
	1891,	-	2	-	2	1	2	1	1	1	2	2	-	14
	1892,	2	1	2	1	1	1	1	-	1	1	-	3	14
	1893,	2	2	-	-	2	1	2	1	-	2	-	2	14
	1894,	-	2	-	-	-	1	-	1	-	-	-	-	4
	1895,	-	2	-	1	-	-	-	-	2	1	-	2	8
Totals for the Months, . .		6	10	4	5	6	7	7	6	5	6	3	9	74

This Table shows that the total number of records of *Sagitta* for the six selected stations, and for the seven years, is four hundred and five, being somewhat fewer than the number of records of *Calanus*, but considerably in excess of the number of records of any other of the species that have been referred to. In making an analysis of the Table, it is found that while there is a considerable difference between the number of records for Station IV. and that for Station VIII., if the three inner stations and the three outer stations be considered separately, the difference in the numbers of each group is much less. These points will be made more evident by arranging the stations and the number of records for each as a formula, thus :—

Formula (1) shows the number of records of *Sagitta* for each of the six selected stations :—

	The Three Inner Stations.			The Three Outer Stations.		
Numbers of Stations, - - - -	I.	III.	IV.	V.	VIII.	IX.
Number of Records for each Station, -	66	58	56	74	77	74

The numbers for the three outer stations are nearly equal, but Station I. in the other three shows a greater divergence. It may also be noticed that Station I. and Station VIII., that have the highest number of records in each of the two groups, are intermediate stations—that is, Station I. is situated between Stations III. and IV., and Station VIII. between Stations V. and IX. Moreover, if a comparison be made of the total number of the records for each of the twelve months, it will be observed that those for February and December are considerably greater than the numbers for the other months; but if the numbers are separated into two groups to correspond with the inner and outer stations, the difference in the numbers for each of the twelve months for the three inner stations is increased, while the difference becomes less in the other group of numbers. This is best seen by arranging the numbers as a formula.

Formula (2) shows the total number of records of *Sagitta* for each of the twelve months, for (*a*) the inner stations, I., III., and IV.; (*b*) the outer stations, V., VIII., and IX.; and (*c*) for all the six stations :—

Names of the Months,		Jan.	Feb.	Mar.	April.	May.	June.	July.	Aug.	Sep.	Oct.	Nov.	Dec.
Total number of Records for each Month, for	(a) the Three Inner Stations,	19	30	10	16	15	7	9	11	6	14	19	24
	(b) the Three Outer Stations,	18	30	10	18	21	17	19	17	15	20	17	26
	(c) all the Six Stations,	37	60	20	34	36	24	28	28	21	34	36	50

There is one point very clearly brought out in connection with the distribution of the *Sagitta*—viz., that, whether a comparison be made of the monthly numbers of records for each station or for the three inner and three outer stations or for them all, the numbers for February and most of those for December are greater than the others. This would seem to indicate that *Sagitta* was usually more common during these months than at other times of the year.

It will be further observed that the highest total number of yearly records, viz. 79, is that for 1892, and the smallest, 31, for 1894, and that if the yearly records be divided into two sets to correspond with the three inner and the three outer stations, the results, as regards the highest and lowest numbers, are somewhat different. In this arrangement of the figures the highest number for the three inner stations is that for 1893, and the lowest that for 1889, while the maximum and minimum numbers for the outer stations are for the same years as in the former arrangement. Formula (3) will show these results more clearly.

Formula (3) shows the total number of records of *Sagitta* for all the seven years; for (*a*) the inner stations, I., III., and IV.; (*b*) the outer stations, V., VIII., and IX.; and (*c*) for all the six stations :—

Numbers of the Years,		1889.	1890.	1891.	1892.	1893.	1894.	1895.
Total number of Records for each Year, for	(a) the Inner Stations,	13	26	34	33	36	18	20
	(b) the Outer Stations,	24	33	40	48	42	13	25
	(c) all the Six Stations,	37	59	74	81	78	31	45

It would thus appear that *Sagitta*, like the majority of the other pelagic organisms of which there are records, are usually of more frequent occurrence in the Firth of Forth and its vicinity during the colder months, and that they were more numerous during 1891, 1892, and 1893 than during the other years. These again coincide with the yearly maxima for *Parathemisto*, and may be the result of a larger migration during these years, as was probably the case with *Parathemisto*, or to circumstances favouring a greater local increase; for though the difference in the number of the yearly records may be partly accounted for by the greater or fewer number of experiments, this does not explain the whole of it. Unfortunately the experiments were not carried on long enough nor so regularly as to indicate whether the difference shown is due to the actual increase or decrease of these organisms, and whether it was temporary and accidental or part of a more or less regular cycle of changes in their migration or development.

(4.) The Cœlenterata.

THE CTENOPHORA AND HYDROMEDUSÆ.

The next groups of organisms that are numerically important and which may be specially noticed are the Ctenophora and Hydromedusæ. These include the well-known and beautiful *Pleurobrachia*, which sometimes occurs in such immense numbers that they seem to monopolise the whole of the water-space where they are. The swarms of these things that are sometimes met in with by the *Garland* are so great that the tow-nets, if unfortunately in use at the time, are found, when hauled on board, to be literally full of these crystal spheres, while little of anything else is found in the nets. The appearance of such a mass of *Pleurobrachia*—each globe, by the rapid movement of its delicate cilia, scintillating in the light—is wonderful, and cannot be accurately described. *Beroë* is another of the Cœlenterates that was sometimes met with in considerable numbers, but not usually in such immense swarms as *Pleurobrachia*. Various other forms, such as *Sarsia, Geryonopsis, Thaumantias,* &c., were also observed during the years referred to, but from want of time no attempt was made to differentiate the species, and they are therefore usually all included under the name of Ctenophora : on a few occasions only are *Pleurobrachiæ* and a few of the others specially mentioned. There are two hundred and ninety-six records of Ctenophora, including Hydromedusæ, among the lists of tow-net gatherings for the seven years, and their distribution for the months and stations is shown by the annexed Table :—

TABLE XI., showing the Distribution of the Ctenophora and Hydromedusæ, indicated by the Tow-net Records for the Years, Months, and Stations.

Three Inner Stations.	Years.	Jan.	Feb.	March.	April.	May.	June.	July.	August.	Sept.	Oct.	Nov.	Dec.	Totals for Years and Stations.
	1889,	2	2	–	–	–	1	2	2	–	–	1	–	10
	1890,	–	–	1	1	–	–	2	2	–	2	1	–	9
	1891,	1	1	–	–	1	1	–	–	–	2	2	–	8
I.	1892,	1	2	1	–	–	2	1	–	–	1	1	2	11
	1893,	2	1	–	1	1	1	1	2	2	2	1	–	14
	1894,	–	2	–	–	–	–	–	2	–	–	–	2	6
	1895,	–	1	–	–	–	–	1	–	–	1	–	1	4
Total number of Records for all the Months,		6	9	2	2	2	5	7	8	2	8	6	5	62
	1889,	1	–	–	–	–	2	2	2	–	–	1	–	8
	1890,	–	2	–	–	2	–	2	1	–	2	1	–	10
	1891,	1	–	–	–	–	1	1	1	1	1	2	–	8
III.	1892,	1	–	1	1	2	2	1	–	1	1	2	2	14
	1893,	2	1	–	–	1	1	–	1	2	1	2	1	12
	1894,	–	1	–	–	–	1	–	2	–	–	–	4	8
	1895,	–	1	–	–	–	–	2	–	–	1	–	2	6
Total number of Records for all the Months,		5	5	1	1	5	7	8	7	4	6	8	9	66

TABLE XI.—*continued.*

Inner Stations—continued.	Years.	Number of the Records for each Month of each Year.												Totals for the Years and Stations.
		Jan.	Feb.	March.	April.	May.	June.	July.	Aug.	Sept.	Oct.	Nov.	Dec.	
	1889, .	1	–	–	–	–	2	2	2	–	–	1	–	8
	1890, .	–	1	–	1	2	–	–	1	–	2	2	1	10
	1891, .	1	1	–	–	–	1	–	1	–	1	–	–	5
IV.	1892, .	2	1	–	–	–	1	2	–	–	–	1	2	9
	1893, .	2	1	–	–	1	–	–	2	2	2	2	1	13
	1894, .	–	–	–	–	–	–	–	1	–	–	–	–	1
	1895, .	1	–	–	–	–	–	–	2	–	2	–	2	7
Total number of Records for all the Months,		7	4	–	1	3	4	4	9	2	7	6	6	53

Three Outer Stations.	Years.	Number of the Records for each Month of each Year.												Totals for the Years and Stations.
		Jan.	Feb.	March.	April.	May.	June.	July.	Aug.	Sept.	Oct.	Nov.	Dec.	
	1889, .	1	1	–	–	–	1	2	1	–	–	1	–	7
	1890, .	–	1	–	–	1	–	2	–	–	2	1	–	7
	1891, .	–	1	–	–	–	–	1	–	1	2	1	–	6
V.	1892, .	1	1	1	–	1	2	1	–	2	–	2	1	12
	1893, .	1	–	–	–	2	–	–	2	1	1	–	1	8
	1894, .	–	1	–	–	–	1	–	1	–	–	–	–	3
	1895, .	–	1	–	–	–	–	–	–	–	–	–	–	1
Total number of Records for all the Months,		3	6	1	–	4	4	6	4	4	5	5	2	44
	1889, .	1	–	–	–	–	–	2	2	–	–	1	–	6
	1890, .	–	–	–	–	1	–	–	1	–	–	–	–	2
	1891, .	–	–	–	–	–	–	–	–	–	2	–	–	2
VIII.	1892, .	1	–	1	–	–	2	1	–	2	1	–	2	10
	1893, .	1	1	–	–	1	1	1	1	1	–	–	–	7
	1894, .	–	1	–	–	–	2	–	–	–	–	–	–	3
	1895, .	–	–	–	–	1	–	–	1	–	1	–	1	4
Total number of Records for all the Months,		3	2	1	–	3	5	4	5	3	4	1	3	34

TABLE XI.—*continued.*

Outer Stations—continued.	Years.	Jan.	Feb.	March.	April.	May.	June.	July.	Aug.	Sept.	Oct.	Nov.	Dec.	Totals for the Years and Stations.
							Number of the Records for each Month of each Year.							
	1889,	1	-	-	-	-	-	-	2	1	-	-	1	-	5
	1890,	-	-	-	-	-	1	-	-	1	2	-	-	-	4
	1891,	-	-	-	-	1	-	2	-	-	1	2	1	-	7
IX.	1892, .	-	-	-	-	-	1	2	2	-	-	2	-	3	10
	1893, .	1	1	-	-	-	-	1	1	-	-	1	-	1	6
	1894, .	-	-	-	-	-	2	-	-	-	-	-	-	-	2
	1895, .	-	1	-	-	-	-	-	1	-	-	-	-	1	3
Total number of Records for all the Months,		2	2	-	1	2	7	5	3	3	5	2	5	37

In 1889 small Cœlenterates were abundant in a bottom tow-net gathering collected at Station IV. in July. In 1890 they were abundant in a surface tow-net gathering at the same station in May. In 1891 they were abundant in a bottom tow-net gathering collected at Station I. in July, and at Station IX. in June ; and they were also abundant in a surface tow-net gathering at Station IX. in October. Cœlenterates were frequently obtained in 1892, but in none of the records for the selected stations are they described as abundant. In 1893 they were abundant in a bottom tow-net gathering at Station I. in August, and also at Station IV. in the same month, but at the latter station they were abundant in the surface as well as in the bottom tow-net ; in the same year they were also abundant in a surface tow-net gathering at Station V. in May. In 1894 Cœlenterata were abundant in bottom tow-net gatherings collected in June at Stations III. and IV., while in August 1895 they were abundant in a surface gathering at Station III., and in a bottom gathering at Station IV.

From an examination of the annexed Table showing the general distribution of Ctenophora and Hydromedusæ, it will be observed that there is a somewhat curious contrast between their distribution and that of *Sagitta.* On comparing the total number of the records of Cœlenterata and *Sagitta* for each of the three inner and outer stations, it is found that while in the case of *Sagitta* the larger numbers are those of the outer stations, it is just the reverse with the Cœlenterates, the largest numbers being those of the inner stations. This contrast is better seen by arranging the numbers of the records of *Sagitta* and Ctenophora for each of the stations, in juxtaposition as a formula, thus :—

Formula (1) shows the total number of records of Cœlenterata for all the six stations, to which the numbers of the records of *Sagitta* are added for comparison :—

	Inner Stations.			Outer Stations.		
Numbers of the Stations,	I.	III.	IV.	V.	VIII.	IX.
Number of Records for each Station (Cœlenterata)	62	66	53	44	34	37
Number of Records for each Station (*Sagitta*)	66	58	56	74	77	74

It would thus appear that while *Sagitta* was comparatively scarce at the inner stations, Cœlenterata were more numerous. This difference does not seem be merely accidental; it is too general and persistent for that being the case, and is possibly due rather to a difference in the habits of the two kinds of organisms, but the reason, whatever it is, does not appear to be very obvious. An examination of the total numbers of the monthly records of Cœlenterata for each station, as well as for all the six stations, shows one or two other points of difference between the distribution of these organisms and that of *Sagitta*. In the case of these Cœlenterates the lowest numbers of records for any of the six stations are those for March and April, whereas, in the case of *Sagitta*, the lowest numbers are, with one exception, those for June and July. It will be further observed that, if the numbers of the monthly records are separated to correspond with the inner and the outer stations, ten of the monthly numbers for the inner stations are greater, and some of them considerably greater, than the corresponding numbers for the outer stations, while one of the other two numbers is equal to, and the second only slightly less than, the numbers for the same months for the outer stations, which emphasises still further the difference in the distribution of these Cœlenterates between the inner and outer stations. This is more clearly shown by the Formula (2).

Formula (2) shows the total number of records of Ctenophora and Hydromedusæ for each of the twelve months, for (*a*) the inner stations, I., III., and IV.; (*b*) the outer stations, V., VIII., and IX.; and (*c*) for all the six stations :—

Names of the Months,		Jan.	Feb.	Mar.	April.	May.	June.	July.	Aug.	Sept.	Oct.	Nov.	Dec.
Total number of Records for each Month, for	(*a*) the Inner Stations,	18	18	3	4	10	16	19	24	8	21	20	20
	(*b*) the Outer Stations,	8	10	2	1	9	16	15	12	10	14	8	10
	(*c*) all the Six Stations,	26	28	5	5	19	32	34	36	18	35	28	30

That Cœlenterates were more numerous at the inner than at the outer stations is still further shown by separating the yearly total numbers so as to correspond with the two groups of stations as in Formula (3).

Formula (3) shows the total number of records of Ctenophora and Hydromedusæ for each of the seven years, for (*a*) the inner stations, I., III., and IV.; (*b*) the outer stations, V., VIII., and IX.; and (*c*) for all the six stations :—

Numbers of the Years,		1889.	1890.	1891.	1892.	1893.	1894.	1895.
Total number of Records for each Year, for	(*a*) the Inner Stations,	26	29	21	34	39	15	17
	(*b*) the Outer Stations,	18	13	15	32	21	8	8
	(*c*) all the Six Stations,	44	42	36	66	60	23	25

This formula shows that the number of yearly records for the outer stations is considerably lower for every year except 1892, when the respective numbers more nearly approximate. The highest number for the inner stations is that for 1893, and for the outer stations the highest number is that for 1892, while the lowest numbers for both are in 1894 and 1895.

(5.) Larval and Young Crustacea.

The next group of organisms described in the tow-net records which are of numerical importance are the young of various species of Crustacea belonging to the Decapoda, Schizopoda, and Cirripedia. Young and larval Crustacea are generally more or less frequent during the summer months, and sometimes they occur in great abundance, and because of this, and also because of their nutritive qualities, they form one of the most important constituents of fish-food existing in the sea; this remark applies, of course, to young Entomostracans as well as to the other orders of this great class of invertebrata. The minute nauplei, the larger megalops, and other stages of crustacean growth, along with embryo mollusca, form, no doubt, the principal food of young fishes of all kinds. Mr H. Dannevig, in his paper * describing his experiments in the rearing of larval and post-larval plaice and other flat fishes, gives some interesting observations bearing on the value of young Crustacea as fish food. The examination of the stomach of large numbers of young fishes at the Marine Laboratory of the Fishery Board for Scotland at Tarbert, Loch Fyne, during 1886-87, some of the results of which are published in the Board's Fifth Annual Report, also show that the importance of young Crustacea as fish food has long been recognised, and, as a matter of fact, their importance in this respect cannot well be overstated. It is evident, therefore, that the study of the distribution of these miniature forms, and of the influences that govern their increase or decrease, as well as the investigation of their life-histories and development, are well within the scope of scientific fishery work.

The records of miniature Crustacea included in the lists of tow-net fauna that form the basis of this paper refer chiefly to the young of various species of the Decapoda. Among the more common of the larval Decapoda included in the tow-net gatherings are those of *Portunus* (or swimming crabs), of *Porcellana longicornis*, of the Eupaguridæ (or soldier or hermit crabs), and of *Nephrops norvegicus* (the "Norway lobster"). The young of *Galathea* (or what some fishermen call the "bastard lobster") and of various shrimps, such as *Crangon* and *Pandalus*, which are at times more or less frequent, are also included among these records under the general name of young or larval Crustacea. Young Schizopoda, as such, are specially referred to in the lists only when they happen to be common or abundant; at other times they are included with the young Decapods. The larvæ and young of the Barnacle are more frequently recorded separately, but they also are sometimes included with the others, and the table of distribution comprises the records of all the three groups. The total number of records referring to young and larval Crustacea is about three hundred and forty, and an enumeration of them is contained in the annexed Table. This Table shows that in some respects the distribution of the immature Crustacea is somewhat similar to that of the Cœlenterata. If, for example, the total number of records of immature Crustacea for each of the three inner stations be compared with those for the three outer stations, they will be found on an average larger than the others, though this difference is not so marked as is the case with Cœlenterata.

It will also be observed that the yearly maxima of records for the various stations are not so regular, and also that the larger of the monthly records occur during the summer months.

---

* Fifteenth Annual Report of Fishery Board for Scotland. Part iii. Pp. 175-193.

[TABLE.

TABLE XII., showing the Distribution of the Larval Crustacea (including Decapoda, Schizopoda, and Cirripedia), as indicated by the Tow-net Records for the Years, Months, and Stations :—

Three Inner Stations.	Years.	Jan.	Feb.	March.	April.	May.	June.	July.	August.	Sept.	Oct.	Nov.	Dec.	Totals for the Years and Stations.
	1889,	-	-	-	-	-	1	2	2	-	2	-	-	7
	1890,	-	1	2	2	1	-	2	2	-	2	2	-	14
	1891,	-	-	1	2	2	2	1	2	-	2	1	-	13
I.	1892,	-	-	-	2	2	1	2	-	2	1	2	-	12
	1893,	-	-	-	2	2	1	2	3	2	2	-	-	13
	1894,	-	-	-	1	-	-	-	2	-	-	-	-	3
	1895,	-	-	-	2	1	-	1	-	2	2	-	-	8
Total number of Records for all the } Months,		-	1	3	11	8	5	10	10	6	11	5	-	70
	1889,	-	-	-	-	-	1	2	2	-	-	1	-	6
	1890,	-	-	-	2	-	-	2	2	-	2	1	-	9
	1891,	-	-	-	-	3	-	1	2	1	2	2	-	11
III.	1892,	-	-	-	2	1	1	2	-	2	1	-	-	9
	1893,	-	-	-	-	2	2	2	2	2	1	1	-	12
	1894,	-	-	-	2	-	1	-	-	-	-	-	1	4
	1895,	-	-	-	2	1	-	-	2	1	2	-	1	9
Total number of Records for all the } Months,		-	-	-	8	7	5	9	10	6	8	5	2	60
	1889,	-	-	-	-	-	1	2	2	-	-	1	-	6
	1890,	-	-	-	1	1	-	2	2	-	2	1	-	9
	1891,	-	-	-	2	1	1	2	2	-	-	1	-	9
IV.	1892,	-	-	1	2	2	-	2	-	2	2	1	-	12
	1893,	-	-	-	2	1	1	-	1	2	1	-	-	8
	1894,	-	-	-	2	-	1	-	1	-	-	-	-	4
	1895,	-	-	-	1	2	-	-	2	-	2	-	-	7
Total number of Records for all the } Months,		-	-	1	10	7	4	8	10	4	7	4	-	55

[TABLE.

TABLE XII.—continued.

Three Outer Stations.	Years.	Jan.	Feb.	March.	April.	May.	June.	July.	August.	Sept.	Oct.	Nov.	Dec.	Totals for the Years and Stations.
	1889,	–	–	–	–	1	1	–	3	–	–	1	–	6
	1890,	–	–	–	1	1	–	2	–	2	2	1		10
	1891,	–	–	1	1	2	–	1	2	1	1	1	–	10
V.	1892,	–	–	–	2	1	1	1	–	2	–	2	–	9
	1893,	–	–	–	–	2	2	1	2	2	1	1	–	11
	1894,	–	–	–	2	–	–	–	1	–	–	–	–	3
	1895,	–	–	–	2	1	–	–	2	2	–	–	–	7
Total number of Records for all the Months,		–	–	1	8	8	4	5	10	9	4	6	1	56
	1889,	–	–	–	–	–	1	1	2	–	–	1	–	5
	1890,	–	–	–	1	2	–	–	2	2	2	–	1	10
	1891,	–	–	–	2	1	2	2	2	2	1	1	–	13
VIII.	1892,	–	–	–	–	2	–	–	–	2	–	–	–	4
	1893,	–	–	–	–	1	–	2	1	1	–	–	1	6
	1894,	–	–	–	2	–	–	–	1	–	–	–	–	3
	1895,	–	–	–	1	1	–	–	2	2	2	–	–	8
Total number of Records for all the Months,-		–	–	–	6	7	3	5	10	9	5	2	2	49
	1889,	–	–	–	–	–	1	2	2	–	–	1	–	6
	1890,	–	–	–	1	1	–	–	2	1	2	–	1	8
	1891,	–	–	–	1	1	1	2	1	2	–	–	–	8
IX.	1892,	–	–	1	–	3	–	1	–	2	1	1	–	9
	1893,	–	–	–	1	2	2	2	1	–	1	–	–	9
	1894,	–	–	–	1	–	1	–	1	–	–	–	–	3
	1895,	–	–	–	2	1	–	–	2	2	–	–	–	7
Total number of Records for all the Months,		–	–	1	6	8	5	7	9	7	4	2	1	50

There is one aspect of the question of the distribution of these young Crustaceans, as well as of that of the various other invertebrates touched upon in this paper, that it is well to keep in view in an inquiry of this kind. I refer to the individual frequency or abundance of these organisms at one time, or at different times, apart from the frequency or rarity of their recorded occurrences. In the Table showing the distribution of the young Crustacea the maximum numbers of the yearly records are, speaking generally, those for 1891, 1892, and 1893 ; while, on the other hand, if the individual frequency of the creatures be taken into account, the results brought out will be somewhat different. For example, young Crustacea are described as abundant in four of the records for 1890

—viz., twice for Station I. in July, and once for Stations V. and VIII. in September. They are described as abundant in four of the records for 1891—viz., once for Station VIII. in May, once for Stations III. and IX. in July, and once for Station I. in August. They are recorded as abundant three times in 1892—once for each of the three Stations V., VIII., and IX. in September. Young Crustacea are described as abundant five times in 1893, once for Station I. and twice for Station III. in July, and also twice for Station V. in June. In 1894 there are over twenty records of young Crustacea, but in none are they described as abundant unless the three April records of young *Balani* (for Stations I., III., and IV.) be counted. On the contrary, in 1895 young Crustacea are described as abundant no fewer than twelve times, once at each of the three outer stations (V., VIII., and IX.) in April, once at Stations III. and V. in August, once at Stations I., III., and VIII. in September, and once at Stations I., III., and IV. in October—thus indicating that, notwithstanding the difference in the number of records when compared with those for 1891, 1892, or 1893, young Crustacea were locally more plentiful in the estuary and its vicinity in 1895 than during the other three years referred to.

The number of separate records of young Schizopoda is thirty-two for the seven years, and their distribution is indicated by the following three formulæ :—

Formula (1) shows the total number of records of Schizopoda for each of the stations :—

Number of the Stations,	I.	III.	IV.	V.	VIII.	IX.
Total number of Records for each Station,	12	4	0	3	7	6

Formula (2) shows the total number of records of Schizopoda for each of the twelve months :—

Names of the Months,	Jan.	Feb.	Mar.	April.	May.	June.	July.	Aug.	Sept.	Oct.	Nov.	Dec.
Total number of Records for each Month,	0	0	2	6	4	3	2	4	4	5	1	1

Formula (3) shows the total number of records of Schizopoda for each of the seven years :—

Numbers of the Years,	1889.	1890.	1891.	1892.	1893.	1894.	1895.
Total number of Records for each Year,	5	16	6	1	4	0	0

The separate records of young *Balani* number fifty-eight. These larval Crustaceans are described as abundant in 1891 at Station I. in April, and in 1892 at Station III. in April. In 1893 they appear to have been plentiful at the inner stations, and are described as abundant in a surface tow-net gathering at Station I. in April, in the bottom tow-net at Station III. in June, and in a surface tow-net gathering at Station IV. in April. In 1894 young *Balani* are described as abundant in the bottom tow-net gatherings collected at Stations I., III., and IV. in April. In all the other records of these immature Crustaceans they are either described as common or few. Their distribution is indicated by the following three formulæ :—

Formula (1) shows the total number of records of young *Balani* for each of the six stations :—

Numbers of the Stations,	I.	III.	IV.	V.	VIII.	IX.
Total number of Records for each Station,	12	12	17	3	8	6

This formula shows that young *Balani* were, as might be expected, more numerous at the three inner stations, and that the highest number of records of these three stations is that for Station IV. This difference in the numbers of the records of these young forms is satisfactorily explained by the fact that the greater number of the Barnacle tribe, in our seas at least, are to be found attached to rocks, stones, and other objects inshore, and that there the young are usually to be found.

Formula (2) shows the total number of records of young *Balani* for each of the twelve months for all of the six stations :—

Names of the Months,	Jan.	Feb.	Mar.	April.	May.	June.	July.	Aug.	Sept.	Oct.	Nov.	Dec.
Total number of Records for each Month,	0	0	1	27	14	10	4	2	0	0	0	0

It is evident from the numbers in this formula that the *Balani* hatch out principally in April, and that after April there is a rapid diminution in the production of larval forms, after which a more or less complete cessation takes place, there being not a single record for all the six months from September to February.

Formula (3) shows the total number of records of young *Balani* for each of the seven years for all the six stations :—

Numbers of the Years,	1889.	1890.	1891.	1892.	1893.	1894.	1895.
Total number of Records for each Year,	4	9	20	7	11	5	2

The number of records for 1891 is greatly in excess of that for any of the other seven years, but the reason for this marked difference does not seem to be very clear.

But the records of young Schizopoda and *Balani* form a comparatively small part of the whole number. The largest proportion of the three hundred and forty records of young and larval Crustacea are those of various species of Decapods, as has already been stated. But resuming consideration of the young Crustacea, we find by an examination of the general Table of Distribution that with one exception there are no records for January and February in all the seven years and for all the stations. After this a sudden increase is observed over all the stations, so much so that in regard to two of the stations the maximum of records is reached in April. From this time the number decreases somewhat till August, when a marked extension occurs over all the stations, the number afterward gradually diminishing till November. In December there is a somewhat sudden drop to one or two records, or, in the case of two of the stations, to none at all. This sudden increase and subsequent continuance of the moderately high numbers of the monthly records till November is an indication that in March and April an extensive and more or less simultaneous hatching of Crustacean ova takes place, and that it continues more or less all through the summer and autumn and on till about the end of the year. Of course it is not suggested that it is the same species that produces these young forms all the year through, but that the ova of different species hatch out at different seasons and thus keep up the supply of larval and young forms. The subjoined formulæ, prepared from the numbers in the Table of Distribution, show more clearly by the arrangement of the numbers in this way some of the special features referred to in the distribution of the young and larval Crustacea.

Formula (1) shows the total number of the records of young Crustacea for each of the six stations, which are divided into inner and outer stations :—

	Inner Stations.			Outer Stations.		
Number of the Stations,	I.	III.	IV.	V.	VIII.	IX.
Total number of Records for each Station,	70	60	55	56	49	50

This formula shows that Station I. has a higher number of records than any others, and that the total number of records for the three outer stations is 155, while the total number for the three inner stations is 185, or fully 16 per cent. more than the outer.

Formula (2) shows the total number of records of young Crustacea for each of the twelve months, for (a) the three inner stations, I., III., and IV.; (b) the three outer stations, V., VIII., and IX.; and (c) for all the six stations :—

Names of the Months,		Jan.	Feb.	Mar.	April.	May.	June.	July.	Aug.	Sept.	Oct.	Nov.	Dec.
Total number of Records for each Month, for	(a) the Inner Stations, -	0	1	4	29	22	14	27	30	16	26	14	2
	(b) the Outer Stations, -	0	0	2	20	23	12	17	29	25	13	10	4
	(c) all the Six Stations.	0	1	6	49	45	26	44	59	41	39	24	6

Formula (3) shows the total number of the records of young Crustacea for each of the seven years, for (a) the inner stations, I., III., and IV.; (b) the outer stations, V., VIII., and IX.; and (c) for all the six stations :—

Numbers of the Years, - - - - -		1889.	1890.	1891.	1892.	1893.	1894.	1895.
Total number of Records for each Year, for	(a) the Inner Stations,	19	32	33	33	33	11	24
	(b) the Outer Stations,	17	28	31	22	26	9	22
	(c) all the Six Stations,	36	60	64	55	59	20	46

But besides young and larval Crustacea there are other immature forms mentioned in the lists of tow-net gatherings, and among them are occasional records of embryo and young Mollusca (including Gasteropoda, Lamellibranchiata, and Cephalopoda), young and larval Annelides, as well as plutei and other immature stages of star-fishes.

Infusoria and microscopic algæ are also described as being occasionally in greater or less abundance, but the records of these forms are few in number.

There are occasional records of larval and post-larval fishes among the lists of tow-net gatherings, but they also are comparatively few.

(6.) PELAGIC FISH OVA.

There are a considerable number of records of pelagic fish ova among the published lists of tow-net gatherings collected during the seven years. These records, which for the sake of clearness are arranged in tabular form similar to the other Tables, show that the total numbers for the inner and outer stations are in inverse ratio to the records of young Crustacea for the same stations, and this difference may probably be accounted for by the outer stations being nearer to the more important spawning grounds situated in the vicinity of the estuary.

[TABLE.

TABLE XIII., showing the Distribution of Fish Ova, as indicated by the Tow-net Records for the Years, Months, and Stations.

Three Inner Stations.	Years.	Jan.	Feb.	March.	April.	May.	June.	July.	Aug.	Sept.	Oct.	Nov.	Dec.	Totals for the Years and Stations.
	1889,	–	–	–	–	–	1	–	–	–	–	–	–	1
	1890,	–	–	–	1	1	–	–	–	–	–	–	–	2
	1891,	–	–	–	1	2	1	–	1	–	–	–	–	5
I.	1892,	–	–	–	1	1	1	1	–	–	–	–	–	4
	1893,	–	–	–	1	1	1	–	–	–	–	–	–	3
	1894,	–	–	–	1	–	–	–	–	–	–	–	–	1
	1895,	–	–	–	–	1	–	–	–	–	–	–	–	1
Total number of Records for all the Months,		–	–	–	5	6	4	1	1	–	–	–	–	17
	1889,	–	–	–	–	–	–	–	–	–	–	–	–	–
	1890,	–	–	–	1	–	–	–	–	–	–	–	–	1
	1891,	–	–	–	–	1	1	1	1	–	–	–	–	4
III.	1892,	–	–	–	1	–	1	–	–	–	–	–	–	2
	1893,	–	–	–	–	1	–	–	–	–	–	–	–	1
	1894,	–	–	–	2	–	–	–	–	–	–	–	–	2
	1895,	–	–	–	–	1	–	–	–	–	–	–	–	1
Total number of Records for all the Months,		–	–	–	4	3	2	1	1	–	–	–	–	11
	1889,	–	–	–	–	–	1	–	–	–	–	–	–	1
	1890,	–	–	–	1	2	–	–	–	–	1	–	–	4
	1891,	–	–	–	–	2	1	1	1	–	–	–	–	5
IV.	1892,	–	–	–	1	–	–	–	–	–	–	–	–	1
	1893,	–	–	–	–	1	–	–	–	–	–	–	–	1
	1894,	–	–	–	1	–	–	–	–	–	–	–	–	1
	1895,	–	–	–	–	1	–	–	–	–	–	–	–	1
Total number of Records for all the Months,		–	–	–	3	6	2	1	1	–	1	–	–	14

[TABLE.

TABLE XIII.—*continued.*

Three Outer Stations.	Years.	Number of the Records for each Month of each Year.												Totals for the Years and Stations.
		Jan.	Feb.	March.	April.	May.	June.	July.	Aug.	Sept.	Oct.	Nov.	Dec.	
	1889,	–	–	–	–	1	–	–	–	–	–	–	–	1
	1890,	–	–	1	1	1	–	–	–	–	–	–	–	3
	1891,	–	–	1	1	1	1	1	1	–	–	–	–	6
V.	1892,	–	–	1	–	2	1	1	–	–	–	–	–	5
	1893,	–	–	–	–	1	–	–	1	–	–	–	–	2
	1894,	–	1	–	1	–	–	–	–	–	–	–	–	2
	1895,	–	–	–	1	1	–	–	1	–	–	–	–	3
Total number of Records for all the Months,		–	1	3	4	7	2	2	3	–	–	–	–	22
	1889,	–	–	–	–	–	1	–	–	–	–	–	–	1
	1890,	–	–	1	1	1	–	–	–	–	–	–	–	3
	1891,	–	1	–	1	1	–	1	–	–	–	–	–	4
VIII.	1892,	–	–	1	–	3	1	1	–	–	–	–	–	6
	1893,	–	–	–	–	1	–	–	1	–	–	–	–	2
	1894,	–	1	–	1	–	–	–	–	–	–	–	–	2
	1895,	–	1	–	1	1	–	–	–	–	–	–	–	3
Total number of Records for all the Months,		–	3	2	4	7	2	2	1	–	–	–	–	21
	1889,	–	–	–	–	–	–	1	–	–	–	–	–	1
	1890,	–	–	–	1	–	–	–	–	–	1	–	–	2
	1891,	–	1	–	1	1	1	1	1	–	–	–	–	6
IX.	1892,	1	–	1	–	2	1	1	–	–	–	–	1	7
	1893,	1	–	–	1	–	2	–	–	–	–	–	–	4
	1894,	–	1	–	1	–	–	–	1	–	–	–	–	3
	1895,	–	–	–	1	1	–	–	1	–	–	–	–	3
Total number of Records for all the Months,		2	2	1	5	4	4	3	3	–	1	–	1	26

It will be seen by a glance at the annexed Table that while the number of records of fish ova for Stations I., III., and IV. are 16, 11, and 14 respectively, those for the outer stations, V., VIII., and IX., are 22, 21, and 26. Moreover, it may be stated in connection with this difference between the inner and outer stations, and as tending to emphasise it, that the inner stations were examined a greater number of times than the outer stations, owing to the outer stations being more exposed, and therefore less easily examined during bad weather.

Formula (1) shows the number of records of pelagic fish ova for each of the six stations, thus :—

	(*a*) Inner Stations.			(*b*) Outer Stations.		
	I.	III.	IV.	V.	VIII.	IX.
Numbers of the Stations,						
Total number of Records for each Station,	16	11	14	22	21	26

This formula exhibits a marked difference between the number of records for the inner and outer stations, those for the outer stations being considerably larger. The total numbers for each of the two groups of stations are 41 and 69 respectively, showing a difference of 28 in favour of the outer stations, which is equal to nearly 41 per cent.—a difference which, as already stated, is doubtless owing to the outer stations being nearer the more important spawning-grounds.

Formula (2) shows the total number of records of pelagic fish ova for each of the twelve months, for (*a*) the inner stations, I., III., and IV.; (*b*) the outer stations, V., VIII., and IX.; and (*c*) for all the six stations :—

Names of the Months,	Jan.	Feb.	Mar.	April.	May.	June.	July.	Aug.	Sep.	Oct.	Nov.	Dec.
(*a*) the Inner Stations,	0	0	0	11	14	9	3	3	0	1	0	0
(*b*) the Outer Stations,	2	6	6	13	18	8	7	7	0	1	0	1
(*c*) all the Six Stations,	2	6	6	24	32	17	10	10	0	2	0	1

The monthly maxima of the records of pelagic fish ova, as might be expected, are to be found in the early spring and summer months, those for April and May especially being considerably higher than for the other months. The arrangement of the numbers to correspond with the inner and outer stations brings out another contrast between them : for the inner stations there are no records of fish ova till April, nor are there any after October, but for the outer stations there are two records in January, six in February and in March, and one in December ; while with the exception of June the numbers for the five central months are considerably above those of the corresponding months for the inner stations. This seems to indicate that spawning outside begins at a comparatively early period of the year, and may continue more or less over all the months : inside the estuary, on the other hand, spawning is later in commencing, and rarely continues beyond the month of August. It is probable that some of the ova collected at the inner stations may have drifted in or been carried in by currents from the outside, but it is also well known that there are a number of fishes, whose eggs float, that spawn within the estuary, so that the majority of the records for the inner stations are those of fishes that have really spawned there.

The next *Formula* (3) shows the total number of records of pelagic fish ova for each of the seven years for (*a*) the inner stations, (*b*) the outer stations, and (*c*) for all the six stations :—

	1889.	1890.	1891.	1892.	1893.	1894.	1895.
(*a*) the Inner Stations,	2	7	14	7	5	4	3
(*b*) the Outer Stations,	3	8	16	18	8	7	9
(*c*) all the Six Stations,	5	15	30	25	13	11	12

The largest number of records are those for 1891 and 1892. Another point in connection with the distribution of the pelagic fish ova, as shown by the published tow-net records, is the occurrence of them in several of the bottom tow-net gatherings. The records show that pelagic fish ova were found in bottom tow-net gatherings on eight different occasions, and it is noteworthy that on a few of these occasions no fish eggs were observed in the surface tow-net. This feature in the distribution of the pelagic fish ova cannot be shown in the Table, but is clearly indicated in the chart showing the comparative frequency of fish ova during the various months and years for all the stations. Pelagic fish ova were found in the bottom tow-net in 1889 at Station IV. in June; in 1890 at Station IV. in April and May, and at Station V. in May; in 1891 at Station I. in May, and in the same month at Station IV.; in 1892 at Station VIII. in May; and in 1894 at Station III. in April. In the records for all the seven years pelagic fish ova are three times described as abundant, and this only in 1894, when they are so described once for each of the three outer stations.

(7.) EXPLANATION OF THE CHARTS (PLATES IV. TO VII.).

Before concluding this paper on the tow-net gatherings collected in the Firth of Forth in connection with the trawling investigations of the Fishery steamer *Garland* during the years 1889 to 1895, and published in the Annual Reports for these years, I desire to say a few words about the charts appended hereto, by way of explaining their contents. These charts are intended to supplement what has already been said concerning the distribution of a few of the organisms that are numerically or otherwise of greater importance than the others in their relation to fishery questions. The charts exhibit certain important features respecting the distribution of these organisms that could not be readily shown in the Tables, by indicating concisely and graphically the relative scarcity or abundance of these organisms as well as their distribution vertically and horizontally. The years and months are arranged in regular sequence along the top of each chart. Where the organisms are present in the tow-net gatherings they are indicated under four degrees of frequency, viz. :—(1) as abundant, represented by the contraction " ab."; (2) as common, represented by " com. "; (3) as frequent, represented by " fr."; and (4) as few or rare, represented by " f. or r."; if the organisms are altogether absent or are not referred to in the tow-net lists, this fact is indicated by the word " none," which is the last in the series. These abbreviations are used for each of the six stations, and are arranged alongside of them. The tow-net records for each month are represented by round dots; the dots are arranged to show whether the species referred to on the chart was abundant, common, frequent, few or rare, or absent in the gathering represented. Moreover, these dots are joined together by thick or thin lines : the thick lines show that the records are for bottom tow-net gatherings, and the thin lines for surface tow-net gatherings. When the distribution of two organisms or groups of organisms are represented separately on the same chart, they are distinguished from each other by using black and red lines. There is one other point in connection with the charts that requires to be explained. It has been stated that, for various reasons, the selected stations were not examined every month during the seven years, and that there are now and again omissions of the monthly records of tow-net fauna, either because no trawling was done, or if done that the usual lists of the contents of the tow-nets could not be prepared. The months when such omissions occurred are indicated on the charts by the thick and thin lines when crossing the column or columns under these months being discontinuous.

The first chart (Plate IV.) illustrates the distribution of *Calanus finmarchicus* and *Temora longicornis*. The first is shown by the black lines, the other by the red lines.

The second chart (Plate V.) illustrates the distribution of *Sagitta bipunctata* and the Cœlenterata. *Sagitta* is represented by the black lines, and the Cœlenterata by the red lines.

o

The third chart (Plate VI.) illustrates the distribution of the larval and young Decapod Crustacea (including the Schizopoda) and that of the larval and young *Balani*.   The *Balani* are represented by the red lines, the others by the black lines.

The fourth chart (Plate VII.) shows the distribution of the pelagic fish ova so far as that is indicated by a study of the tow-net records for the seven years.

The following list contains the names of all the species of Mollusca and Crustacea referred to in the preceding notes, and of *Sagitta* and *Tomopteris*, and the stations where each was obtained :—

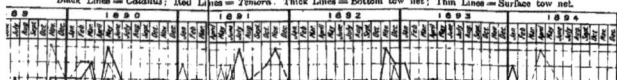

CHART showing the distribution and relative abundance of *Calanus finmarchicus* and *Temora longicornis* in the Firth of Forth. Black Lines = *Calanus*; Red Lines = *Temora*. Thick Lines = Bottom tow net; Thin Lines = Surface tow net.

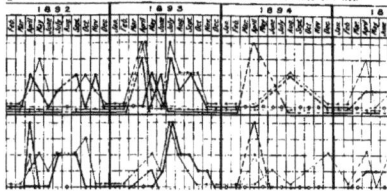

g *Balani*. Thick Lines = Bottom tow net; Thin Lines = Surface tow net.

(8. LIST OF THE SPECIES OF MOLLUSCA AND CRUSTACEA REFERRED TO IN THE PRECEDING NOTES, AND SHOWING THE STATIONS WHERE THEY WERE OBTAINED.

Names of the Species.	Station I.	Station III.	Station IV.	Station V.	Station VIII.	Station IX.	Remarks.
MOLLUSCA.							
NUDIBRANCHIATA.							
Doris sp. (? *repanda,* A & H.), -			×				1890, in November.
Doto coronata, Gm., - - - -				×			1890, in December.
Eolis sp. (1), - - - - -					×		1891, in April.
Eolis sp. (2), - - - - -	×						1895, in May.
LAMELLIBRANCHIATA.							
Cuspidaria cuspidata (Olivi), -						×	1893, in April.
PTEROPODA.							
Limacina retroversa (Flem.), -	×	×		×		×	Not uncommon.
Clione borealis (Brug.), -		×			×		1889, Jan.; 1893, Dec.
CRUSTACEA.							
DECAPODA.							
Macropodia rostrata (Lin.), -			×				1890 & '91, Oct. & Aug.
Crangon allmanni (Kin.), -		×	×	×	×	×	Frequent.
Egeon fasciatus, Risso, -		×					1891, in September.
Cheraphilus nanus (Kroyer), -		×	×				Not very rare.
Hippolyte varians, Leach, -				×			1890, December.
Hippolyte fascigera, Gosse, -					×		1894 & '95, Jan., Feb.
Spirontocaris pusiola (Kroyer), -			×				1895, February.
Pandalus montagui, Leach, -		×		×	×		Not rare in tow-nets.
Pandalus brevirostris, Rathke, -		×		×			1881, March, Sept.
SCHIZOPODA.							
Boreophausia raschii (M. Sars), -	×	×	×	×	×	×	Mod. com., 1889-90.
Thysanoessa neglecta (Kr.), -	×	×	×	×	×	×	Mod. com., 1891-92.
Siriella jaltensis, Czern., -				×			1893, March.
Siriella armata (M'Edw.), -			×				1892, March.
Gastrosaccus spinifer (Goes.), -	×	×		×	×	×	Not uncommon.
Heteromysis formosa, S. L. Smith, -	×	×	×				1894-95, Aug., Feb.
Erythrops goesii, G. O. Sars, -	×	×		×	×	×	Frequent.
Mysidopsis didelphys (Norm.), -	×	×			×	×	1890-91, March.
Mysidopsis gibbosa, G. O. Sars, -	×	×	×				Not uncommon.
Mysidopsis angusta, G. O. Sars, -	×	×					1891, March.
Leptomysis gracilis, G. O. Sars, -	×	×		×			1890-91, Oct.-Dec.
Leptomysis lingura, G. O. Sars, -	×	×					1892, March.
Hemimysis lamornæ (Couch), -	×	×			×	×	Not very rare.
Macropsis slabberi (Van Ben.), -	×	×	×	×			Locally frequent.
Praunus flexuosus (Müller), -			×	×			Locally frequent.
Praunus inermis (Rathke), -				×	×	×	1895, February.
Schistomysis spiritus, Norm., -	×	×	×	×	×		Not uncommon.
Schistomysis ornatus, G. O. Sars, -	×	×	×	×	×	×	Frequent.
Neomysis vulgaris (J. V. Thomp.), -			×				Locally frequent.
CUMACEA.							
Iphinoë trispinosa (Goodsir), -	×						1889-90, July, Feb.
Cumopsis goodsiri (V. Ben.), -		×					1891, September.
Leucon nasicus, Kroyer, -	×			×	×		Not very rare.
Eudorella truncatula (Spence Bate), -	×			×	×		Not very rare.
Diastylus rugosa, G. O. Sars, -	×			×	×		Not very rare.
Pseudocuma cercaria (V. Ben.), -	×			×	×	×	Frequent.
Campylaspis rubicunda (Lillj.), -						×	1893, April; 1894, Aug.
Petalosarsia declivis (G. O. Sars), -						×	1893, April.

Names of the Species.	Station I.	Station III.	Station IV.	Station V.	Station VIII.	Station IX.	Remarks.
ISOPODA.							
Gnathia (?) *maxillaris* (M'Edw.), -					×		1890, March.
Eurydice pulchra, Leach, -		×					1890, November.
Idothea baltica (Pallas), -	×	×	×	×	×	×	Frequent.
Idothea emarginata (Fabr.), -	×						1891, May.
Idothea linearis (Lin.)				×			1893, January.
Janira maculosa, Leach, -						×	1890, April.
AMPHIPODA.							
Hyperia galba (Mont.), -	×	×	×	×	×	×	Frequent.
Hyperoche tauriformis (Bate), -	×	×	×	×	×	×	Frequent.
Parathemisto oblivia (Kr.), -	×	×		×	×	×	Sometimes common.
Euthemisto compressa (Goes), -					×		1892, Feb., Nov.
Callisoma crenata (Bate), -	×				×	×	1890, Oct.; 1893, Nov.; 1895, Dec.
Hippomedon denticulatus (Bate), -						×	1894, February.
Bathyporeia sp., -				×			1894, February.
Ampelisca sp., -		×					1891, October.
Stenothoë marina (Bate), -				×			1890, April.
Argissa hamatipes (Norm.), -	×			×	×	×	Not uncom. at Outer St'ns.
Metopa alderi, Bate, -	×	×		×	×	×	Not rare in bot. gatherings.
Periocalodes longimanus (Bate), -				×	×	×	Frequent.
Iphimedia obesa (Rathke), -					×		1892, January.
Apherusa bispinosa (Bate), -	×	×	×	×	×	×	Frequent.
Apherusa borealis (Boeck), -	×	×					Not very rare.
Paratylus swammerdami (M'Edw.), -	×	×	×	×			Frequent.
Dexamine sp., -		×					Rare.
Melphidipella macera (Norm.), -					×		1892, March.
Amathilla homari (Fabr.), -				×			1892, December.
Gammarus sp., -				×			Rare.
Gammaropsis erythrophthalmus, Lillj., -					×		1890, December.
Dulichia sp., -						×	Rare.
Pariambus typicus (Kr.), -		(?)	(?)		×	×	Not uncommon.
CLADOCERA.							
Evadne nordmanni, -		×	×	×	×	×	Frequent.
Podon (?) *polyphemoides*, -		×	×		×	×	Not uncommon.
COPEPODA.							
Calanus finmarchicus (Gun.), -	×	×	×	×	×	×	Common.
Pseudocalanus elongatus, Boeck, -	×	×	×	×	×	×	Frequent.
Temora longicornis (Müll.)	×	×	×	×	×	×	Common.
Centropages hamatus, Lillj., -	×				×		1889, June; '95, May, Dec.
Centropages typicus, Kr., -				×	×	×	1889, Aug.; 1890, Oct.
Metridia hibernica (B. & R.), -						×	1892, May.
Candace pectinata, Brady, -	×		×	×	×	×	Not very rare.
Acartia sp., -	×	×	×	×	×	×	Common.
Anomalocera patersoni (Temp.), -	×	×	×	×	×	×	Frequent.
Parapontella brevicornis (Lub.), -	×						1893, June.
Oithona (?) *similis*, Claus, -	×		×				Occasionally frequent.
Thalestris longimana, Claus, -		×					1891, June.
Thalestris serrulatus, Brady, -	(?)					×	1891, June.
Alteutha depressa (?) (Baird), -	×	×		×		(?)	Not very rare.
Dermatomyzon nigripes (B. & R.), -	×						1894, August.
Monstrilla sp., -	×	×	×	×			1890-92, Aug., Oct.; 1893,
Caligus rapax (Müller), -	×	×	×		×	×	Frequent. [Oct.
ANELLIDA and MEDUSA.							
Sagitta bipunctata, G. & G., -	×	×	×	×	×	×	Common.
Tomopteris oniscifrucis, Esch., -	×	×	×	×	×	×	Frequent.
Aurelia aurita, Lam., -	×			×			Sometimes common.

VII. NOTES ON RECENT GATHERINGS OF MICRO-
CRUSTACEA FROM THE CLYDE AND THE MORAY
FIRTH. By Thomas Scott, F.L.S., Mem. Soc. Zool. de France.

Pub. July 28th 1899

(Plates X–XIII.)

In the following notes my remarks refer chiefly to the rarer forms that
have been observed in gatherings of Microcrustacea submitted for
examination during 1898. The gatherings examined have been collected
in the Moray Firth and the Firth of Clyde, and therefore the notes refer
chiefly to these localities. I am indebted to Mr. F. G. Pearcey,
naturalist on board the s.s. "Garland," for most of the gatherings
forwarded for examination.

COPEPODA.

Paracalanus parvus (Claus).

>   1863. *Calanus parvus*, Claus. Die frei-lebenden Copopoden,
>   p. 173, Pl. XXVI., figs. 10-14; Pl. XXVII., figs. 1-4.

This Copepod, which was observed in the Firth of Clyde for the first
time in September 1897, has occurred in several of the tow-net gatherings
collected during 1898, both at the surface and bottom. It was obtained
at Stations IX. and XIII. in August, and at Stations VII. and VIII. in
September. Station XIII. is one of the Upper Loch Fyne stations, and
its occurrence there makes it an addition to the Loch Fyne fauna. It is
a small species, and may therefore have been previously passed over as a
young *Calanus*. It has also been observed in the Moray Firth and in
the Firth of Forth.

*Bradyidius armatus (Brady).

>   1878. *Pseudocalanus armatus*, G. S. Brady, Mon. Brit. Copep.,
>   vol. i., p. 46 (non *P. armatus*, Boeck—see Giesbrecht, *Zool.
>   Anzeiger*, 1897, p. 25).

This species is referred to because its distribution appears to be
somewhat restricted. Though not very plentiful, it is one of the more
widely diffused of the Clyde Copepoda, and it has been known for
many years as a Clyde species. It is usually obtained in gatherings
collected with the bottom tow-net, and much less frequently in surface
gatherings. Though the species has been recorded from the East Coast
of Scotland, it seems to be of rare occurrence there. I do not find a
single reference to it in any of the descriptions of tow-net gatherings
collected on the East Coast during the past year.

Euchæta norvegica, Boeck.

>   1864. *Euchæta prestandreæ*, Boeck, Overs. Norg. Kyster iagt.
>   Copep., *Forh. Vid. Selsk. Christiania*, p. 236.
>   1872. *Euchæta norvegica*, Boeck, Nye Slaegt. og Art. af Salt-
>   vandscopep, *Forh. Vid. Selsk. Christiania*, p. 40.

I have no record of this species for the Moray Firth; but it has, as
in previous years, been obtained in tow-net gatherings from various

---

* See " Additional Notes ' at the end of this paper.

parts of the Clyde and Loch Fyne.   *Euchœta* appears to be even more restricted in its distribution than *Bradyidius*.   According to Sars ("Norw. North Sea Exped.," Crust., Part I., p. 284), this species was at first ascribed by Boeck to *Euchæta prestandreæ*, Philippi, but was afterwards described by him under the name which it now bears.

Scolecithrix hibernica, A. Scott.

1896.   *Scolecithrix hibernica*, A. Scott, Ann. and Mag. Nat. Hist., (6) vol. xviii., p. 362, Pl. XVII. and XVIII.

This species, though only recently discovered, is at times not very rare in the Clyde and Loch Fyne tow-net gatherings.   The following records of its occurrence will indicate sufficiently the wide distribution of the species in the Clyde district.   During August last year it was obtained at Stations III. and IV. (in Kilbrennan Sound), and at Stations XIII., XIV., XV., and XVII. (Upper Loch Fyne), and in September at Stations VII. and VIII. (4 or 5 miles south of Ailsa Craig).

I have now to record its occurrence for the first time in the Moray Firth, having obtained it in a tow-net gathering collected in June last year at Station XVI. (vicinity of Smith Bank) from a depth of about forty fathoms, but it was apparently rare in this gathering.   The opinion expressed by the describer of the species that *Scolecithrix hibernica* was really a deep-water form, and that its being so would partly account for its having been so long overlooked, is more or less confirmed by what is observed regarding its distribution in the Clyde and in the Moray Firth.

Scolecithrix pygmæa, sp. n. (Pl. X., figs. 1–9).

Description of the Female.—Somewhat like *Scolecithrix hibernica*, A. Scott, in general appearance but smaller, the length of the specimen figured is, exclusive of tail setæ, ·95 mm. (about $\frac{1}{27}$ of an inch).   The rostrum is small.   The last segment of the thorax is produced on each side into a hook-like process (fig. 1).   The antennules are scarcely as long as the thorax; they are twenty-four jointed; the first two joints are moderately large; the third to the seventh are smaller; but the eighth is about twice as long as the preceding joint, and sub-equal in length to the first and second.   The joints that immediately follow the eighth are shorter, but the others gradually increase in length, so that several of the last joints are about as long as the eighth.   The end joint is very small. The antennules are only sparingly setiferous, but the terminal joints are furnished with a few plumose hairs as shown by the figure.   The formula gives approximately the proportional lengths of all the joints, as follows :—

Numbers of the joints,	1	2	3	4	5	6	7	8	9	10	11	12	13	14	15	16	17	18	19	20	21	22	23	24
Lengths of the joints,	12	15	8	5	6	6	12	8	7	·8	·9	·9	·9	·9	11	11	11	11	11	11	10	14	12	·5

The antennæ, mandibles, and maxillæ are all somewhat similar to those of *Scolecithrix hibernica*.   The anterior foot-jaws are furnished with several lobes on the inner aspect as in *Scolecithrix dubia*, Giesbrecht. The distal lobe is armed with a long slender spine, but the others are setiferous.   The special joint of the anterior foot-jaws carries a number of the long slender worm-like hairs which form one of the principal characters of the genus (fig. 3).   The posterior foot-jaws are elongate, and somewhat like those of *Scolecithrix hibernica* (fig. 4). The first four pairs of swimming feet are also somewhat similar to those of *Scolecithrix hibernica*, except that the marginal spines of the outer branches of the fourth pair are stouter than those of the outer branches of the same pair in that species. The terminal spines are also slightly different (figs. 5, 6). The fifth pair appears to be wanting in the female. The abdomen is, proportionally, scarcely so long as that of *Scolecithrix hibernica*. The first segment is about equal to the combined length of the next two, and is rather more dilated ; the second segment is somewhat shorter than the third ; but the length of the third and fourth is about equal. The caudal furcæ, which are about as long as broad, are somewhat longer than the segment to which they are articulated, and the furcal setæ are long and plumose (fig. 8).

Description of the Male.—The male of *Scolecithrix pygmœa* resembles that of *Scolecithrix hibernica* in several aspects, but differs particularly in the structure of the fifth pair of thoracic feet.. In this pair the basal joint is stout, and armed with several curved spines as in *Scolecithrix hibernica*, but the right branch is more slender, and the first joint of that branch is not so irregular in outline. In the present species the first joint of the right branch becomes gradually but only slightly dilated towards the distal end, and is not produced into a large lobe as in *Scolecithrix hibernica* ; the second joint is smaller, and proportionally much more slender than in that species ; the marginal thumb-like process is small, and situated near the middle of the joint. The left branch of the present form (fig. 7 l.) is also rather more slender than that of *Scolecithrix hibernica*, but the process at the distal end of the second joint is somewhat more produced and attenuated. The abdomen consists of five segments. The second, which is rather longer than the first, is about one and a half times the length of the following segment ; the third and fourth segments are sub-equal, but the last is very small ; the furcæ are about equal in length to the penultimate segments of the abdomen (fig. 9). Figure 10 represents the fifth thoracic feet of the male of *Scolecithrix hibernica* for comparison with those of the species now described. The figures of both are of the same magnification.

Habitat.—Firth of Clyde and Loch Fyne. Not very rare.

Remarks.—This *Scolecithrix* has been under observation for a considerable time. At first I was inclined to regard it simply as a form of *Scolecithrix hibernica*, but as it continues to turn up both alone and in company with that species, and as all of the specimens are characterised by the same distinctive features, I think it will be more satisfactory to describe it under a separate name. It is distinctly a smaller species than *Scolecithrix hibernica*, being scarcely a millemetre in length. If male and female specimens of the two species be placed side by side—the males together and the females together—the difference in size is readily noticed. The structure of the fifth thoracic feet of the male, and the structure and armature of the anterior foot-jaws of the female, are characters by which the species may be distinguished ; the lengths of the abdominal segments in both male and female are also proportionally different.

Centropages typicus, Kroyer.

        1849.   *Centropages typicus*, Kroyer. Nat. Tidskr., (2) ii.,   p. 588, t. 6.

This species appears to be much rarer in the Clyde than *Centropages hamatus*, for while the latter form occurs in nearly all gatherings collected in August and September last year, I have only three records for *Centropages typicus*. On the East Coast of Scotland *Centropages typicus* appears to be more common. In a series of gatherings from the

Moray Firth collected during May and June last year, both species were nearly equally frequent, *Centropages hamatus*, however, was even here rather the more common of the two species.

Isias clavipes, Boeck.

> 1864. *Isias clavipes*, Boeck.   Overs. Norg. Copep., Forh. Vid. Selsk, Christiania, p. 18.

This fine species has been obtained in several of the Clyde tow-net gatherings, both surface and bottom.   But though occurring now and again in various parts of the Clyde, it seems to be always a scarce species. The following are a few of the more recent records of *Isias* from the Clyde—at Station II. (Kilbrennan Sound), in surface and bottom tow-nets, at Station XII. (between Arran and the Heads of Ayr), in the bottom tow-net in August, and in the surface tow-net at Station VII. in September 1898.

**Eurytemora lacinulata* (Fischer).

> 1853. *Cyclopsina lacinulata*, Fischer, Beitr. z. Kenntn. d. Cyclo-piden, Bull. Soc. Imp. Natur., Moscow, XXVI., p. 86–90, Pl. II., figs. 4–17, 34.

This species has been obtained during the past year in brackish water-pools at Hunterston, Firth of Clyde.   *Eurytemora lacinulata* has been recorded from the Clyde district on one or two former occasions, but not previously from Hunterston.

**Metridia hibernica* (Brady and Robertson).

> 1873. *Paracalanus hibernicus*, Brady and Robertson, Ann. and Mag. Nat. Hist. (4), vol. xii., p. 126, Pl. VIII., figs. 1–3.

This species has already been recorded from Loch Fyne, and seems to be distributed, though very sparingly, all over the Clyde estuary. *Metridia hibernica* somewhat resembles *Metridia longa* (Lubbock), but is smaller than that species.   It may be difficult to discriminate between the young of the two species; but there need be little difficulty in distinguishing the adult forms, especially if males are present.   *Metridia hibernica* has also been observed both in the Moray Firth and in the Firth of Forth.

**Candace pectinata*, Brady.

> 1878. *Candace pectinata*, Brady, Mon. Brit. Copep., vol. i. p. 49, Pl. VIII. figs. 14, 15 ; Pl. X. figs. 1–12.

Though this species has been obtained both in the Firth of Clyde and the Moray Firth, as well as in the Firth of Forth, it has not been observed within recent months.

Labidocera wollastoni (Lubbock).

> 1857. *Pontella wollastoni*, Lubbock, Ann. and Mag. Nat. Hist. (2), vol. xx., p. 406, Pl. X. fig. 13 ; and Pl. XI. figs. 9–11, 18.

The only records of this species I have for the past year are two, and both are for the Clyde.   They are as follows : --In a bottom tow-net gathering collected at Station IX. on August 31st, rare ; and in a surface gathering collected at Station VIII. on September 23rd (only one specimen was observed in this gathering).

* See '' Additional Notes '' at the end of this paper.

Anomalocera patersonii, Templeton.

> 1837. *Anomalocera patersonii*, Templeton, Trans. Entom. Soc.,
> vol. ii. p. 35, Pl. V., figs. 1–3.

This species, which is one of the most richly coloured of the British Copepoda, was occasionally observed during the past year in tow-net gatherings both from the Firth of Clyde and the Moray Firth, but it seldom occurred in any quantity.

Parapontella brevicornis (Lubbock).

> 1857. *Pontella brevicornis*, Lubbock, Ann. and Mag. Nat. Hist.
> (2), vol. xx., Pl. XI., figs. 4–8.

Though *Parapontella* may occasionally be found moderately common more frequently one or a few specimens only are obtained in any single gathering—such, at least, is my experience in regard to the distribution of this species in the Scottish seas. *Parapontella brevicornis* has during the past year occurred sparingly both in the Firth of Clyde and in the Moray Firth. Living specimens are readily distinguished, even amongst crowds of *Calanus, Pseudocalanus, Temora*, etc., by their peculiar dark or blackish colour, but much of this colour is lost when the specimens are preserved in spirit.

Acartia clausii, Giesbrecht.

> 1889. *Acartia clausii*, Giesbrecht, *Rendiconti R. Accad. d. Lincei*,
> vol. v., fasc. 11.

This is the only species of *Acartia* I have hitherto observed in the Clyde district. The spines, with which the fifth pair of feet of the female are armed, are short and very stout, and therefore very different from those of *Acartia longiremis*, Lilljeborg. In the Moray Firth district both *Acartia clausii* and *Acartia longiremis* are met with ; the first is frequent in the open sea, but it has also been observed inshore. On June 6th, 1898, both species occurred in a gathering collected at Station III. (Cromarty Firth), where there is usually a more or less admixture of fresh water, and also at Stations I. and II. (off the Nairn Coast) on the 7th of the same month. Neither *Acartia bifilosus*, Giesbrecht, nor *Acartia discaudata*, Giesbrecht, have been observed in the Moray Firth district, but it is quite possible that they may yet be found there—especially in that part of the district known as the Beauly and Cromarty Firths, where the conditions seem to be favourable for these two species.

Cervinia bradyi, Norman.

> 1878. *Cervinia bradyi*, Norman ; Brady, Mon. Brit. Copep.,
> vol. i., p. 86, Pl. XXIVᴀ., figs. 3–13.

A single specimen of this curious species was obtained in a small gathering of Microcrustacea washed from a quantity of mud brought up in the bottom tow-net at Station XII. (Firth of Clyde) on 29th August 1898, from a depth of from forty to forty-three fathoms. *Cervinia* was discovered at Oban by the Rev. A. M. Norman in 1877. It has also been recorded from the Irish Sea by I. C. Thompson, of Liverpool ; but this appears to be the first time the species has been observed in the Clyde. It is quite distinct from any other species of the British Copepoda.

Ectinosoma curticorne, Boeck.

> 1864. *Ectinosoma curticorne*, Boeck, Overs. Norg. Copep. Forh. Vid. Selsk. Christiania.

This species has been obtained during the past year at Hunterston, Firth of Clyde.

Ectinosoma herdmani, T. and A. Scott.

> 1896. *Ectinosoma herdmani*, T. and A. Scott, Rev. Brit. Copep. belonging to the gen. *Bradya* and *Ectinosoma*, p. 432, Pl. XXXVI., &c.

This species was obtained during the past year at Hunterston, Firth of Clyde, and also in a gathering from Cromarty Firth, collected 4th November 1897. The last is a new record for this species.

Ectinosoma gracile, T. and A. Scott.

> 1896. *Ectinosoma gracile*, T. and A. Scott, Rev. Brit. Copep. belonging to the gen. *Bradya* and *Ectinosoma*, p. 429., Pl. XXXVI., &c.

Several specimens of this apparently rare species were obtained in a shore gathering collected on 4th November 1897, a little to the east of Invergordon, Cromarty Firth. This is the only time I have obtained *Ectinosoma gracile* since it was discovered in the Firth of Forth in 1896. It is a very small species and easily overlooked. One of the Cromarty Firth specimens with ova measured only 0·43mm. ($\frac{1}{58}$th of an inch).

Bradya typica, Boeck.

> 1872. *Bradya typica*, Boeck, Nye Slægt. og Art. of Saltv. Copep., Forh. Vid. Selsk. Christiania, p. 14.

This species occurred in a gathering collected at Station XVI. (Moray Firth), 10th July 1898, but only one or two specimens were obtained. It has also been obtained in a gathering from Station XII. (Firth of Clyde), collected 29th August 1898; in one from Station XV., collected 22nd August; and in another from Station XVII. (both in Upper Loch Fyne), collected on the 24th of the same month.

Bradya hirsuta, T. and A. Scott.

> 1896. *Bradya hirsuta*, T. and A. Scott, Rev. Brit. Copep. belonging to the gen. *Bradya* and *Ectinosoma*, p. 423, Pl. XXXV., &c.

This species was obtained in a gathering collected at Station XVI. (Moray Firth), 10th July, 1898, but only a few specimens were observed.

Misophria pallida, Boeck.

> 1864. *Misophria pallida*, Boeck, Overs. Norg. Copep., p. 24.

This is a somewhat rare species. It has already been recorded for Loch Fyne, and has also been found in Kilbrennan Sound, though not previously recorded. It occurred during the previous year in a bottom tow-net gathering from Station XVII. (Upper Loch Fyne), collected on 7th December, but only two specimens, however, were obtained in this gathering.

Robertsonia tenuis (G. S. Brady and Robertson).

> 1875. *Ectinosoma tenue,* B. and R., "Proceed. of the Brit. Ass.," p. 196.

A few specimens of this well-marked species occurred in a small gathering of Microcrustacea from Station XVI. (Moray Firth), depth 30 to 40 fathoms, collected on 10th June 1898. The gathering consisted of the washings of some dredged material.

Delavalia mimica, T. Scott.

> 1897. *Delavalia mimica,* T. S., Fifteenth Ann. Rep. Fish. Board Scot., Pt. III., p. 150, Pl. I., figs. 1–9.

This species was obtained in a gathering from Station XVI. (Moray Firth), collected on 10th June 1898, but was somewhat rare.

Delavalia æmula, T. Scott.

> 1893. *Delavalia æmula,* T. S., Eleventh Ann. Rep. Fish. Board Scot., Pt. III., p. 204, Pl. IV., figs. 36–47.

This species was obtained in some dredged material collected a little to the west of Invergordon, and was also somewhat rare.

Delavalia giesbrechti, T. and A. Scott, var. (Pl. XIII., figs. 20–22).

> 1890. *Delavalia giesbrechti,* T. and A. S., Ann. Scot. Nat. Hist., p. 225, Pl. IV., figs. 1–10.

A form of *Delavalia* has been obtained at Hunterston, Firth of Clyde, which resembles *Delavalia giesbrechti* very closely, but it wants the peculiar broad tail setæ that constitute such a marked feature in that species (fig. 22); there is also a slight difference in the arrangement of the four marginal setæ on the inner portion of the basal joints of the fifth thoracic feet (fig. 21). In other respects the two forms appear to be similar. Fig. 20, which represents the first pair of swimming feet, shows the second joint of the inner branches to be rather more slender than the same joint in typical *D. giesbrechti.*

Psyllocamptus, gen. nov.

Similar to *Canthocamptus,* but the inner branches of the first pair of swimming feet, which are about equal in length to the outer branches, are two-jointed, while the inner branches of the next three pairs are all three-jointed. Moreover, the antennules in spirit specimens are distinctly bent at the second joint as in *Nitochra.*

Psyllocamptus fairliensis, sp. n. (Pl. XIII., figs. 12–19).

Description of the Female.—Length about ·6 mm. ($\frac{1}{41}$ of an inch). Body slender and elongate, and resembling *Canthocamptus* in general appearance (fig. 12). Antennules moderately short and setiferous, eight-jointed; the first two robust, the penultimate joint smaller than the others (fig. 13); the formula shows approximately the proportional lengths of all the joints—

Number of the joints,	1 · 2 · 3 · 4 · 5 · 6 · 7 · 8
Proportional lengths of the joints,	24 · 24 · 7 · 9 · 6 · 7 · 4 · 6

Antennæ short, three-jointed; secondary branch small, one-jointed, and furnished with three setæ (fig. 14).

Mandibles as in *Canthocamptus*, mandible-palp one-branched, and furnished with a few setæ; the basal joint is armed with a moderately stout apical spine (fig. 15). Other mouth organs as in *Canthocamptus*. First pair of swimming feet with both branches moderately short, and of nearly the same length; the outer branches are three-jointed, but the inner are composed of only two joints, as in *Attheyella*; the first joint of the inner branch is about equal in length to the first two joints of the outer branches, and is also somewhat stouter; the second joint is little more than half the length of the first one (fig. 16).

The second, third, and fourth pairs of swimming feet have both branches three-jointed; the outer branches are considerably longer than the inner; in the fourth pair the inner branches are only about a third of the length of the outer branches (fig. 17); the first joint of the inner branches of the fourth pair are very small; the outer branches are furnished with long terminal setæ. The fifth pair are small and foliaceous; and the inner portion of the basal joint is considerably produced, and is subcylindrical in outline; the apex is subtruncate and bears four setæ, the two inner ones being small, while the other two are elongate; the secondary joint is ovate, and furnished with several setæ on the outer margin and apex. All the setæ are of considerable length, except a small one near the base of the outer margin (fig. 18). Caudal furca short, and about as broad as long (fig. 19).

Habitat.—Shore between Fairlie and Hunterston, Firth of Clyde. Rather rare. No males have been observed.

Remarks.—The Copepod just described resembles more or less closely not only *Canthocamptus*, but also *Attheyella* and *Mesochra*. It differs from the typical *Canthocamptus* in having the inner branches of the first pair of swimming feet only two-jointed, while the inner branches of all the next three pairs are three-jointed. It also differs from *Attheyella* and *Mesochra* in having the inner branches of the second, third, and fourth pairs three-jointed, though agreeing with these two genera in the structure of the first pair. Moreover, it somewhat resembles *Nitochra* in the antennules being distinctly bent at the second joint; but in that genus all the first four pairs of swimming feet have the inner branches three-jointed.

Psyllocamptus fairliensis seems to form one of the links in a chain of Copepods that at the one end terminates in *Mesochra*, which has the inner branches of all the four pairs of swimming feet two-jointed, and at the other end in *Nitochra*, which has the same branches all three-jointed. The generic name is derived from the two Greek words *psylla*—a flea, and *kamptos*—flexible.

Huntemannia jadensis, S. A. Poppe.

> 1884. *Huntemannia jadensis*, Poppe, Abhandl. d. nat. Ver. Bremen, Bd. IX., p. 59.
>
> 1885. *Huntemannia jadensis*, Poppe, Die freilebenden Copep. des Jadebusens, op. cit., Bd. XI., p. 167, Pl. VII., figs. 10-20
>
> 1895. *Huntemannia jadensis*, T. and A. Scott, Ann. and Mag. Nat. Hist. (6), vol. xv., p. 57, Pl. VI., figs. 21, 22.

This species was described by Dr. Poppe in 1884. It was first detected in Scotland in 1894, in brackish water pools, at the head of West Loch Tarbert (Cantyre), and a record of its occurrence there was published in the "Annals and Magazine of Natural History" for 1895, but up till the present time this appears to be the only record of its occurrence in Scotland. On 4th November 1897, Mr. F. G. Pearcey collected a gathering of small crustacea, on the shore near low-water mark, to the east of Inver-gordon, in the Cromarty Firth ; this he afterwards sent to me for exa-mination. Several rare Copepods have been obtained in this gathering, and one of them—*Ectinosoma gracile*—has already been referred to ; another of these rare forms is the species under consideration. Only four specimens of *Huntemannia* were obtained in this gathering from Cromarty Firth, so that the species, which is very well marked, is probably rare. The first pair of thoracic feet are stout, the outer branches are three, and the inner one-jointed. They are armed with strong marginal spines. The basal joint of the first feet carry each, interiorly, a comparatively large thumb-like process, instead of a spine. This process was quite conspicuous, even without dissection, in each of the four speci-mens obtained. Probably the species is local as well as rare.

Pseudotachidius coronatus, T. Scott.

> 1898. *Pseudotachidius coronatus,* T. Scott, Sixteenth Ann. Rep. Fish. Board for Scot, Pt. III., p. 267, Pl. XIII., figs. 12-26 ; Pl. XV., figs. 1-4.

This somewhat remarkable species was described in 1898 from one or two specimens obtained amongst some small Crustacea sent to me from Lower Loch Fyne by Mr. F. G. Pearcey. They had been dredged from 105 fathoms. I have now to record the species from other two localities in the Clyde district, and from moderately deep water—viz., from Station XII., 40-43 fathoms, washed from mud brought up in the tow-net, 29th August 1898 ; and from Station XVII., Upper Loch Fyne, washed from trawl refuse, 7th December 1898. Only one specimen was obtained in each of these two gatherings.

Tetragoniceps macronyx, T. Scott.

> 1892. *Tetragoniceps macronyx,* T. Scott, Tenth Ann. Rep. Fish. Board for Scot., Pt. III., p. 253, Pl. X., figs. 19, 28.

This well-marked and somewhat rare species was described from specimens obtained in the Firth of Forth. I have now to record its occurrence in the Cromarty Firth, having obtained it in a gathering of material dredged in the vicinity of Invergordon in 1896, but only recently examined.

Laophonte thoracica, Boeck.

> 1863. *Laophonte thoracica,* Boeck, Overs. Norg. Copep., p. 54.

What appears to be two forms of this species have been observed in the Moray Firth district, one a deep-water form, which appears to be the typical one. This was obtained in a gathering from Station XV. (vicinity of Smith Bank), depth 24-49 fathoms, collected 20th November 1897. The other was obtained in the Cromarty Firth, and a little to the west of Invergordon, where the depth of water is only a few fathoms, and where there is usually a certain admixture of fresh water.

Laophonte serrata (Claus).

> 1863. *Cleta serrata,* Claus, Die frei-lebenden Copep., p. 123, Pl. XV., figs. 13-20.

Laophonte serrata, which appears to be a rare species, occurred in the same gathering in which the *Huntemannia* was obtained, and is now for the first time recorded for the Cromarty Firth district.

Cletodes tenuipes, T. Scott.

> 1897. *Cletodes tenuipes*, T. Scott, Fifteenth Ann. Rep. Fish,
> Board for Scot., Pt. III., p. 170, Pl. I., figs. 19–27.

This species, which is apparently rare, was also obtained in the same gathering from Cromarty Firth in which the *Huntemannia* occurred. *Cletodes tenuipes* was described from Clyde specimens, and it is interesting now to find it also on the East Coast.

Cletodes perplexa, sp. n. (Pl. XI., figs. 12–20; Pl. XII., fig. 1).

Description of the Female.—Length of the specimen figured ·67mm. ($\frac{1}{37}$ of an inch). The body is stout anteriorly but tapers gradually towards the posterior end; in spirit specimens the tail is generally incurved as shown in the figure (fig. 12, Pl. XI.). Rostrum short and slightly recurved. Antennules very short, moderately stout, and composed of five joints; the first and second joints are large; the third is about half the size of the second; the fourth is very small; and the last is about one and a half times the length of the third (fig. 13, Pl. XI.). The approximate proportional lengths of the various joints are shown in the formula—

Numbers of the joints,　　$1 \cdot 2 \cdot 3 \cdot 4 \cdot 5$

Proportional lengths of the joints,　$\overline{21 \cdot 25 \cdot 12 \cdot 4 \cdot 19}$

There are a number of coarsely plumose setæ on the distal half of each antennule, and a small asthetask springs from the end of the third joint. Antennæ three-jointed; secondary branches small, each furnished with two coarsely plumose setæ and a small hair (fig. 14, Pl. XI.). Mouth organs nearly as in *Cletodes linearis* (Claus); figs. 15 and 16, Pl. XI., show the form of the anterior and posterior foot-jaws. The inner branches of the first four pairs of swimming feet, which are all two-jointed, have the first joint small, while the second is slender and elongate. The inner branches of the first pair have the first joint somewhat dilated, while the principal terminal seta of each is fully three times the entire length of the inner branches; these branches are also somewhat shorter than the three-jointed outer branches (fig. 17, Pl. XI.). The inner branches of the second, third, and fourth pairs are comparatively shorter than those of the first pair, and the terminal setæ of both the inner and outer branches of these three pairs are long and plumose (fig. 18, Pl. XI.). The fifth pair of feet differ from those usually observed in *Cletodes;* the basal joint, which is proportionally much dilated, is produced into a large and strong spine-like process which is slightly curved at the end and bordered with minute spinules; the secondary joint is rudimentary, and bears three small setæ at its truncate apex; two setæ spring from the opposite margin of the large basal joint (fig. 19, Pl. XI.). The caudal furcæ are long and slender; two small setæ spring from near the middle of the outer margin of each of the furcæ, and they each bear a long spiniform terminal seta (fig. 1, Pl. XII.).

The male differs little from the female, except that the antennules are modified in the usual way. The fifth pair of thoracic feet are nearly the same as those of the female (fig. 20, Pl. XI.).

Habitat.—Vicinity of Smith Bank, Moray Firth. Rare.

Remarks.—This very distinct species was obtained amongst some dredged material collected on the 6th October 1898, and sent to me by Mr. F. G. Pearcey. The fifth thoracic feet form one of the most striking characters of this species, not only because of their remarkable form, but also because in all the specimens examined they projected nearly straight out from the body of the animal instead of being adpressed, as is usually the case. The incurved position of the posterior portion of the abdomen and caudal furca is also a more or less constant feature in this species so far as regards all the specimens examined. Except for the somewhat abnormal form of the fifth pair of feet, the species appears to be a typical *Cletodes*.

Dactylopus tenuiremis, Brady and Robertson.

> 1895. *Dactylopus tenuiremis*, Brady and Robertson, Brit. Assoc. Report, p. 197.

This apparently rare species occurred in the shore gathering collected to the east of Invergordon, Cromarty Firth, in November 1897. There is no previous record of *Dactylopus tenuiremis* from the Moray Firth district. It is a somewhat critical species, but appears to be distinct.

Dactylopus minutus, Claus.

> 1863. *Dactylopus minutus*, Claus, Die frei-lebenden Copep., p. 126, Pl. XVI., figs. 14-15.

This *Dactylopus* occurred amongst a number of other things in a gathering from Station VI. (Firth of Clyde) collected 1st September 1898. It is a comparatively small species and appears to be rare.

Thalestris helgolandica, Claus.

> 1863. *Thalestris helgolandica*, Claus, Die frei-lebenden Copep., p. 131, Pl. XVII., figs. 12-21.

This rare species was obtained in a bottom tow-net gathering from Station IV. (Kilbrennan Sound, Firth of Clyde) on 24th August 1898. The Rev. A. M. Norman has also obtained *Thalestris helgolandica* in the Firth of Clyde. This species of *Thalestris*, as well as *Thalestris hibernica*, has been found recently in some material dredged in 1886 a little to the west of Invergordon, Cromarty Firth, but both appear to be scarce.

Cylindropsyllus fairliensis, sp. n. (Pl. X., figs. 11-14; Pl. XI., figs. 1-4).

Description of the Female.—The body is elongate, slender, and cylindrical. The length of the specimen figured is 1·73mm. (nearly $\frac{1}{15}$ of an inch). The thorax is composed of five distinct segments, the first of which is rather longer than the combined lengths of the next two, but the second to the fifth are sub-equal. The abdomen is also composed of five distinct segments; the first to the fourth are of nearly the same length as the posterior thoracic segments, the last being about one and a half times the length of the penultimate segment. The caudal furcæ are short and broad; the interior half of the apex of each is somewhat produced and bears a long and moderately stout seta and two or three small hairs; the exterior portion of the apex is abruptly concave, the concavity being bounded externally by an acute angle, and interiorly by the produced setiferous portion just referred to and as shown in the figure (fig. 6, Pl. XI.). The rostrum is short. The antennules are moderately short and stout, eight-jointed; the first four joints are larger than the last four, a stout asthetask springs from the produced upper angle of the fourth joint; the fifth joint, which is smaller than any of the others, is only about half the length of the preceding one; the next three are sub-equal and somewhat longer than the fifth (fig. 1, Pl. XI.). The formula shows approximately the proportional lengths of all the joints—

Numbers of the joints,	1 · 2 · 3 · 4 · 5 · 6 · 7 · 8
Proportional lengths of the joints,	22 · 15 · 16 · 12 · 6 · 10 · 11 · 10

Antennæ stout, three-jointed, with a very small secondary branch bearing a single seta (fig. 2, Pl. XI.). The mandibles are small and elongate; the palp is small and consists of a single slender branch bearing two short apical setæ (fig. 12, Pl. X.). Maxilla short and moderately stout, with the apex broadly truncate and armed with a number of strong teeth. The palp small, two-jointed, and furnished with a few terminal and subterminal hairs (fig. 13, Pl. X.). Posterior foot-jaws stout, armed with a stout terminal claw and two stout marginal processes (fig. 14, Pl. X.). The inner branches of the first four pairs of swimming feet are all two-jointed; those of the first pair are nearly equal in length to the three-jointed outer branches; but in the second, third, and fourth pairs the inner branches are considerably shorter than the three-jointed outer branches. All the four pairs are moderately stout and are furnished with elongate marginal spines, while the terminal setæ of both the outer and inner branches are long and plumose (figs. 3 and 4, Pl. XI.). The fifth pair of thoracic feet are small and provided with about two moderately long spiniform setæ and one or two small hairs (fig. 5, Pl. XI.).

Habitat.—Brackish water-pools near Fairlie, Firth of Clyde. Apparently rare.

Remarks.—This Copepod at first sight closely resembles *Cylindropsyllus lævis*, Brady, though of somewhat larger size; but even without dissection the caudal furcæ are seen to be distinctly different from those of that species, and if a specimen be dissected several other differences are noticed.

The antennæ, for example, are three-jointed, while in the typical *Cylindropsyllus* they are only two-jointed: the inner branches of the swimming feet are also more fully developed than they are in *Cylindropsyllus*. Such differences may yet render it necessary to remove this Copepod to another genus; but, meantime, as no males have yet been observed, I prefer to leave it in the genus to which for the present it is doubtfully ascribed.

Leptocaris, gen. nov.

The Female.—Body slender, somewhat resembling *Cylindropsyllu s* Secondary branches of the antennæ very small, one-jointed. Mandibl e palp obsolete—in this respect, the mandibles are somewhat similar to those of *Maraenobiotus*. Maxillæ also somewhat similar to those of *Maraenobiotus*, but the palp is a small cylindrical process with a dilated base. Footjaws similar to those of *Cylindropsyllus*. Inner branches of first, second, third, and fourth pairs of swimming feet two-jointed, and considerably shorter than the three-jointed outer branches; fifth pair very small, one-branched.

The Male.—The male is similar to the female, except that the antennules are modified and hinged for grasping, and that each of the fifth pair of thoracic feet is armed with a stout spine on its inner aspect, in addition to a few small setæ.

Leptocaris minutus, sp. n. (Pl. X., figs. 15-21; Pl. XI., figs. 7-11).

Description of the Female.—Body elongate and slender. No distinction between the thorax and abdomen. Thorax composed of five, and the abdomen of four segments (fig. 15. Pl. X.). The first thoracic segment is somewhat longer than the entire length of the next two, the second to the third are subequal, the fourth and fifth—which are also subequal—are rather longer than the second and third. The first abdominal segment is about one and a half times longer than the next, the second and third are subequal, while the ultimate segment is rather longer than the anterior one. Rostrum small. The antennules are very short, and moderately setiferous—seven jointed. The first joint is considerably dilated ; the second, which is only about half the length of the first, is also somewhat dilated. The third joint is nearly as long as the first. The fourth joint—which is furnished with a moderately long asthetask— and the last are of equal lengths, and are each as long as the second. The fifth and sixth joints are somewhat smaller than the others (fig. 16, Pl. X.). These differences are more clearly shown by the formula annexed—

Numbers of the joints,	1 · 2 · 3 · 4 . 5 . 6 . 7
Proportional lengths of the joints,	12 · 6 . 10 . 6 . 4 . 5 . 6

Antennæ small, three-jointed ; secondary branches very small, one-jointed (fig. 18, Pl. X.). The mandibles are also small ; the mandible palp is obsolete, being represented by a single small hair (fig. 19, Pl. X.). The maxillæ are very small, the biting part is moderately broad, and armed with a few comparatively elongate teeth. The palp is a small cylindrical process, arising from a moderately broad base, and furnished with a few hairs (fig. 20, Pl. X.). Posterior foot-jaws small. They somewhat resemble those of *Cylindropsyllus lævis* (fig. 21, Pl. X.). The inner branches of the first four pairs of thoracic feet are all two-jointed, and shorter than the three-jointed outer branches. The two joints that compose the inner branches are, in each of the four pairs, more or less subequal, but those of the first pair are rather stouter than the others. All the four pairs of feet are small (figs. 7 and 8, Pl. XI.). The fifth pair are minute. Each consists of a small semicircular appendage bearing three or four small setæ (fig. 9, Pl. XI.). The caudal furcæ, which are small and cylindrical, are scarcely twice as long as broad ; but each carries a long terminal spiniform seta, and also a few small hairs (fig. 11, Pl. XI.).

The Male.—So far as can be made out, the male does not differ much from the female, except that the antennules are modified for grasping, as in other Harpactids. The fifth thoracic feet are also each provided with an elongate and stout spine, in addition to the setæ observed on the fifth pair of the female (fig. 10, Pl. XI.).

Habitat.—Brackish water-pools on the shore near Hunterston, Firth of Clyde. Rather rare.

Remarks.—This Copepod is somewhat like a *Moraria* or a *Maraenobiotus* in general appearance, as well as in some of the structural details ; but it differs from these two genera, not only in the form of some of the mouth organs, and of the fifth pair of thoracic feet, but in other details of structure as well. Neither does it agree with *Cylindropsyllus*, although it has a general resemblance to the members of that genus. For these and other reasons, I have instituted for its reception the genus *Leptocaris* (Greek—*leptos*, slender ; *karis*, a shrimp).

Idya cluthæ, sp. n. (Pl. XII., figs. 2-6).

Description of the Female.—Length of the specimen figured, 1·17mm. ($\frac{1}{22}$ of an inch). Somewhat like *Idya furcata* in general appearance, but rather more slender (fig. 2). The antennules are moderately short ; being only about two-fifths of the length of the thorax, they resemble generally the antennules of *Idya furcata*, but the entire length of the first four joints is proportionally shorter. The first four joints are sub-equal in length ; the fifth is rather shorter than the one next to it ; while the seventh is distinctly smaller than either the fifth or sixth (fig. 3). The proportional lengths of all the joints are nearly as in the annexed formula—

Number of the joints,	$1 \cdot 2 \cdot 3 \cdot 4 \cdot 5 \cdot 6 \cdot 7 \cdot 8$
Proportional length of the joints,	$\overline{22 \cdot 28 \cdot 25 \cdot 24 \cdot 6 \cdot 9 \cdot 4 \cdot 18}$

The antennæ and mouth organs are similar to those of *Idya furcata*. The first pair of thoracic feet resemble those of *Idya furcata*, but the seta that springs from the end of the first joint of the outer branches is short and straight, and none of the terminal or sub-terminal setæ bear secondary spine-like apical cilia so characteristic of *Idya furcata* and one or two other members of the genus. The second joint of the inner branches is proportionally stouter than the same joint in *Idya furcata*, and the terminal claws are long and slender instead of being short and moderately stout. Moreover, the spines on the inner and outer aspects of the second basal joint are also small and slender (fig. 4). The second, third, and fourth pairs of swimming feet are rather more slender than the same appendages in *Idya furcata* (fig. 5). The fifth pair of feet have the second joint long and slender, and the margins do not appear to be ciliated ; the seta which springs from the inner angle, and also that which springs from the outer angle, of the basal joint are long and slender, while the terminal setæ of the secondary joint are also elongate (fig. 6). The abdomen is elongate, being equal to nearly two-thirds of the length of the thorax ; the first and second segments appear to be, at least partly, coalescent ; their combined length is equal to half the entire length of the abdomen ; the last abdominal segment is very small. The caudal furca are short, and about as long as broad.

Habitat.—Loch Fyne and Firth of Clyde. Generally distributed, and apparently not very rare.

Remarks.—This distinct species of *Idya* appears to be unlike any previously described member of this genus. The two most prominent characters by which it may be distinguished from all closely allied species are—(1) The armature of the first pair of thoracic feet, and especially the long terminal spines of the inner branches, and (2) the long slender fifth feet. *Idya cluthæ* may by these two characters be distinguished at a glance even without dissection. Like other forms of *Idya*, this one bears a comparatively large ovisac. Both males and females have been obtained, and both are equally distinct. Hitherto this species has occurred only in moderately deep water.

Monstrilla danæ (?), Claparède.

Several specimens of *Monstrilla*, all of which appear to belong to the same species—viz., *Monstrilla danæ*, Claparède—have been obtained during the past year. They are all from the Clyde district, chiefly Upper Loch Fyne and Kilbrennan Sound. Usually one or two, rarely three or four, specimens were obtained in a single gathering.

The gatherings in which *Monstrilla* was observed were from the following stations :—Station II. (three specimens), Station III. (one specimen), Station IV. (two specimens), Station VI. (three specimens), Station XIII. (two specimens), Station XIV. (four specimens), Station XVII., two gatherings (one specimen each). These gatherings were all collected in August and November 1898.

Dermatomyzon nigripes (Brady and Robertson).

> 1875. *Cyclopicera nigripes*, Brady and Robertson, Brit. Assoc.
> Report p. 197.

This fine species occurred in only one of the gatherings at present under consideration—viz., in a bottom tow-net gathering from Station XV. (Moray Firth), collected 20th November 1897.

Rhynchomyzon purpurocinctum (T. Scott).

> 1893. *Cyclopicera purpurocinctum*, T. Scott, Eleventh Ann. Rep.
> Fish. Board for Scot., Part III., p. 209, Pl. III., figs. 29-40.

This well-marked species was obtained in the gathering from Station XV. (Moray Firth), in which *Dermatomyzon nigripes* occurred, and in another collected at Station II., also in the Moray Firth, 5th November 1897. In this species the last three thoracic segments are of a dark purple colour. Dr. W. Giesbrecht has found *Rhynchomyzon purpurocinctum* in Naples Bay.

Neopontius angularis (T. Scott).

> 1898. *Neopontius angularis*, T. Scott, Sixteenth Ann. Rep. Fish.
> Board for Scot., Part III., p. 271, Pl. XIV., figs. 1-11.

This species was described in 1898 from specimens obtained at Otter Spit, Upper Loch Fyne. I have now to record it from a bottom tow-net gathering from Station IV. (Kilbrennan Sound), Firth of Clyde, collected 24th August 1898 (27-29 fathoms).

Bradypontius papillatus (T. Scott) (Pl. XI., fig. 21; Pl. XII., figs. 7-15).

> 1888. *Artotrogus papillatus*, T. Scott, Sixth Annual Report of
> the Fishery Board for Scotland (Appendix), p. 232, Pl. VIII.,
> figs. 7-12.
>
> 1895. (?) *Bradypontius chelifer*, Giesbrecht, Ann. and Mag. Nat.
> Hist., ser. 6, vol. xvi. p. 183 (August 1895).

This species, described in 1888 in the Sixth Annual Report of the Fishery Board for Scotland, has recently been re-examined, and some further details of structure have been elucidated which I now propose to notice by way of supplementing the original description.

The length of the specimen figured is 1·2mm. ($\frac{1}{2\text{T}}$ of an inch). The first thoracic segment is equal to rather more than half the length of the thorax and abdomen combined; the abdomen is moderately elongate, and the furcæ are rather longer than broad. In general appearance this species somewhat resembles *Cribropontius normani* (B. and R.) (fig. 7, Pl. XII.).

The antennules are eight-jointed. The first and second joints are elongate; the third to the seventh are comparatively short; while the last is about twice the length of the penultimate joint (fig. 8, Pl. XII.). The proportional lengths of all the joints are approximately as shown by the formula—

Numbers of the joints,	1 · 2 · 3 · 4 · 5 · 6 · 7 · 8
Proportional lengths of the joints,	36 · 50 · 18 · 10 · 14 · 12 · 16 · 31

A moderately long asthetask springs from the end joint, as shown in the figure.

The antennæ are apparently four-jointed, and a very small secondary branch bearing two minute hairs springs from the end of the second joint (fig. 9, Pl. XII.). The mandibles are long and slender (fig. 10, Pl. XII.). Figure 10A represents the apical portion of one of the mandibles greatly enlarged, which somewhat resembles the apical portion of the mandibles in *Bradypontius magniceps* (Brady). The maxillæ resemble very closely those of *Dyspontius striatus*, Thorell, but the inner lobe is slightly longer than the outer, and the terminal setæ appear to be shorter than those of the maxillæ of that species (fig. 11, Pl. XII.). The first joint of the anterior foot-jaws is large and robust, but the second is elongate and slender, somewhat dilated at the extremity, and armed with a short, stout, and finger-like subapical claw (fig. 12, Pl. XII.). The posterior foot-jaws have the first and second joints moderately robust, but the end joints are somewhat slender, and terminate in a short, stout claw, as shown in the figure (fig. 13, Pl. XII.)

In the first pair of swimming feet, which are moderately stout, the outer and inner branches are nearly of equal length. The outer branches are armed exteriorly with short, stout, dagger-like marginal spines, and the interior marginal setæ are one on the second and five on the last joint. The inner branches are furnished with one seta on the inner margin of the first joint, and two on the second joint ; while the third joint has five setæ on the inner margin and apex, and a small one on the outer margin (fig. 14, Pl. XII.).

In the fourth pair the outer branches are stout and elongate ; the first and second joints have each one marginal seta ; the third joint bears five marginal setæ, and is also armed with a moderately large sabre-like terminal spine in addition to the small spines on the outer margin ; the inner branches, which are three-jointed, and scarcely reach to the end of the second joint of the outer branches, are slender, and provided with only a few minute hairs on the margins and two small apical spines (fig. 15, Pl. XII.).

Fifth feet small, one-jointed, subquadrate, each of them furnished with one small marginal and two short apical setæ. There is also close to each foot exteriorly a long seta with a slightly dilated base which springs from the edge of the segment to which the fifth feet are attached (fig. 21, Pl. XI.).

Habitat.—Firth of Forth. Rare. No males observed.

Remarks.—As already stated, this specimen was first partly described and figured in the Appendix to the Sixth Annual Report of the Fishery Board for Scotland, published in 1888. It was described under the name of *Artotrogus papillatus*, but some doubt was expressed as to its being a true *Artotrogus*. No more specimens having been observed, the interest in the species passed away, and it was practically forgotten. Recently, however, my son got hold of the original specimen from which the species was described, and with the assistance of Dr. W. Giesbrecht's work on the "Diagnosis, Synonymy, and Distribution of the Ascomy-zontidæ" made a careful examination of the characters by which the species is distinguished, as well as a series of delineations illustrating its principal appendages. The description given above is the result of this extra research. From the additional information that has been obtained by this re-examination, there can be no doubt that our species is a true *Bradypontius.* It agrees perfectly with Dr. Giesbrecht's definition of that genus. It may also be identical with the species described by Dr. Giesbrecht under the name of *Bradypontius chelifer* from the Bay of Naples, and, if so, the distribution of the species will be very considerably extended.

AMPHIPODA.

A few of the Amphipods observed in the tow-net and other gatherings forwarded to me from the "Garland" may now be referred to.

The Hyperiidæ were of rare occurrence in the tow-net gatherings forwarded from the Clyde or Moray Firth during the past year. *Hyperia galba* (Mont.), *Hyperoche tauriformis* (Bate), and *Parathemisto* were observed in one or two of the Moray Firth gatherings, but in those from the Clyde only *Hyperoche* and *Parathemisto* were observed, the one from Stations I. and VIII. in both the surface and bottom tow-net gatherings, and the other from Station VII.

The Orchestiidæ observed include *Orchestia mediterranea*, a species that appears to be of rare occurrence in the Clyde district. One or two specimens were obtained amongst decaying sea-weed, on the shore between Fairlie and Hunterston in September. *Orchestia mediterranea* is readily distinguished from the more common *Orchestia littorea* by the form of the hands of the second gnathopoda in the male; in these appendages the propodos are triangular instead of ovate; the palm, which is almost straight, extends from near the base of the propodos, and has a triangular tooth-like projection anteriorly near the origin of the claw. The claw is long and somewhat sinuate, and nearly of the same length as the palm (figs. 9-11, Pl. XIII., represent the anterior and posterior gnathopods and one of the posterior pereiopods).

Only two specimens of this species have been recorded from the Clyde district by the late Dr. Robertson in Part I. of his Catalogue of Clyde Amphipoda and Isopoda. One of these he discovered at the west end of Cumbrae; the other was sent to him by Mr. John Smith, Kilwinning, who obtained it at the mouth of the Garnock.

A number of Amphipods belonging to the Lysianassidæ have been observed. I will, however, refer to only one of them—viz., the curious *Normanion quadrimanus* (Bate and Westwood), a single specimen of which was obtained in a bottom tow-net gathering from Station I., Firth of Clyde (near Davaar Island), collected 15th December 1898. In Part II. of the late Dr. Robertson's Catalogue of the Clyde Amphipoda, that author records having, along with the Rev. Dr Norman, captured *Normanion* off Farland Point, Cumbrae, which seems to be the only previous record of its occurrence in the Clyde. Professor Sars has shewn that *N. quadrimanus* is parasitic in its habits, and states that he has found it in great abundance clinging to the skin of fishes (both living and dead) caught on a fishing line set in deep water. It may, therefore, be found to be more common in the Clyde than it has hitherto appeared to be if a careful examination were to be made of the fishes caught in the deeper parts of the estuary.

The Ampeliscidæ were represented in recent tow-net gatherings from the Clyde by one or two moderately rare forms, such as *Ampelisca lævigata*, Lilljeborg; *Ampelisca spinipes*, Boeck; and *Haploops tubicola*, Lilljeborg.

Amongst the Phoxocephalidæ the only species that need be referred to is *Harpina crenulata*, Boeck. Four specimens of this Amphipod were obtained in a gathering of Crustacea dredged in Campbeltown Loch (Cantyre) in 1897. This appears to be the first record of *H. crenulata* for the Clyde.

Argissa hamatipes (Norman)—one of the Pontoporeiidæ—occurred sparingly in tow-net gatherings from the Clyde district as well as from the Moray Firth. Both male and female specimens were observed.

The curious *Stegocephaloides christianensis* (Boeck) occurred in a bottom tow-net gathering collected in the Clyde at Station XVII. (Upper Loch Fyne) on 7th December 1898. This species appears to be somewhat rare in the Clyde district.

Amongst the Amphilochidæ the somewhat rare *Amphilochus tenuimanus,* Boeck, was obtained in a bottom tow-net gathering from Station VII., Firth of Clyde, collected at night on 23rd September 1898. *A. tenuimanus* is recorded in Dr. Robertson's Catalogue, Part I., page 28.

Amphilochoides odontonyx (Boeck).=*Amphilochoides pusillus,* Sars, was obtained in the Clyde at Stations III., IV., VII., and IX., but only in bottom tow-net gatherings; while *Gitana sarsi,* Boeck (another of the same family of Amphipoda), occurred in a gathering from Station VI. collected on 25th August 1898.

The family Stenothoidæ was represented in the tow-net gatherings from the Clyde by *Stenothoë marina* (Spence Bate), *Cressa dubia* (Spence Bate), and one or two other forms; and in a gathering from Station XV. Moray Firth, *Stenothoë marina, Metopa robusta,* G. O. Sars (a species recorded from the Moray Firth a few years ago in "Ann. and Mag. Nat. Hist.," ser. 6, vol. xiii., p. 148, February 1894), and *Cressa dubia* (Spence Bate). In *Metopa robusta*—the record of which in 1894 appears to be the first for Britain—the form of the first gnathopoda is quite distinct from that of the same appendages of any other British species except, perhaps, *Metopa polexiana.*

Monoculodes packardi, Boeck, occurred in a gathering from Station XVII. (Upper Loch Fyne) collected on 7th December 1898; and *Synchelidium brevicarpum* (Spence Bate) at Station IV. (Kilbrennan Sound, Firth of Clyde) on 24th August 1898.

Epimeria tuberculata, G. O. Sars. Two specimens of this fine species were obtained in a gathering from Station IV., Firth of Clyde, on 1st September 1898. *Epimeria tuberculata,* which appears to be a rare species in the Clyde, was added to the British fauna about two years ago.

Eusirus longipes, Boeck, was obtained from the Clyde in gatherings Stations VI., VII., and VIII. It occurred in one gathering from Station VI., collected on 25th August 1890; in two other gatherings from Station VII., collected on 3rd September and 24th September; and in one from Station VIII., collected on the 29th of the same month.

Guernea coalita (Norman). This curious little species was observed in washings of material dredged at Station VI., Firth of Clyde. The same species has been captured off Millport by the late Dr. Robertson, and is recorded in Part II. of his catalogue of Clyde Amphipoda and Isopoda; and his appears to be the only previous record of the occurrence of *G. coalita* for the Clyde.

Melphidippella macera (Norman). This Amphipod, though obtained occasionally in tow-net gatherings, is apparently scarce, as only one or a very few specimens are found in any single gathering. It occurred in four of the series of gatherings at present under consideration, all the four being from the Clyde, viz. :—In a gathering from Stations III. and IV., collected on 24th August 1898 ; in a gathering from Station VII. collected on 24th September; and in one from Station VIII., collected on 29th of the same month.

Photis longicaudatus (Spence Bate) occurred sparingly in a gathering from Station IV., Firth of Clyde.

Amongst the Podoceridæ *Podocerus palmatus*, Stebbing and Robertson, and *Podocerus pusillus*, G. O. Sars, have been obtained in tow-net gatherings from the Clyde—the first in a gathering from Station VII. collected on 24th September 1898, and the other in a gathering from Station VI., collected on 1st September.

Unciola planiceps, Norman, was obtained in the same gathering from Station VI. in which *Podocerus pusillus* just recorded occurred. *Unciola* appears to be an addition to the Amphipod fauna of the Clyde.

ISOPODA.

The following are a few of the more interesting of the Isopoda that have been observed in tow-net and other gatherings recently examined. Those from the Clyde are, *Leptognathia breviremis*, Lilljeborg, collected at Station XII., between the south end of Arran and the Ayrshire coast ; *Leptognathia brevimana* (Lilljeborg), at Station VI., 1st September 1898. Another closely allied species—*Tanaopsis laticaudata*, G. O. Sars—was moderately frequent amongst a gathering of material from Campbeltown Loch (Cantyre) collected in May 1897. *Munna boecki*, Kroyer, which is larger than *Munna kroyeri*, Goodsir, and apparently more frequent in the Clyde, has been obtained at Whiting Bay, and one or two other places. Both *Munna boecki* and *Munna kroyeri* have been observed in the Moray Firth district.

Paramunna bilobata, G. O. Sars, recorded for the Clyde in 1894, by the Rev. A. M. Norman (Ann. and Mag. Nat. Hist. (6), vol. xii., p. 280, footnote), was obtained at Station IV. (Kilbrennan Sound). It has already been recorded for Loch Fyne in Part III. of the Sixteenth Annual Report of the Fishery Board for Scotland.

Pleurogonium rubicundum, G. O. Sars, was recorded by the Rev. A. M. Norman, for the Clyde, in 1894, in the same footnote in which *Paramunna* is recorded (see reference under that species), and is, so far as I know, the only record for the Clyde hitherto. *Pleurogonium inerme*, G. O. Sars, has been obtained in a gathering from Campbeltown Loch (Cantyre), collected in 1897, and in one from Station IV. (Kilbrennan Sound), collected on 24th August 1898. *Pleurogonium spinosissimum*, G. O. Sars, another of these minute Isopods, has been recorded for the Clyde by the late Dr. Robertson, of Millport.

Pseudione crenulata, G. O. Sars (a parasite of *Munida rugosa*), apparently new to Britain, and *Pseudione affinis*, G. O. Sars (parasite on *Pandalus montagui*), have both been observed in the Clyde, while *Bopyroides hippolytis* (Kroyer) has been obtained in the Moray Firth.

Microniscus. The little parasite known by this name, and which, in the tow-net gatherings of the " Garland," is sometimes found clinging to *Calanus, Pseudocalanus,* and other Copepods, has just been shown by Prof. Sars ("Crustacea of Norway," vol. ii., p. 218-220, 1898) to be one of the post-larval stages of a species of *Phryxus.* In my paper on the "Marine Fishes and Invertebrates of Loch Fyne," published in the Fifteenth Annual Report of the Fishery Board for Scotland (p. 136, 1897), attention was directed to the close resemblance between *Microniscus* and the young of *Phryxus fusticaudatus,* Spence Bate, but no definite opinion was expressed as to the relationship between them. Probably more than one kind of *Bopyrus* is represented by these *Micronisci,* but it may, at this stage, be difficult to distinguish the one kind from the other.

CUMACEA.

Several interesting Cumaceans have been observed during the examination of tow-net gatherings recently collected, the following of which may be referred to :—

Lamprops fasciata, G. O. Sars, has been obtained sparingly in a gathering from the Cromarty Firth. *Hemilamprops rosea,* Norman, occurred in a gathering from Station IV. (Kilbrennan Sound), Firth of Clyde. *Leucon nasicus,* Kroyer, was obtained in a gathering from Station XII. (Firth of Clyde), depth 40-43 fathoms. *Eudorellopsis deformis* (Kroyer)—a curious little Cumacean—was taken at Stations VII. and VIII., Firth of Clyde, in moderately deep water. *Eudorella truncatula* (Spence Bate) occurred in gatherings from Clyde Stations VII. and VIII., and *Eudorella marginata* (Kroyer) in a gathering from Station XII.

Campylaspis rubicunda, Lillejeborg, was obtained in gatherings from Clyde Stations XII. and XVII. *Cumella pygmæa,* G. O. Sars, occurred in a gathering from Station IV. (Kilbrennan Sound), Firth of Clyde, 24th August 1898, and in one from Station XV. Moray Firth, 15th November 1897.

Cuma pulchella, G. O. Sars, though only recognised within recent years as a member of the British fauna, has apparently a wide distribution around our shores. It was obtained in the Firth of Forth in 1889-90, and recorded in Part III. of the Eighth Annual Report of the Fishery Board for Scotland, p. 329, and afterwards in the Liverpool Bay District (Eighth Annual Report of the Liverpool Marine Biological Committee, p. 25). I have now to record its occurrence in the Clyde, having obtained one or two specimens in some washings of dredged material from Station VI. As pointed out by Dr. Norman, the first joint of the seventh foot is furnished with a series of backward-directed tooth-like processes, by which character *C. pulchella* may be distinguished from its congeners.

SCHIZOPODA.

The Schizopoda, though plentiful in some of the gatherings, were usually limited to a few species, amongst which the Euphausiidæ were the most numerous. The Schizopod usually of most frequent occurrence in the Clyde and Loch Fyne gatherings is *Boreophausia raschii,* but in the Moray Firth *Thysanoessa neglecta*, Kröyer, is frequently the prevailing form. * *Nyctiphanes norvegicus*, which is also of frequent occurrence, and of large size, in Loch Fyne, is generally comparatively rare in the Moray Firth and in the Firth of Forth. This *Nyctiphanes*, though occasionally met with in considerable numbers in other parts of the Clyde district, seldom attains such a large size as it does in Loch Fyne.

Erythrops serratus and *Erythrops elegans* have been obtained in gatherings collected during the past year both in the Clyde and in Loch Fyne, but neither have been observed in the Moray Firth. *Erythrops serrata* formed part of the contents of a hake's stomach captured at Station XIV. (Loch Fyne). The Epicarid parasite *Aspidoecia Normanni*, Giard and Bonnier, was obtained on *Erythrops elegans* at Station VII., September 1898.

Siriella norvegica, G. O. Sars, was obtained at Station VII. (Firth of Clyde). The *Siriellæ*, though represented in the Clyde by at least four species, are usually of rare occurrence, so that even the commoner forms are worth recording. *Siriella norvegica* is one of the less familiar of the British *Siriellæ*.

Schistomysis ornatus, G. O. Sars, and *Hemimysis lamornæ* (R. Couch) has been occasionally met with in the Clyde tow-net gatherings during the past year. The first has been obtained in gatherings from Stations VII. and VIII., near the mouth of the estuary, and from Stations XIV. and XVII. in Upper Loch Fyne ; the other was obtained in a gathering from Station VII.

Leptomysis gracilis, G. O. Sars, and *Leptomysis lingaura*, G. O. Sars, were also obtained in Clyde tow-net gatherings recently examined. *Leptomysis gracilis* occurred in a gathering from Station VIII., and *Leptomysis lingvura* in two different gatherings from Station VII., and in gatherings from Stations XIV. and XVII.

Mysidopsis gibbosa, G. O. Sars, *Mysidopsis didelphys* (Norman), and *Mysidopsis angusta*, G. O. Sars, have all occurred in gatherings recently collected in the Clyde and Upper Loch Fyne. *Mysidopsis gibbosa* was obtained in gatherings from three stations, viz., Stations VII., VIII., and XVII.; *Mysidopsis didelphys* in a gathering also from Station XVII., and *Mysidopsis angusta* in one from Station VIII.

Before concluding these notes on the tow-net gatherings collected on board the " Garland " and forwarded for examination, it may be of interest to refer to a young form of crustacean which is sometimes met with in these gatherings.

The study of the changes of form to be met with in the life-history of the Crustacea is a profoundly interesting one. The variations of form observed in the different species are sometimes so perplexing that they have occasionally puzzled even experienced students. Numerous larval and young forms are now and again captured in the tow-nets, but usually they belong to species that are fairly well known. It sometimes happens, however, that specimens are obtained which are not so easily disposed of, and I now draw attention to a curious form which is occasionally noticed

---

* *Thysanoessa neglecta* has recently been observed in the Clyde. It occurred in a tow-net gathering from Station X.. collected on the 16th of January last. An exposition of the characters which distinguished this from the closely allied species, *T. longicaudata*. will be found in Rev. Dr. A. M. Norman's excellent " Revision of the British Schizopoda," published in the " Annals and Magazine of Natural History," June-September 1892.

in tow-net gatherings from the Clyde—the only Scottish locality where it has as yet been observed. The form referred to, which in general appearance is not very unlike the widely distributed crustacean known as *Lucifer*, has been known for a considerable period; it was described and figured under the name of *Trachelifer*, by the late Mr. George Brook in 1888*, but is evidently immature, and there is still some doubt as to the species to which it really belongs. The neck of this young form is long and slender; the thorax is small, and is furnished with more or less rudimentary appendages; the slender abdomen is more than twice the length of the neck, and the last abdominal segment is as long as all the other segments of the abdomen put together; the telson and uropods are comparatively short, and more or less rudimentary. Figure 16, Pl. XII., represents one of the Clyde specimens, which measured over all about 16·5 millemetres ($\frac{2}{3}$ of an inch). The specimens that have been observed vary in length to a small extent, but all possess the same slender *Lucifer*-like form.

Habitat.—Station VII., Firth of Clyde. (I have also a specimen from Loch Fyne collected in 1886.)

Additional Remarks.—Figures 17 to 20, Plate XII., exhibit on a somewhat enlarged scale portions of the Lucifer-like crustacean referred to above. In fig. 17, which represents the front part of the cephalic segment, the eyes are large and somewhat divergent; the triangular rostrum is slightly shorter than the eyes; both pairs of antennæ are slender and elongate; the antennal scales are also slender and rather shorter than the basal part of the antennules. Fig. 18 represents what appears to be one of the first peræopods. In the specimen dissected this was the only pair that had the extremities of the principal branches chelate; all the other trunk legs appear to be simple. Fig. 19 represents one of the first pair of abdominal appendages, which are all more or less rudimentary; each appendage consists of a single unjointed branch, with a furcated extremity. Fig. 20 represents the posterior end of the last abdominal segment, together with the uropods and telson; the appendages of the last abdominal segment form tapering and slightly curved processes; the uropods are foliaceous, and little more than half the length of the telson; the telson is comparatively of large size. In the specimen dissected for drawing, the telson was somewhat imperfect. The extremity is therefore indicated by dotted lines, but in another specimen in which the telson was fairly perfect the following characters were observed :—The terminal lateral processes had each two small teeth on the inner margin, and the part between the lateral processes was furnished with twenty-two short and slender marginal spines; the two middle spines were rather shorter than the others; and there was a slight but perceptible gradation in the length of those on each side of the two central ones, the spines nearer the centre being somewhat shorter than those more distant. The larva above referred to was described by Claus as *eine in vieler Hinsicht merkwürdigen Larve* in a paper *Zur Kenntniss der Kreislaufsorgane der Schizopoden und Decapoden*, in Arb. d. z. Inst. Univ. Wien. V. 1884, p. 302 (32), Pl. VIII., figs. 48-50. The same writer subsequently described a somewhat more advanced specimen as the larva of *Calliaxis adriatica*, Heller, ibid. VI. 1886, p. 63, Pl. V., fig. 45. The identity of Brook's *Trachelifer* with Claus's *Calliaxis*—larva is pointed out by Korschelt and Heider, Lehrb. d. vergl. Entwicklungsgesch. d. wirbell. Th., I., p. 471. *Calliaxis* is not yet known as a British form, being only known from the Adriatic and from Naples : at the latter station the larva is met with in the surface-fauna, but the adult has only been found once in 25 years (S. Lo Bianco, Mitth. Zool. Stat. Neapel. XIII., p. 503, 1899).

ADDITIONAL NOTES.

Shortly after the MS. of this paper had been sent to the printer I had the privilege of perusing the sixth portion of *Das Tierreich*, which has recently been published. This portion contains a revision of the Copepoda belonging to the sub-order Gymnoplea by Giesbrecht and Schmeil; several of the species of Copepoda mentioned in this paper belong to the same sub-order, and on comparing these with the revision referred to I find that the names of four of them have been somewhat altered, as follows:—

Bradyidius armatus (Brady).—Corrected name, *Bradyidius armatus* (Vanhöffen).

Eurytemora lacinulata (Fischer).—Corrected name, *Eurytemora velox* (Lillj), Brady.

Metridia hibernica (Brady and Robertson).—Corrected name, *Metridia lucens*, Boeck.

Candace pectinata, Brady.—Corrected name, *Candacia pectinata*, Brady

---

EXPLANATION OF THE PLATES.

PLATE X.

Scolecithrix pygmœa, sp. n.

Fig. 1.	Female, lateral view	× 53.
Fig. 2.	Antennule	× 126.
Fig. 3.	Anterior foot-jaw	× 380.
Fig. 4.	Posterior foot-jaw	× 380.
Fig. 5.	Foot of first pair of swimming-feet	× 250.
Fig. 6.	Foot of fourth pair	× 190.
Fig. 7.	Fifth pair of feet, male (r., right; l., left branch)	× 127.
Fig. 8.	Abdomen and caudal furca, female, dorsal view	× 190.
Fig. 9.	Abdomen and caudal furca, male, dorsal view	× 190.

Scolecithrix hibernica, A. Scott.

Fig. 10.	Fifth pair of feet, male	× 127.

Cylindropsyllus fairliensis, sp. n.

Fig. 11.	Female, lateral view	× 53.
Fig. 12.	Mandible	× 760.
Fig. 13.	Maxilla	× 760.
Fig. 14.	Posterior foot-jaw	× 760.

Leptocaris minutus, gen. et sp. n.

Fig. 15. Female, lateral view	× 107
Fig. 16. Antennule, female	× 760.
Fig. 17. Antennule, male .	× 760.
Fig. 18. Antenna .	× 760.
Fig. 19. Mandible .	× 760.
Fig. 20. Maxilla .	× 760.
Fig. 21. Posterior foot-jaw	× 760.

PLATE XI.

Cylindropsyllus fairliensis, sp. n.

Fig. 1. Antennule	× 380.
Fig. 2. Antenna . .	× 380.
Fig. 3. Foot of first pair of swimming-feet	× 250.
Fig. 4. Foot of fourth pair	× 190.
Fig. 5. Foot of fifth pair . .	× 380.
Fig. 6. Last abdominal segment, and caudal furca .	× 380.

Leptocaris minutus, gen. et sp. n.

Fig. 7. Foot of first pair of swimming-feet	× 760.
Fig. 8. Foot of fourth pair .	× 760.
Fig. 9. Foot of fifth pair, female . .	× 760.
Fig. 10. Foot of fifth pair, male . . .	× 760.
Fig. 11. Last abdominal segment, and caudal furca .	× 253.

Cletodes perplexa, sp. n.

Fig. 12. Female, lateral view	× 760.
Fig. 13. Antennule	× 760.
Fig. 14. Mandible .	× 760.
Fig. 15. Anterior foot-jaw .	× 760.
Fig. 16. Posterior foot-jaw	× 760.
Fig. 17. Foot of first pair of swimming-feet .	× 760.
Fig. 18. Foot of fourth pair	× 380
Fig. 19. Foot of fifth pair, female . .	× 760.
Fig. 20. Foot of fifth pair, male	× 760.

PLATE X.

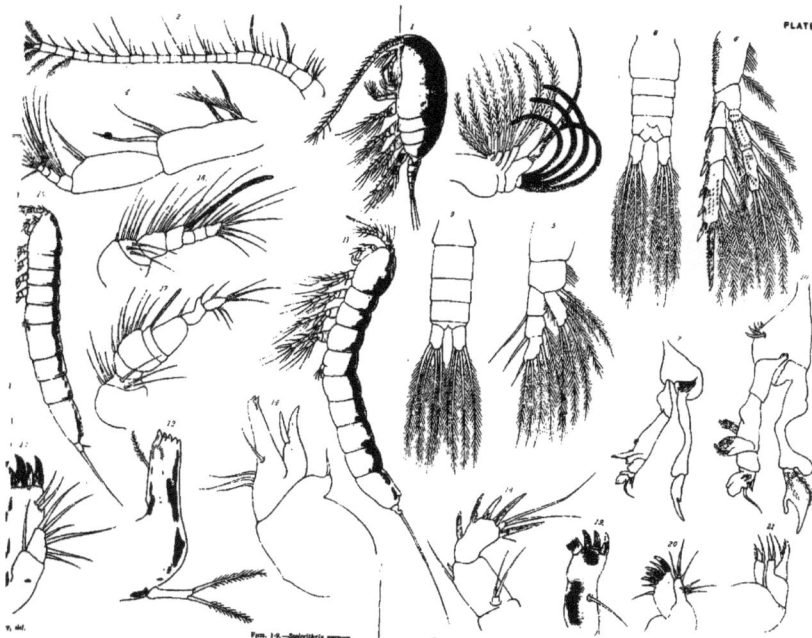

v, del.

Figs. 1-9.—*Scutellidium pygmæum.*
Figs. 11-14.—*Cylindropsyllus frisbiensis.*

Figs. 10.—*Scutellidium hibernica.*
Figs. 15-21.—*Leptocaris minutus.*

PLATE XI.

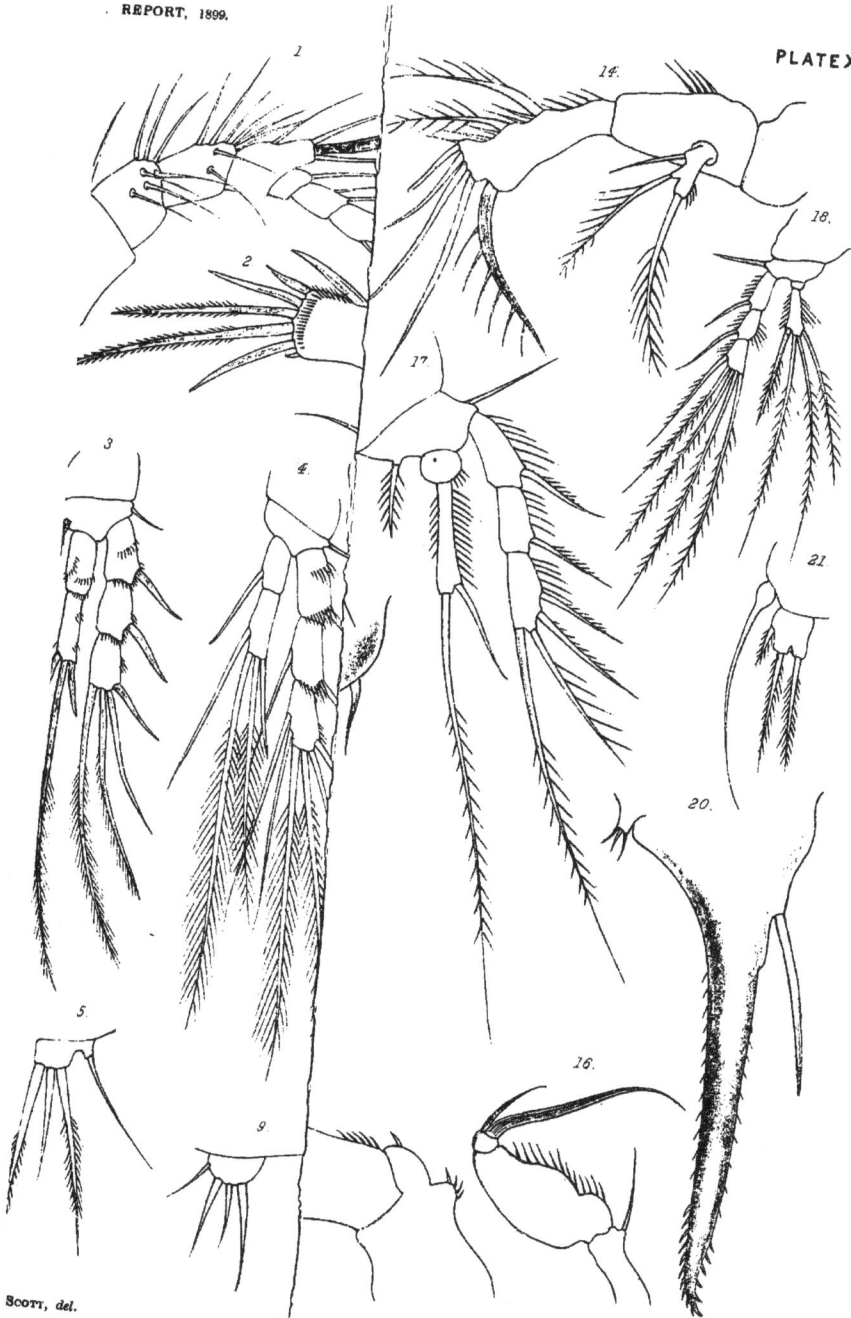

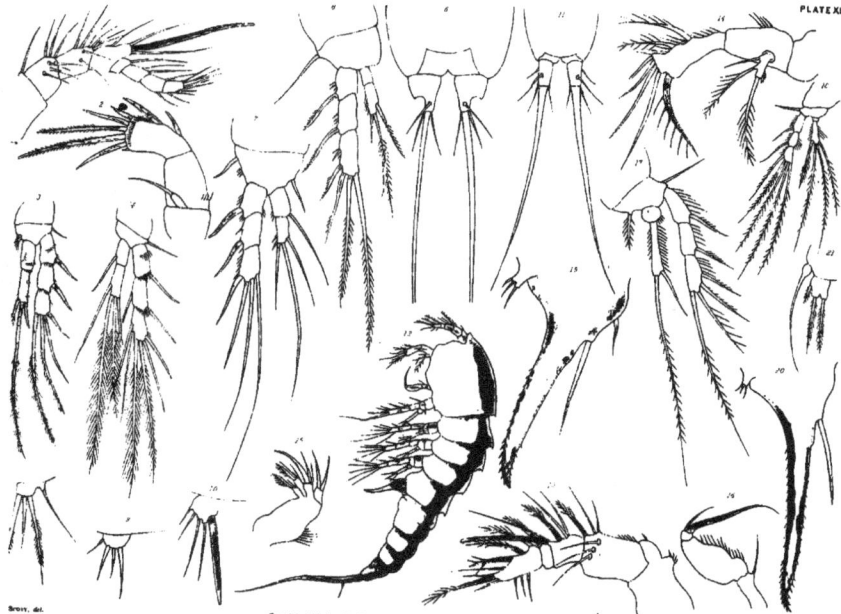

PLATE XI

Brady, del.

Figs. 1-6.—*Cylindropsyllus fairliensis*.
Figs. 15-20.—*Cletodes propinqua*.

Figs. 7-13.—*Laophonte minuta*.
Fig. 21.—*Bradyepsilus papillosus*.

PLATE XIII.

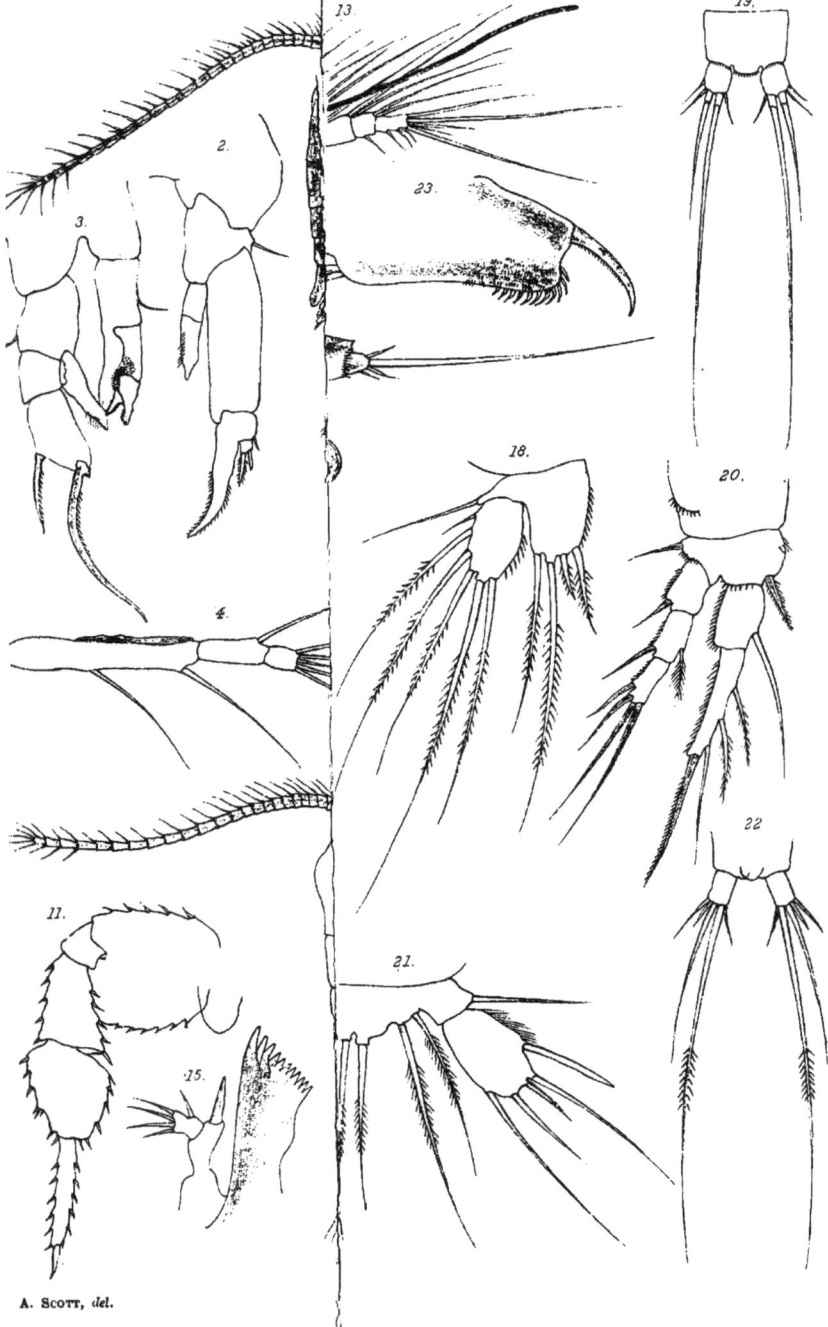

PLATE XII.

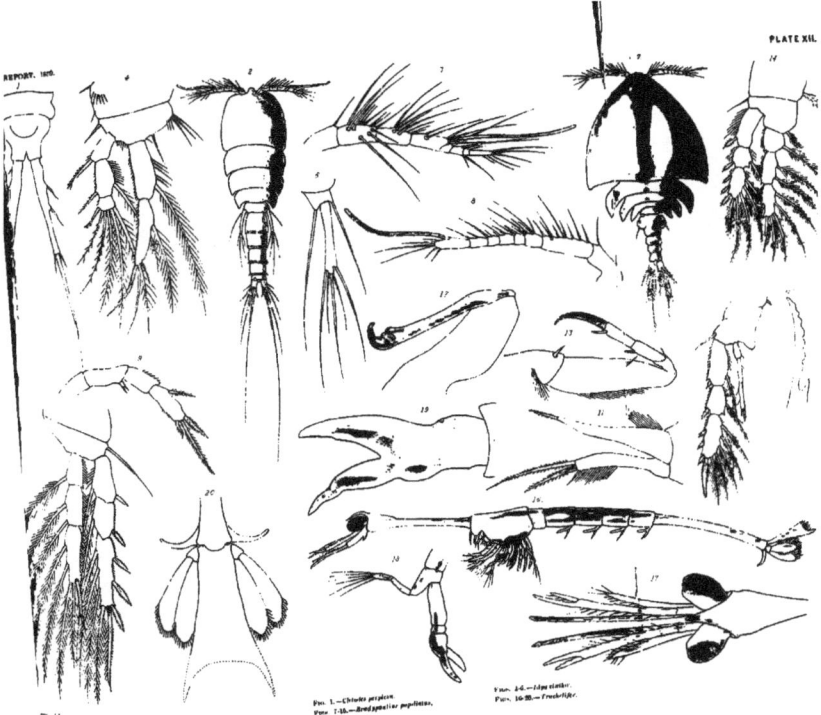

Fig. 1.—*Chlorus perplexa*.
Figs. 7-16.—*Bradyaemellus pupillatus*.

Figs. 5-6.—*Idya elaclini*.
Figs. 16-20.—*Frechtlift*.

Bradypontius papillatus (T. Scott).

Fig. 21. Foot of fifth pair . . ×260.

PLATE XII.

Cletodes perplexa, sp. n,

Fig. 1. Caudal furca . . . . ×250.

Idya cluthæ, sp. n.

Fig. 2. Female, dorsal view . . . . . . × 53.
Fig. 3. Antennule . . . . . ×190.
Fig. 4. Foot of first pair of swimming-feet . ×190.
Fig. 5. Foot of fourth pair . . ×190.
Fig. 6. Foot of fifth pair . . . . ×190.

Bradypontius papillatus (T. Scott).

Fig. 7. Female, dorsal view . × 53.
Fig. 8. Antennule . ×190.
Fig. 9. Antenna . ×380.
Fig. 10. Mandible . . ×253.
Fig. 10A. Extremity of mandible . . ×1520.
Fig. 11. Maxilla . . ×380.
Fig. 12. Anterior foot-jaw ×190.
Fig. 13. Posterior foot-jaw . ×190.
Fig. 14. Foot of first pair of swimming-feet ×190.
Fig. 15. Foot of fourth pair . ×190.

Larva of *Calliaxis (Trachelifer).*

Fig. 16. Specimen, showing side view . × 10.
Fig. 17. Head, showing antennæ, etc. . ×26½.
Fig. 18. One of first peræopods . ×26½
Fig. 19. One of the abdominal appendages ×160.
Fig. 20. Telson and uropods ×26½.

PLATE XIII.

* *Diaptomus laciniatus.*

Fig. 1. Female, dorsal view   .   × 32.

Fig. 2. Foot of fifth pair, female   × 190.

Fig. 3. Foot of fifth pair, male .   × 127.

Fig. 4. Last three joints of right antennule, male   × 190.

---

* *Diaptomus hircus,* Brady.

Fig. 5. Female, dorsal view   .   .   × 42.

Fig. 6. Foot of fifth pair, female   × 253.

Fig. 7. Foot of fifth pair, male .   × 127.

Fig. 8. Last three joints of right antennule, male   × 170.

Orchestia mediterranea.

Fig. 9. One of the anterior gnathopods   × 18.

Fig. 10. One of the posterior gnathopods   × 18.

Fig. 11. One of the last pair of peræopods   × 9.

Pysllocamptus fairliensis, T. Scott.

Fig. 12. Female, side view   .   × 160.

Fig. 13. One of the antennules, female .   × 456.

Fig. 14. One of the antennæ   × 456.

Fig. 15. One of the mandibles   .   × 760.

Fig. 16. Foot of first pair of swimming-feet   × 456.

Fig. 17. Foot of fourth pair of swimming-feet .   × 456.

Fig. 18. Foot of fifth pair .   .   × 380.

Fig. 19. Last abdominal segment, and caudal furca .   × 190.

Delavalia giesbrechti, var.

Fig. 20. Foot of first pair of swimming-feet   .   .   × 380.

Fig. 21. Foot of fifth pair .   .   _   × 380.

Fig. 22. Last abdominal segment, and caudal furca   × 190.

---

* *Ceriodaphnia quadrangula,* var.

Fig. 23. Post-abdomen of female .   .   .   × 152

* Reference to these species will be found in my paper on the Fresh-Water Lochs of Scotland in
this Report.

VI.—THE FISHES OF THE FIRTH OF CLYDE.

By Thomas Scott, F.L.S., Mem. Soc. Zool. de France.

It has been considered desirable that a list should be prepared of the fishes which have from time to time been observed within the limits of the Clyde estuary. The list is not a descriptive one, but is merely an enumeration of the various species which have come under the notice of the writer, together with those which have been recorded by different authors who have written on the fauna of the Clyde, or which have been seen by persons whose accuracy may be relied on. The following are the chief sources from which my information concerning Clyde fishes has been obtained :—

(1.) A List of Loch Fyne Fishes, prepared for the most part by the late George Brook, Esq., F.L.S., and published in the *Fourth Annual Report of the Fishery Board for Scotland* (1886). (A Revised List of Loch Fyne Fishes is published in the Board's *Fifteenth Annual Report*, Part III., 1897).*

(2.) A paper by Dr. Albert C. L. G. Günther, F.R.S., entitled, "Report on the Fishes obtained by Mr. (now Sir) J. Murray in deep water on the North-West Coast of Scotland, between April 1887 and February 1888," published in the Proceedings of the Royal Society, Edinburgh, Vol. XV. (1888). In this report 31 species are recorded, one of which was new to science, while two were additions to the British fauna. The species recorded in Dr. Günther's paper were mostly from the Clyde area, and there are some interesting notes on the bathymetrical distribution of the species.

(3.) *A Vertebrate Fauna of Argyle and the Inner Hebrides*, by J. A. Harvey-Brown and Thomas E. Buckley, published in 1892. The part of this work which is referred to in the present paper is the separate reprint of the list of fishes. In the list of fishes given in this work, the authors include Dr. Günther's records mentioned above, in the form of foot-notes under each of the species to which the records specially refer ; additional Clyde records are also given in this work.

(4.) The statistics published in the *Annual Reports of the Fishery Board for Scotland*, in so far as they relate to the fishes of the Firth of Clyde, and especially to those fishes which have been captured by the Fishery steamer "Garland."

I have also to acknowledge my indebtedness to Mr. Duthie, the Fishery Officer at Girvan, and to Mr. Gray of the Marine Station, Millport, Cumbrae, for interesting information relating to Clyde fishes. Other sources of information are duly acknowledged in the sequel.

Before proceeding with the enumeration of species, it may be of interest to refer to some points concerning the distribution of one or two of the forms which are usually considered to be more or less rare in the Clyde area. Previous to the investigations carried on by Dr. (now Sir) John Murray, by means of the s.s. "Medusa," the fish usually called the Sharp-tailed Lumpenus, *Lumpenus lampetræformis*, was not known to occur within the limits of the Clyde; but now, by the use of a small-mesh trawl-net, this fish is found to be comparatively frequent

---

* The compiler of this revised list had the privilege of consulting the records of Loch Fyne fauna obtained by the s.s. "Medusa" as the result of the investigations carried on by Sir John Murray on the West Coast.

and more or less generally distributed, especially in the deep-water area. The same may be said of the Argentine, *Argentina sphyræna*. This fish, which used to be considered a rare species in the Clyde, is really not very uncommon when sought for with suitable appliances. It may be pointed out further that if anyone were to look through the lists of Clyde fishes captured by the "Garland," he would fail to find a single record of a Halibut, which is only occasionally taken by a trawl. Moreover, the halibut is not mentioned for the Clyde in Dr. Günther's paper, nor in the work published by Harvey-Brown and T. E. Buckley already cited, yet the Fishery Officer at Girvan informs me that halibut are not very rare in the seaward part of the Clyde estuary, and that they are sometimes caught in the deep water off Ayr; while in a letter, dated the 28th December last, he states that he had seen, a few days before, a young halibut landed at Girvan, which had been caught between that place and Ailsa Craig, and which weighed about a stone.

In looking through the statistics of the steamer "Garland," it will also have been observed that Turbot and Brill are not very frequently mentioned in the lists of fishes captured by the steamer in the Clyde estuary, yet there is a more or less regular turbot fishery carried on off Girvan, and sometimes a considerable number of these fishes, captured in the Clyde by gill-nets, are brought to market. It is obvious from facts such as these that, though one mode of fishing may yield negative results in respect of certain species, it does not necessarily follow that these species are absent or even rare. The results obtained by the use of the ordinary beam or otter trawl are usually very different from those obtained by the use of special nets or lines, therefore a kind of fish that may be seldom or never captured by one set of appliances, may by the use of a different set be found comparatively frequent.

The present enumeration comprises 113 species of Clyde fishes, but the occurrence of one or two of these appears to be somewhat doubtful; and it is also very desirable in the case of one or two others which, though their presence in the Clyde seems to be fairly well attested, further information should be obtained concerning them. I have indicated such species by enclosing their names within square brackets, and also by the notes referring to them.

There are several kinds of fishes which appear to be equally at home in the sea and in brackish water, and in some cases even in water that is fresh or nearly so; and there is a considerable divergence of opinion as to which, and how many, of these should be included in lists of *marine* species, and of those which should more properly be regarded as *fresh-water* forms. One has only to compare Professor H. G. Seeley's interesting work on *The Fresh-Water Fishes of Europe* with that of *The British Marine Food-Fishes*, by Professor M'Intosh and Mr. A. T. Masterman, to find examples of this difference of opinion. In the present list there will probably be found species which, in the opinion of some people, should have been excluded as belonging more properly to the fresh-water group; but when we find distinguished writers failing to agree on such a point as this, I may be excused if unable to prepare a list perfect in this respect. The basis of this list is the enumeration of Loch Fyne fishes, already referred to, prepared by the late George Brook, Esq., and published in 1886.

I have followed as far as possible the nomenclature used by Professor M'Intosh and Masterman in their *British Marine Food-Fishes*, while the arrangement of the species is in accordance with that of the *History of British Fishes* by Dr. Francis Day.

The following abbreviations are also used :—

(B. & S.) Refers to the list of Loch Fyne fishes published in the *Fourth Annual Report of the Fishery Board for Scotland.*
(Günther) Refers to Dr. Günther's paper published in the Proceedings of the Royal Society, Edinburgh, 1888, Vol. XV., pp. 205-220.
(H.B.) Refers to the separate reprint of the List of Fishes published in Harvey-Brown and Buckley's work on the Vertebrate Fauna of Argyle and the Inner Hebrides.
(G.) Refers to the "Garland's" statistics.
(M.) Refers to MS. records for Loch Fyne of the s.s. "Medusa," showing some of the records obtained by Sir John Murray while investigating the marine fauna and flora of the West Coast.

SUB-CLASS TELEOSTEI.

Order ACANTHOPTERYGII.

Fam. PERCIDÆ.

Polyprion americanus (Bl. Schneider). The Stone Basse.

This is the *Polyprion cernium* of Day's *British Fishes.* "One was taken years ago at the mouth of the Clyde" (Aflalo). "It was also taken off Little Cumbrae in 1870 by Dr. J. Young" (H.B. p. 184). The Stone Basse appears to be only an occasional visitor to the Clyde, for although it is included in the *Vertebrate Fauna of Argyle and the Inner Hebrides* the above is the only locality for the species given by the authors.

Fam. MULLIDÆ, Swainson.

Mullus barbatus, Linné. Surmullet, Red Mullet.

Loch Fyne (Captain Campbell of Inverneil). Captain Campbell in referring to this species says:—"I caught one in Loch Fyne . . . in a trammel net in 10 fathoms, and though I tried in many places for them I never got another" (*see* H.B., p. 185).

Fam. SPARIDÆ, Cuvier.

Pagellus centrodontus (De la Roche). The Common Sea Bream.

One specimen of *Pagellus centrodontus* was taken by Mr. M. P. Bell at Cumbrae, on July 12th, 1885 (*see* Robertson in Proc. N.H.S. Glasgow, Vol. I., p. 119). When at Tarbert in 1885–86, I remember seeing a specimen which was washed ashore in East Loch Tarbert, and which appeared to have been but a short time dead. It was probably captured in the nets of the herring boats fishing in the vicinity of East Tarbert and having been thrown back into the sea had drifted ashore. In the *Vertebrate Fauna of Argyle and the Inner Hebrides, Pagellus centrodontus* is described as "locally numerous throughout the West of Scotland." The species is sometimes brought in in large numbers to the fish market at Aberdeen.

Fam. COTTIDÆ.

Cottus scorpius, Linné. The Sea Scorpion.

Cottus scorpius is generally distributed throughout the Clyde and Loch Fyne, but it seems to be more frequent inshore than in deep water. "Two immature specimens were taken between Cloch Lighthouse and

Whiting Bay in 15 to 30 fathoms in July, and one immature specimen between French and Kilbrennan Sound in 10 to 14 fathoms in March" (Günther). The variety *grœlandicus* is also occasionally obtained in Loch Fyne; a specimen of the variety in the Fishery Board's collection at Bay of Nigg measures slightly over one foot in length. Dr. Day remarks that the sea scorpion is said to attain to six feet in length in the Greenland seas, while the largest recorded example in Great Britain is fifteen inches.

Cottus bubalis, Euphrasén. The Father Lasher.

Moderately common in Loch Fyne (B. & S., 1886). "A very young specimen was obtained at the Mull of Cantyre in 60 fathoms in February, and another immature one in the Sound of Sanda in 20 fathoms in March, 1885" (Günther). Little Harbour, Upper Loch Fyne, December 1896. Off Inveraray, 1897. Young specimens are moderately frequent amongst seaweed in the shallow inshore bays of Loch Fyne.

Cottus lilljeborgii, Collett. The Norway Bullhead.

"Off Ardrossan, 15 to 30 fathoms; Sound of Sanda, 20 fathoms" (Günther). This is one of the species added to the British fauna by Dr. (now Sir) John Murray. It seems to be a small species. The specimen caught off Ardrossan measured $2\frac{1}{4}$ inches in length, while that from the Sound of Sanda was only $1\frac{3}{4}$ inches long.

Trigla lineata, Gmelin. The Streaked Gurnard.

East Loch Tarbert, Loch Fyne, not common (B. & S.). One was obtained in the trawl of the s.s. "Garland" at trawling Station VI.* (near the mouth of the Clyde) on November 25th, 1895. Two were captured at trawling Station XI. on April 21st, 1897, and another at Station VIII. in September of the same year. Mr. Pearcey records that five were caught last year (1899) at Stations I., VI., VII., and IX. Dr. Day refers to one specimen of *T. lineata* having been procured near Ayr, and a second in October 1844 at Glasgow.† It appears to be a rare species in the Clyde.

Trigla cuculus, Linné. The Red Gurnard.

Red Gurnards are occasionally captured by the "Garland" in the seaward part of the Clyde estuary, but do not seem to be very common. They appear to have been more frequently recorded from the Clyde last year (where the "Garland" was chiefly employed) than for some years previously. In looking over the returns of the "Garland" for 1899, it is interesting to observe that while these fishes were captured in every haul made at Station VI., at Station VIII. they occurred in three of the hauls, and in only two of the hauls made at each of the Stations I., VII., IX., and X. At Stations III. and XII. they were only once captured, and none are recorded from any of the hauls made at Stations II., IV., V., and XI.

Ninety hauls were made by the "Garland" last year at Stations I. to XII., and Red Gurnards were obtained in twenty-one of them; the number of specimens recorded from these twenty-one hauls is 73, and the number obtained in each haul may be summarised as follows:—

---

* The situation of the Stations is described on p. 20 of this Report.

† *British Fishes*, Vol. I., p. 58.

7	hauls	contained	1	Red	Gurnard	each	=	7.
3	,,	,,	2	Red	Gurnards	each	=	6.
3	,,	,,	3	,,		,,	=	9.
3	,,	,,	4	,,		,,	=	12.
1	,,	,,	5	,,		,,	=	5.
1	,,	,,	7	,,		,,	=	7.
2	,,	,,	8	,,		,,	=	16.
1	,,	,,	11	,,		,,	=	11.

21   73

Fifteen specimens of Red Gurnards were obtained in one haul at Station VIII. in December 1898, but such a large number in one haul is exceptional. They are also recorded by Landsborough in his *Natural History of Arran*, p. 492.

Trigla lucerna, Linné. The Sapphirine Gurnard.

This, which is the *Trigla hirundo* of Day's *British Fishes*, appears also to be moderately rare in the Clyde. Thompson remarks that he has seen this species in autumn captured in salmon nets at Ballantrae, in Ayrshire, and on different parts of the coast in the country (Day).* I do not find it recorded amongst the captures of Clyde fishes made by the Fishery steamer "Garland."

Trigla gurnardus, Linné. The Grey Gurnard.

Common, and generally distributed throughout the seaward part of the Clyde and in Loch Fyne, especially during the summer months. This seems to be a common species throughout the Scottish seas. Large quantities of gurnards are sometimes landed at the fish market at Aberdeen, and they nearly all belong to this species.

Triglops murrayi, Günther. Murray's *Triglops*.

"Several specimens from $2\frac{1}{2}$ to 4 inches long were obtained at the Mull of Cantyre at a depth of 64 fathoms in the months of February and March, and 4 miles south-east of the Island of Sanda in 35 fathoms in the middle of March 1887" (Günther). One specimen $2\frac{1}{2}$ inches long was captured in the shrimp-trawl of the "Garland," in the vicinity of Sanda, November 1896. *Triglops murrayi* is described as closely allied to *Trigla pengellii*; it is one of the species added to the British fauna by Dr. (now Sir) John Murray.

Fam. CATAPHRACTI, Müller.

Agonus cataphractus, Linné. The Pogge.

Ardrishaig (Dr. Scouler). Occasionally in other parts of Loch Fyne and the Clyde estuary. "Five specimens from the Mull of Cantyre, 49 to 50 fathoms, February. Two specimens from Kilbrennan Sound, 10 to 20 fathoms, March 1888" (Günther). It has also been got by the "Garland" at Stations I. and VI.

Fam. PEDICULATI, Cuvier.

Lophius piscatorius, Linné. The Angler.

The Angler is frequently taken in the trawl-net of the Fishery steamer "Garland," and especially at those stations where the water is

* *British Fishes*, Vol. I., p. 62.

moderately deep. One or more are got in most hauls of the trawl at all the Stations, except in Loch Fyne, where it is much less common. No specimens have been caught at Stations XV. or XVII.; one was taken at Station XVI. in over 60 fathoms. The largest obtained was 49 inches (Station I., 30th May 1899). *Lophius* is sometimes obtained larger than any of the Clyde specimens; in December 1841 Thompson examined one that was four and a half feet in length.*

Fam. TRACHINIDÆ, Risso.

Trachinus vipera, Cuvier.   The Lesser Weaver.

Recorded from Arran by Dr. Landsborough in his *Natural History of Arran*, p. 318. He states that the lesser weaver is known to the people at Lamlash as the "Stangster" or "Stang-fish."

Fam. SCOMBRIDÆ, Cuvier.

Scomber scombrus, Linné.   The Mackerel.

Shoals of mackerel are occasionally observed both in the Clyde estuary and in Loch Fyne, and considerably over a thousand cwts. are usually landed by fishermen each year. Dr. Day, though he gives a very full description of the habits, habitats, etc., of the mackerel, makes no allusion to its occurrence in the Clyde or Loch Fyne, neither is any notice taken of its occurrence in these places in the *Vertebrate Fauna of Argyle and the Inner Hebrides*.

Orcynus thynnus, Linné.   The Tunny.

Pennant records a specimen weighing 460 lbs. taken at Inveraray in 1769. An example, 9 feet long, was captured in the Gareloch, nearly opposite Greenock, in July 1831.† This last specimen is said to have been in pursuit of herrings.

Thynnus pelamys, Linné.   The Bonito.

One was captured in the Clyde in July 1832,‡ and purchased for the Andersonian Institute of Glasgow.§ Jenyns is doubtful if the fish recorded by Dr. Scouler is the true *Thynnus pelamys*, but suggests that it is more likely to be the next one.

Pelamys sarda, Bloch.   The Belted Bonito.

One was captured in the Clyde in 1859, and is now in the Hunterian Museum, Glasgow.

Fam. CORYPHÆNIDÆ (part), Swainson.

Brama raii, Bloch.   Ray's Bream.

One specimen in the Hunterian Museum was taken near Ayr (H.B., p. 188).

Lampris luna, Gmelin.   The Opah or King Fish.

One was taken in the Clyde in 1833 (Scouler). A specimen of the king-fish in the Hunterian Museum, Glasgow, was captured in the Clyde in 1864.

* Day, *British Fishes* Vol. I., p. 76.
† Day, *British Fishes*, Vol. I., p. 97.
‡ Scouler, *Mag. Nat. History*, Vol. VI., p. 529 (1833).
§ Day, *British Fishes*, Vol. I., p. 101.

Fam. CYTTIDÆ, Kaup.

Zeus faber, Linné.   The John Dory.

The John Dory has been taken off Rothesay (H.B., p. 186); in Loch Fyne (B. & S.); in Kilbrennan Sound in the shrimp-trawl of the Fishery steamer "Garland," and in other parts of the Clyde estuary (Stations I., II., III., VI., and VII.), but most of the Clyde specimens I have seen are small and immature. Those taken by the "Garland" usually range from 10 to 14 inches; the largest was 19 inches and the smallest three inches.

Fam. GOBIIDÆ, Cuvier.

Gobius flavescens, Fabricius.   The Two-Spotted Goby.

Loch Fyne (Dr. Scouler); common amongst *Zostera* in East Loch Tarbert (B. & S.); Upper Loch Fyne (M.); also in various parts of the Clyde estuary (G.). Though *Gobius flavescens* is widely distributed throughout the Clyde area it does not appear to be very common. This is the *Gobius ruthensparri* of Day's *British Fishes*.

Gobius niger, Linné.   The Black Goby.

Taken in the Clyde area (H.B., p. 193). Moderately common, especially off shore; but most of the specimens captured by the "Garland" are of comparatively small size, few of them reaching a length of four inches.

Gobius minutus, Gmelin.   The Spotted or Speckled Goby.

Common, and generally distributed throughout the estuary, including Loch Fyne and the other sea-lochs within the Clyde area. Günther records *Gobius minutus* from Loch Goil in 45 fathoms, and in the vicinity of Cloch Lighthouse in 43 fathoms. It has also been taken by the "Garland" at the head of Upper Loch Fyne in from 17 to 35 fathoms.

Gobius jeffreysii, Günther.   Jeffrey's Goby.

"Three specimens were obtained in April, viz., in Lamlash Bay (6 to 18 fathoms), off Whiting Bay (20 fathoms), and off Cumbrae Light (56 fathoms). Five specimens in August off the Cloch Lighthouse in 43 fathoms. Two adult specimens, male and female, obtained in Kilbrennan Sound at an uncertain depth (10 to 45 fathoms) on March 22nd, 1888" (Günther). This species has not yet been obtained by the Fishery steamer "Garland."

Several of the species of *Gobius* are somewhat difficult to make out, and it is just possible that other species in addition to those named may also occur in the Clyde.

Fam. CALLIONYMIDÆ, Richardson.

Callionymus lyra, Linné.   The Dragonet or Gemmeous Dragonet.

The dragonet is not uncommon in the Clyde and Loch Fyne; it is frequently mentioned amongst the captures of fishes made by the Fishery steamer "Garland," and especially amongst the fishes taken with the shrimp-trawl. As many as 47 specimens were recorded from one haul made on November 11th, 1896, and Mr. Pearcey records 125 specimens captured in a single haul on September 5th 1897, between Cantyre and Ailsa Craig.

Callionymus maculatus (Bonaparte).   The Spotted Dragonet.

Kilbrennan Sound, rather abundant at 26 fathoms ; Sound of Sanda, 24 to 28 fathoms (H.B., p. 194).   Occasionally captured in the shrimp-trawl of the Fishery steamer " Garland."

Fam. Discoboli, Cuvier.

Cyclopterus lumpus, Linné.   The Lumpsucker or Cock-Paidle.

The lumpsucker is occasionally captured in the Clyde by the "Garland." Young specimens are sometimes observed in the tow-net gatherings collected in Loch Fyne as well as in other parts of the Clyde estuary.   Thompson mentions its existence in the vicinity of Ayrshire.*

Cyclogaster liparis, Fleming.   The Sea Snail or Sucker.

The sucker is not uncommon in Loch Fyne, and also in the Clyde, but usually inshore where the water is shallow and the bottom weedy. It is, however, recorded from water that is also of considerable depth. Dr. Günther in his report to the Royal Society of Edinburgh † records many specimens from 49 and 64 fathoms in the vicinity of the Mull of Cantyre, and three specimens from 30 to 40 fathoms between Cumbrae and Wemyss Point.   It has been got by the "Garland" in 66 fathoms, 8 miles W. by N. of Corsewall Point, and in the same depth S.E. of the Mull of Cantyre.

Cyclogaster montagui, Donovan.   Montagu's Sucker.

Not so common as the last, but the habitat is somewhat similar.

Fam. Gobiesocidæ, Bleeker.

Lepadogaster gouanii (Lacep.).   The Cornish Sucker.

This species has been taken at Millport (H.B., p. 198).   I have not yet seen a Clyde specimen of *Lepadogaster gouanii*; but probably it may not be very rare if sought for in the right places.

Lepadogaster bimaculatus (Cuvier).   The Doubly-spotted Sucker.

Recorded by Dr. Landsborough from Lamlash Bay, Arran.   He describes it as "far from being rare."‡   Taken in the Clyde area (H.B.).

Anarrhichas lupus, Linné.   The Wolf Fish, Cat fish.

Not uncommon amongst the captures of fishes made by the s.s. "Garland," both in the Clyde and Loch Fyne.

Fam. Blenniidæ, Swainson.

Blennius pholis, Linné.   The Shanny.

In Clyde area (H.B., p. 185).   Mr. Alex. Gray, of the Marine Station, Millport, Cumbrae, in a note dated December 11th, 1899, says:— "Last spring I caught a specimen of *Blennius pholis* here, and kept it alive for about four months.   Unfortunately it managed somehow to jump out of the tank, and in the morning I found it on the floor, dead. It is now in our collection, preserved in formaline."

---

* Day, *British Fishes,* Vol. I.. p. 182.
† Proc. Vol. XV., p. 41.
‡ *Natural History of Arran* (1875), p. 328.

Chirolophis galerita, Linné. Yarrell's Blenny.

One specimen was taken amongst boulders at low water in East Loch Tarbert in 1885 (B. & S.). Ballantrae Bank, one specimen in January 1899 (G.).

Pholis gunnellus, Linné. The Gunnel or Butter-fish.

The butter-fish is common and generally distributed, and especially inshore amongst weed and stones, where it may frequently be found when the tide recedes.

Euchelyopus (Zoarces) viviparus, Linné. The Viviparous Blenny.

A specimen of this species, captured in the vicinity of Row, near Helensburgh, is in the collection of the late Dr. Robertson of Millport at the Marine Station there. I am indebted to Mr. Gray, the Curator, for this record.

In the *Vertebrate Fauna of Argyle and the Inner Hebrides*, the viviparous blenny is recorded from Loch Creran, the Sound of Jura, and from Glenshiel, "but it seems to be rare on the West Coast." There is no Clyde record for the species in this work.

Lumpenus lampetræformis (Walbaum). The Sharp-tailed Lumpenus.

" Three adult specimens were found between Cumbrae and Skelmorlie Lighthouse in 20 fathoms in April 1887, and at Cumbrae Lighthouse in 60 fathoms in February 1888 " (Günther). Occasionally captured off shore in the shrimp trawl-net of the Fishery steamer "Garland." As many as a dozen specimens of this rare species have been taken at one time by the Fishery steamer in the lower part of the Clyde estuary, whilst in 1896 four specimens were captured near the head of Upper Loch Fyne.

It requires a net with moderately close meshes to capture this species— an ordinary trawl-net is usually ineffective.

In a work on British Natural History published in 1898, the author, F. G. Aflalo, has the following note within square brackets at p. 371 :— " The sharp-tailed lumpenus was once trawled (1884) off the Carr Light-ship," and he adds that he gives this record " on the authority of M'Intosh and Masterman"; evidently he was unaware of Dr. Günther's records for the Clyde, and the Fishery Board's for the Firth of Forth.

Fam. CEPOLIDÆ, Bleeker.

Cepola rubescens, Linné. The Red Band Fish.

One 15½ inches long was taken on a whiting line 7 miles south of Ayr (Harvey). Another 19½ inches long was found on the beach at Ballantrae after a storm (Thompson). "Two have been taken on the Ayrshire coast just inside the Clyde area " (H.B., p. 195). Perhaps this last record refers to the same specimens mentioned by Harvey and Thompson.

Fam. ATHERINIDÆ, Günther.

Atherina presbyter, Cuvier. The Sand Smelt or Atherine.

Frequent amongst zostera in East Loch Tarbert in the spring of 1885, but not met with later (B. & S.). Mr. Gray, of the Millport Marine Station, states :—" On two occasions I saw shoals of these little fishes in Campbeltown Loch—once at the Kilbrennan shore, where they were pursued by a number of guillemots, which chased them so keenly that they jumped clean up on the gravelly beach, and I collected quite a number of living specimens. Many of the fishes were fully two feet from the water's edge."

Fam. MUGILIDÆ.

Mugil capito, Cuvier.   The Grey Mullet.

The "ear-stones" (*otolites*) of a grey mullet (*Mugil capito,* Cuv.), captured in the Clyde off Fairlie, are in the collection of the late Dr. Robertson of Millport, at the Marine Station there. I am indebted to Mr. Gray, the Curator, for this record.

In the *Vertebrate Fauna of Argyle and the Inner Hebrides, Mugil capito* is recorded from several localities on the West Coast of Scotland, but there is no record of it from the Clyde.

Fam. GASTEROSTEIDÆ.

Gasterosteus aculeatus, Linné.   The Three-spined Stickleback.

Captured occasionally with the push-net at various parts of the shores of Loch Fyne ; we have it also from other parts of the Clyde estuary.

Gasterosteus spinachia, Linné.   The Fifteen-spined Stickleback.

The fifteen-spined stickleback is more or less common amongst sea-weed in shallow water; it is generally distributed along the shore line, both of the Clyde estuary and the lochs branching from it.

Fam. LABRIDÆ, Cuvier.

Labrus bergylta, Ascanius.   The Ballan Wrasse.

The Ayrshire and Wigtown coasts (Thompson). In the Clyde area (Dr. J. Young, *vide* H.B., p. 199). Frequent in Loch Fyne in the autumn (B. & S.).

Labrus mixtus, Linné.   The Striped Wrasse.

Taken occasionally at the mouth of East Loch Tarbert, Loch Fyne (B. & S.). It is also recorded in Harvey-Brown's *Vertebrate Fauna of Argyle and the Inner Hebrides,* and on the authority of Dr. J. Young as occurring in the Clyde area.

A male and a female of this species were captured in Ayr Bay by a fisherman on January 30th this year (1900); the specimens were secured by Mr. Duthie, Fishery Officer, Girvan, who happened to be in Ayr at the time, and were forwarded by him to the Laboratory at Bay of Nigg.

Crenilabrus melops, Linné.   The Goldsinny or Corkwing.

A small specimen of the goldsinny was captured by the "Garland" near Cairndow, at the head of the Loch Fyne, in August 1899. One specimen from Lamlash Bay in 6 to 18 fathoms, April (Günther).

Ctenolabrus rupestris, Linné.   Jago's Goldsinny.

Common, especially near Skate Island, Loch Fyne (B. & S.). Occasionally obtained in the trawl-net of the Fishery steamer "Garland." One specimen was obtained in Lamlash Bay in 6 to 18 fathoms, in April, and another between Cumbrae and Skelmorlie Buoy in 20 fathoms, in the same month, 1887 (Günther).

Centrolabrus exoletus, Linné. Small-mouthed Wrasse, Rock Cook.

Taken occasionally in Loch Fyne (B. & S.).

Coris julis (Linné). The Rainbow Wrasse.

Recorded for the Clyde area on the authority of Dr. J. Young (H.B., p. 200).

Order ANACANTHINI.

Fam. GADIDÆ, Cuvier.

Gadus callarius, Linné. The Cod.

Common and generally distributed. The natural habitat of the cod is in moderately deep water; this is shown by the fact that if they are exposed for a lengthened period to daylight their eyes become more or less diseased. When at Rothesay Aquarium in 1886–87 I observed that the eyes of almost all the specimens of cod kept there were diseased, and one specimen was entirely blind and had to be fed separately. This disease, which at first took the form of an opaque white spot, but which gradually extended all over the eye, was ascribed to the fish being kept exposed to the daylight. The cod were the only fishes in the Aquarium that were affected in this way.

Gadus æglefinus, Linné. The Haddock.

More or less frequent, and generally distributed throughout the Clyde area, but much less abundant than on the East Coast. Upper Loch Fyne in the centre in 65 to 70 fathoms (M.).; between Pennymore and Inveraray (G.).

Gadus luscus, Linné. The Bib, or Whiting Pout.

Occasionally in Tarbert harbour (B. & S.); Tarbert Bank, Lower Loch Fyne; and between Dunderawe and Cairndow, Upper Loch Fyne (G.). Generally distributed throughout the Clyde estuary, but seldom more than 8 to 10 inches in length. Two, 13 inches in length, were obtained near Sanda Island in May 1897. A specimen in the Fishery Board's collection in the Laboratory at Bay of Nigg measures fully 15 inches in length.

Gadus minutus, Linné. The Poor or Power Cod.

Dr. Günther in his paper on Clyde Fishes states that "the specimens obtained on March 10th and 17th were ready to spawn, and had fed on *Nyctiphanes*, sand eels, and *Aphrodite*." This *Gadus* is usually more or less in evidence amongst the contents of the shrimp-trawl of the Fishery steamer "Garland."

Gadus esmarkii (Nilsson). The Norway Pout.

According to Dr. Günther, the distribution of this species in the Clyde extends from Kilbrennan Sound to Lower Loch Fyne, where young specimens were found in tolerable abundance at 80 fathoms. He also remarks that "the species does not appear to be unfrequent in Kilbrennan Sound." He states further that "the characteristics by which *Gadus esmarkii* can be distinguished from its British congeners are :—The lower jaw, which projects beyond the upper; the dentition, the teeth of the outer series in the upper jaw being a little larger than the inner ones; the length of the snout, which is almost equal to the length of the diameter of the eye, which is a little less than one-third the length of the head; the slender barbel, which is about half as long as the eye; and, finally the fin formula—it being D 15–16, 23–25, 22–25; A 27–29, 23–25" (H.B., p. 203). *Gadus esmarkii* is, in some respects, not unlike *Gadus minutus*, and may have occasionally been mistaken for that species. It is occasionally taken in the shrimp trawl-net of the "Garland."

Gadus merlangus, Linné. The Whiting.

This is one of the more common and generally distributed of the *Gadi*; but though extending into Upper Loch Fyne it is described as "not abundant" there. The whiting appears to be more frequent in the seaward portion of the Clyde estuary than it is in the more inland parts.

Gadus poutassou, Risso. Couch's Whiting.

This is readily distinguished from other British *Gadi* by the position of the dorsal fins, and especially by the distance between the second and third dorsals being greater than the distance between the same fins in the whiting, which it somewhat resembles. Three specimens of *Gadus poutassou* were captured in the shrimp-trawl of the "Garland" at the mouth of the Clyde estuary, in 54 fathoms, on September 22nd 1897. These specimens are now amongst the collection of fishes in the Fishery Board's Laboratory at Bay of Nigg.

Gadus virens, Linné. The Coal-fish or Saithe.

Moderately common and generally distributed, but mostly of small size. Full grown coal-fish do not appear to be very plentiful in the Clyde, but the young "podlies" sometimes occur in great numbers in inshore localities, as at East Loch Tarbert harbour and wharf.

Gadus pollachius, Linné. The Pollack or Lythe.

More or less frequent, but usually of small size; large specimens are occasionally brought to East Tarbert, Loch Fyne (B. & S.). The "Garland" also has records of lythe from Loch Fyne as well as off Sanda Island. The specimens from Sanda measured from 23 to 35 inches in length. *Gadus pollachius* has no barbel on the lower jaw.

Merluccius vulgaris, Cuvier. The Hake.

The hake is frequently captured in the Clyde by the "Garland" at all the Stations, as many as 108 having been caught in a single haul of the net; most of the specimens are more or less immature, but sometimes range from 30 to 40 inches.

Phycis blennioides (Brun.). The Greater Fork-beard.

The late Dr. Robertson of Millport, Cumbrae, recorded this species as having been taken near Cumbrae in April 1890.[*] Mr. Alex. Gray, of the Millport Marine Station, informs me that "the palate and forked fins of a specimen of *Phycis* are in the Robertson collections," and are probably portions of the fish referred to by Dr. Robertson, as they agree with the description he gave of the specimen which he recorded. Several specimens of the greater fork-beard have occurred on the coasts of the Solway Firth[†], but it seems to be rare in the Clyde.

---

[*] *Proc. Nat. Hist. Soc. Glasgow*, Vol. III. (N.S.), p. 276 (1892). This specimen measured 24 inches in length.

[†] Day, *British Fishes*, Vol. I., p. 304.

Molua molva, Linné.   The Ling.

The ling, though not uncommon in the Clyde, is not very often captured by the "Garland," and rarely more than one or two specimens are obtained in any single haul made by the beam-trawl.

Onos mustela, Linné.   The Five-bearded Rockling.

Taken between tide marks in East Loch Tarbert; not common (B. & S.).

Onos cimbrius, Linné.   The Four-bearded Rockling.

"Very common, and generally distributed in the Clyde area at depths varying from 6 to 65 fathoms in April, to 70 and 90 fathoms in July and August, to 100 fathoms in November at Upper Loch Fyne and Kilbrennan Sound, at 37 to 46 fathoms in December," &c. (Günther). *Motella cimbria* is frequently captured in the shrimp-trawl of the Fishery steamer "Garland," but rarely in quantities, usually only one or a very few specimens being taken at one time.

Onos tricirratus, Bloch.   The Three-bearded Rockling.

The three-bearded rockling has been taken in Rothesay Bay and other parts of the Clyde area, but appears to be rare.   Part III. of the *Sixteenth Annual Report of the Fishery Board for Scotland* contains records of the capture of a few specimens in Upper Loch Fyne.

Onos maculatus, Risso.   The Spotted Rockling.

"Loch Fyne, 40 fathoms ; Mull of Cantyre, 65 fathoms " (Günther). This species somewhat resembles the last, but the front teeth are large. If *Onos tricirratus* and *maculatus* be really distinct species, the Clyde specimens should perhaps be all included under the latter name.

Raniceps raninus, Collet.   The Lesser Fork-Beard, Tadpole Hake.

A specimen of the lesser fork-beard was recorded from Loch Fyne by the late Dr. Scoular.*

Fam. OPHIDIIDÆ, Müller.

Ammodytes lanceolatus, Le Sauvage.   Greater Sand Launce or Sand Eel.

Taken occasionally in the neighbourhood of East Loch Tarbert (B. & S.).   "Taken in the vicinity of Sanda Island, Sound of Sanda Cantyre.   The young numerous about the end of March " (Günther).

Fam. PLEURONECTIDÆ, Risso.

Hippoglossus vulgaris (Fleming).   The Halibut.

The halibut is occasionally taken in Loch Fyne (B. & S.), but it is usually of small size.

Mr. Duthie, Fishery Officer at Girvan, states *in lit.* :—" Halibut are frequently got here in spring, though they cannot be called plentiful.   On the 25th of March this year (1899) one boat landed four cwts., caught between Ailsa Craig and the Mull of Cantyre ; the fish were of all sizes up to 112 lbs. or more—just what would have been a fair sample of Shetland-caught fish.   I believe they are more plentiful in the neighbourhood of the 'Mull' than here, but an occasional fish is got all round this part of the Firth in the spring months when great-lines are

Proc. Nat. Hist. Soc. Glasgow, Vol. I., p. 8 (1868).

being used. Halibut are occasionally recorded at Ballantrae caught by cod or turbot nets. I have no doubt a stray specimen is also got between Arran and Ayr, although I do not at present remember of more than one cwt. being landed at Ayr this year."

Drepanopsetta plattessoides, Fabr.   The Long Rough Dab.

"Many adults and young (2 inches in length) were obtained in 26 to 46 fathoms in Kilbrennan Sound in December. Many adults and young were also taken between Cumbrae and Wemyss Point in 30 to 40 fathoms in February 1888" (Günther). The long rough dab is very common and generally distributed throughout the Clyde area, especially off shore and in moderately deep water.

Bothus maximus (Linné).   The Turbot.

"One captured in the Clyde area, August 28th, 1888; formerly numerous there" (Günther). Turbots have been taken in the trawl-net of the Fishery steamer "Garland" in the vicinity of Ailsa Craig, in the vicinity of Sanda Island, and in Kilbrennan Sound. Occasionally taken in Loch Fyne (B. & S.). There is a regular turbot fishery carried on off Girvan, a special net called the "turbot net" being used for the capture of the fish.

Bothus rhombus (Linné).   The Brill.

The brill is rather more frequently recorded amongst the lists of fishes captured by the Fishery steamer "Garland" than the turbot, but mention is not often made either of the one or the other. The largest specimen of the brill referred to by Day is one recorded by Thompson, which measured 24 inches in length. The "Garland" has, however, obtained even larger specimens than that; one 25 inches was captured after nightfall at Station VII. on May 21st, 1897. A few measuring 21 and 23 inches have also been taken at different times.

Zeugopterus unimaculatus, Risso.   One-Spotted Topknot.

"One specimen in 10 fathoms off Ardrossan, Clyde area, in April 1888" (Günther). Several specimens were captured in a shallow sandy bay at Barmore, Loch Fyne, in 1885 (B. & S.). Two of these specimens are still in the collection of fishes at the Fishery Board Laboratory, Bay of Nigg, Aberdeen.

Zeugopterus punctatus (Blainville).   Müller's Topknot.

"Clyde area, 60 fathoms" (Günther). Upper Loch Fyne in 10 to 25 fathoms (M.). *Zeugopterus punctatus* has been taken in the trawl-net of the "Garland" in the vicinity of Ailsa Craig and in Kilbrennan Sound.

Rhombus norvegicus, Günther.   The Norway Topknot.

Dr. Günther records the following specimens from the Clyde:—One dredged in Lamlash Bay at a depth of 6 to 18 fathoms, measuring $3\frac{1}{2}$ inches in length, and in excellent condition; one smaller than the last, caught off the Cloch Lighthouse in 43 fathoms; and one $3\frac{1}{2}$ inches long, caught in Kilbrennan Sound in 45 fathoms. (*See* also Dr. Günther's description of the larger of these specimens.*) The Norway topknot is occasionally taken in the shrimp-trawl of the "Garland," but it appears to be a rare species in the Clyde.

Proc. Roy. Soc. Edin., Vol. XV., p. 47 (1888).

Lepidorhombus megastoma, Donovan. The Sail Fluke or Whiff.

The sail-fluke is recorded for the Clyde area by Dr. Günther (1886). It is a species that does not appear to be very rare in the Clyde area, especially in the seaward portion of it, and is more or less frequently mentioned amongst the captures of fishes made by the "Garland"; it is especially frequent at Stations VIII., IX., and X.

Platophrys laterna, Walbaum. The Scald Fish.

Recorded for the Clyde area by Dr. Günther (1888). Taken in the trawl-net of the "Garland" between Sanda Island and Bennan Head. This species appears to be a comparatively rare one in the Clyde estuary; a few specimens were taken last year (1899) at Stations II., V., VI., VII., VIII., and X.

Pleuronectes platessa, Linné. The Plaice.

This is a moderately common and generally distributed species from the head of Loch Fyne to the seaward limits of the Clyde estuary.

Pleuronectes microcephalus, Donovan. The Lemon Dab; also called Lemon Sole and Smooth Dab.

Clyde area, feeding on annelids and *solens* (Günther). The lemon sole is more or less frequent and generally distributed throughout the estuary of the Clyde, especially off shore. Annelids appear to constitute the principal food of the lemon sole, at least in Scottish waters, as shown by the investigations that have been made concerning fish-food on board the Fishery steamer "Garland."

Pleuronectes cynoglossus, Linné. The Witch Sole, Craig Fluke, Pole Dab.

This appears to be the most common of the flat fishes in the Firth of Clyde, especially in the deep water. Dr. Günther records its occurrence at a depth of 80 to 100 fathoms in Loch Fyne. It is most abundant at Stations III., VII., VIII., IX., XI., XV., and XVII.; nearly 400 have been got in a single haul in Loch Fyne. Its food, like that of the lemon sole, consists chiefly of annelids, and it is probably on this account that the species is usually more common on a muddy bottom.

Pleuronectes limanda, Linné. The Common Dab.

Pleuronectes limanda is also one of the more common of the Clyde fishes, and seems to be distributed all over the estuary to the head of Loch Fyne, and probably also throughout the other sea-lochs within the Clyde area (Günther). It is got at all the "Garland" Stations.

Pleuronectes flesus, Linné. The Flounder or Fluke.

Common in Tarbert Harbour, Lower Loch Fyne (B. & S.). Occasionally obtained in Upper Loch Fyne (G.). One was captured off Largymore, Upper Loch Fyne, on April 28th, 1896, and a few have been captured in other parts of the loch with the push-net; but it seems to be a rare fish in the more seaward portion of the Clyde.

Solea vulgaris, Quensel. The Sole or Black Sole.

The black sole is occasionally taken in the trawl-net of the "Garland" at all the Stations, except in Loch Fyne; there are seldom more than one or two specimens, but occasionally ten or twelve, in a haul. It is more difficult to capture black soles during daylight than after darkness sets in. The food of the black sole appears to consist largely of annelids.

Solea variegata, Donovan.   The Variegated Sole or Thickback.

"Two immature specimens were obtained in the vicinity of the Mull of Cantyre in 65 fathoms on March 21st, 1888" (Günther, p. 220).   Two specimens, one about $3\frac{1}{2}$ inches long, the other slightly less, were captured near the mouth of the Clyde estuary at Station VI., on June 15th, 1899.   They were taken with the shrimp-trawl of the "Garland," and are now in the collection in the Laboratory at Bay of Nigg.

Solea lutea, Risso.   The Solenette or Little Sole.

The solenette has been recorded amongst the fishes captured by the "Garland," but appears to be rare in the Clyde.

Order PHYSOSTOMI, Müller.

Fam. SALMONIDÆ, Müller.

Salmo trutta (Fleming).   The Sea Trout.

The sea trout is regularly taken in small quantities just outside East Loch Tarbert (B. & S.).   A fine specimen, which, I understand, had been captured in the Clyde, was kept in one of the tanks in Rothesay Aquarium when I was there in 1886-87.

Salmo salar, Linné.   The Salmon.

There are regular fisheries for salmon at various parts of the Clyde estuary, including the several sea lochs.   The salmon is taken occasionally in herring-nets between Tarbert and Barmore, and probably also in other parts of the Clyde.

Osmerus eperlanus, Linné.   The Smelt.

"One was taken with mussel-bait at Brodick, Arran, October 1888. Not uncommon" (H.B., p. 218).

Argentina sphyræna, Linné.   The Hebridean Smelt.

"Three specimens were obtained in 32 fathoms between Little Cumbrae and Brigaird Point on February 7th, and five obtained in 37 fathoms in Loch Striven on February 15th, 1888" (Günther).   The Hebridean smelt is not very rare in the Clyde estuary, when sought for with suitable apparatus; it is frequently taken with the shrimp trawl-net of the Fishery steamer "Garland," as many as seven specimens having been captured in a single haul; usually, however, only one or a few are obtained at one time.

Fam. SCOMBRESOCIDÆ.

Rhamphostoma belone.   The Garfish.

A specimen of the garfish, 28 inches in length, was taken off Dunoon on May 25th, 1877 (Dr. F. P. Fleming).*   It is taken in the Clyde area occasionally in shoals (H.B., p. 212).   A young specimen was captured in the surface tow-net of the "Garland" in Loch Fyne.

Scombresox saurus, Walbaum.   The Saury Pike.

Taken in the Clyde area (H.B., p. 212).   We have not yet had the good fortune to capture this species, either in the immature or adult stages.

* This record is referred to in Dr. Day's *British Fishes*, Vol. II., p. 150.

[(?) *Exocetus volitans*, Linné. The Flying-fish.

The Rev. David Landsborough, in his *Natural History of Arran*, p. 386, states that the flying-fish has been seen in Ayr Harbour. Two or three kinds of fishes are spoken of as "flying-fish," but probably it is the species named above that he refers to.]

Fam. CLUPEIDÆ, Cuvier.

Clupea harengus, Linné. The Herring.

Common in the Clyde and Loch Fyne, where a great herring fishing is carried on. The movements and habits of the Clyde and Loch Fyne herring have been for a long time subjects of much interest to naturalists, and even yet are not very clearly understood.

Clupea sprattus, Linné. The Sprat.

Sprats are not uncommon in the Clyde, but owing to their having been frequently mistaken for young herring, there is some uncertainty as to their numbers and distribution.

Clupea alosa, Linné. The Allis Shad.

Very large specimens were taken in Loch Fyne in 1888 (H.B., p. 221). Mr. Gray, of the Millport Marine Station, says *in lit.*:—"A fine specimen of this fish was taken in a seine-net off Isle of Ross in the winter of 1894, and was put in spirits by me and left in the old museum in Kirk Street, Campbeltown, and is probably still preserved in the new museum there."

Fam. MURÆNIDÆ, Müller.

Anguilla vulgaris, Cuvier. The Common Eel.

The Common Eel affords a small but regular fishery in Tarbert harbour, Loch Fyne (B. & S.). It was also taken in the shrimp trawl-net of the "Garland" off Inveraray in May 1896.

Conger vulgaris, Cuvier. The Conger.

Moderately large specimens of conger are at times captured in the "Garland's" beam trawl-net, one 50 inches in length was taken near the head of Loch Fyne on May 5th, 1896 ; another of the same length was secured in the vicinity of Ailsa Craig on April 28th, 1897. A specimen 45 inches long is recorded from Whiting Bay, and others of somewhat smaller size from various parts of the Clyde area. I have taken a fairly big specimen between tide marks at Lunderston Bay after the tide had ebbed.

Order LOPHOBRANCHII, Cuvier.

Fam. SYNGNATHIDÆ.

Siphonostoma typhle (Linné). The Deep-nosed or Broad-nosed Pipe-fish.

A specimen of this species was captured amongst *zostera* in East Loch Tarbert (B. & S.). A specimen, which I believe to be the one now recorded here, is in the collection of fishes in the Fishery Board's Laboratory at Bay of Nigg.

Syngnathus acus (Linné). The Great Pipe-fish.

This species is not uncommon in the Clyde and Loch Fyne, especially inshore where the water is shallow.

Nerophis æquoreus (Linné).   The Straight-Nosed Pipe-fish.

East Loch Tarbert, amongst *zostera* (B. & S.).   A fine specimen is in the Robertson collection in the Marine Station at Millport, which was captured at Cumbrae.

Nerophis lumbriciformis (Linné).   The Worm Pipe-fish.

Taken in East Loch Tarbert amongst *zostera* (B. & S.).   I have taken this little fish amongst weed between tide marks in Lunderston Bay, near Inverkip, and it has also been taken by the "Garland" in Campbeltown Loch, Cantyre.

Hippocampus antiquorum, Leach.   The Sea-horse.

Mr. Gray, of the Marine Station at Millport, writes to me concerning this species as follows :—" I picked up a specimen of *Hippocampus* on the Kinloch Park, Campbeltown, in the autumn of 1894, from a spot where herring seine-nets had been spread out to dry, and from the fresh condition of the fish I had no doubt whatever that it had been caught in the nets in the Sound of Kilbrennan, and been shaken out on the spot where I found it.   Still I did not see it alive, so there is room for the element of doubt to creep in ; though this evidence, therefore, may not be quite conclusive, I have very little doubt in my own mind of the little fish having been caught in the Kilbrennan Sound.   Moreover, I may mention, by way of confirming what has just been said, that during the past summer a gentleman visitor to the Marine Station told me that many years ago his own children caught two live *Hippocampus* in a pool on the Fairlie sands, and kept them alive for about a week in a vessel of sea-water.   I took the gentleman to Mrs. Robertson of Fernbank, who happened at the time to be in the museum connected with the station, and to her he repeated the story also.   Since then I feel more firmly convinced that the Campbeltown specimen was a genuinely local one."

Hippocampus has been recorded from Belfast Lough, which is not very far distant from the mouth of the Clyde ; and the same species is also recorded for the West Coast of Scotland in Harvey-Brown and Buckley's work on the *Vertebrate Fauna of Argyle and the Inner Helvrides*.

Order PLECTOGNATHI.

Fam. GYMNODONTES.

Orthagoriscus mola, Linné.   The Short Sun-fish.

A specimen of the short sun-fish was captured off the Esplanade at Greenock on September 10th, 1881 (J. M. Campbell).*   This specimen was said to weigh about a ton; it measured 7 feet 9 inches in length and 3 feet 9 inches in height.   Mr. Campbell states that "the liver was absolutely crammed with a scolecid worm, *Tetrarhynchus reptans*."

SUB-CLASS CHONDROPTERYGII.

Order GANOIDEI.

Fam. ACIPENSERIDÆ.

Acipenser sturio, Linné.   The Sturgeon.

Fine specimens are often noticed in Loch Fyne during the herring fishery, but are seldom captured (B. & S.).   Mr. Alexander Gray, of the

---

* *Proc. Nat. Hist. Soc., Glasgow*, Vol. V., pt. II., p. 177 (1883).

Marine Station, Millport, in a letter to me concerning some Clyde fishes, says that "a sturgeon was caught near Carradale in 10 to 12 fathoms of water; it was about 7 or 8 feet long, but in poor condition, and was sold in the Glasgow market by Mr. Lagan of Paisley for £1 15s. So far as I can recollect, this was in the late autumn of 1895. A few years previous to this I remember seeing another sturgeon, perhaps 5 or 6 feet long, that was caught in a salmon-net at Penimore Bay ; I had no opportunity to examine this specimen, as it was shipped on the steamer ready for conveyance to the Glasgow market when I saw it."

Order ELASMOBRANCHII.

Fam. CARCHARIIDÆ.

Carcharias glaucus, Linné.   The Blue Shark.

This shark has been taken in Loch Fyne (H.B., p. 177). Mr. A. B. Watt of Glasgow saw a specimen of *Carcharias* which had come ashore at Ayr Bay on the 11th or 12th September last year (1899).

Galeus canis, Rondel.   The Tope or Toper.

Several specimens of this species were landed at Girvan in December last. They were at the time mistaken for another species, but were subsequently correctly identified as the tope. I am not aware of this species having been previously recorded for the Clyde; probably it is only an occasional visitor to the estuary, and may on former occasions have been passed over as one of the more common species. A fine specimen of toper was brought to Girvan by a local fisherman on December 28th. It was examined by Mr. Duthie, the Fishery Officer at that place, who was able to identify the species; the specimen measured 5 feet in length and 24 inches in circumference at the thickest part. Mr. Duthie was kind enough to secure the specimen and forward it to the Laboratory at Bay of Nigg, and I was thus enabled to verify Mr. Duthie's identification. The specimen was a female.

Fam. LAMNIDÆ.

[*Alopias vulpes* (Gmel.).   The Thrasher or Fox Shark.

Captain Campbell, of the Fishery steamer "Garland," informs me that a year or two ago he was proceeding down Loch Fyne, and, having reached a point somewhere between Crarae and Castle Lachlan, but rather nearer the former place, he saw a thrasher making an attack on what was probably a cetacean of some kind ; the thrasher was leaping clear out of the water in its usual way when engaged at this kind of work, and was coming down with a slap, the sound of which could be heard distinctly on board the vessel. Captain Campbell knows the thrasher quite well, and has frequently, when voyaging across the Atlantic, been an eye-witness of its tactics, so that there is little likelihood of his having been mistaken about the fish he saw in Loch Fyne.

My friend Mr. Gray of Millport tells me that in July 1895 a thrasher shark got into shallow water in Machrihanish Bay, and that some men there put off in a boat and, with ropes, managed to entangle it and haul it ashore close to the Pans Hotel. This specimen, which measured 15$\frac{1}{2}$ feet in length, was seen by Mr. Gray soon after it was captured. Machrihanish, though not within the Clyde area, is only a few miles west from Campbeltown, and comparatively near the mouth of the Clyde, and tends to show that the *Alopias* may occasionally find its way into the Clyde estuary.]

Selache maxima (Gunner). The Basking Shark.

A young specimen measuring 8 feet in length was captured in Maidens Bay, near Turnberry, Ayrshire (*Scotsman* newspaper, September 26th, 1898). This record was confirmed by the Fishery Officer at Girvan. Dr. Landsborough, in his *Natural History of Arran*, refers to the occurrence of this fish in the Clyde, and says (pp. 95, 96):—" A century ago . . . it was a frequent visitor to the Clyde, and was taken at Ballantrae, in Ayrshire, as well as in Arran. It made its appearance in the first or second week in June, and generally remained only for three or four weeks, though occasionally seen considerably later. . . . Now it is seldom seen." Dr. Landsborough also describes the means by which the fishermen of Arran usually captured these sharks.

Fam. SCYLLIIDÆ.

Scyllium canicula (Linné). The Lesser Spotted Dog-fish, Rough Hound.

" A very young specimen, 8 inches long, was obtained in the Sound of Sanda, at a depth of 20 fathoms" (Günther). Two specimens, each about 25 inches in length, were captured by the "Garland" at Station I, Firth of Clyde, on May 25th, 1897, and three specimens, measuring respectively 25, 28, and 29 inches in length, were captured on the same day in the vicinity of Sanda Island. In a male specimen $12\frac{3}{4}$ inches long taken in 60 fathoms between the Mull of Cantyre and Corsewall Point on May 15th, the spots are distinctly larger than they are on the adult specimens, especially on the anterior portion of the body. This specimen is now in the collection of fishes in the Fishery Board's Laboratory, Bay of Nigg, Aberdeen.

[*Scyllium catulus*, Cuvier. The Larger Spotted Dog-fish, Nurse Hound.

The following notice appeared in the *North British Daily Mail* of December 11th, 1899 :—

" One of the Girvan great-line boats on Saturday, the 9th inst., had an unusual catch of dog or hound sharks. No fewer than seven were caught in the great-lines, set in Lendal Bay, south of Girvan. Some of them measured about 5 feet in length and 24 inches in girth. The species is known as *S. catulus*, or rock shark. The females were full of well-developed eggs. Where they abound they are most destructive to the fishermen by taking their bait and hooks off the lines. The fish were cut up for bait for the crab and lobster creels."]*

Pristiurus melanostomus (Bonaparte). The Black-Mouthed Dog-fish.

" Two adult males in Upper Loch Fyne, 37 fathoms " (Günther). One captured by a Cumbrae fisherman was exhibited at a meeting of the Natural History Society of Glasgow on January 26th, 1869. The black-mouthed dog-fish is occasionally brought into East Tarbert amongst the *Acanthii* in the winter fishing (B. & S.). Upper Loch Fyne, in 10 to 20 fathoms (M.). Captured by the "Garland" in the deep water between Arran and Turnberry Point.

Fam. SPINACIDÆ.

Acanthias vulgaris, Risso. Picked Dog-fish.

Sometimes of common occurrence in the Clyde and Loch Fyne,

---

*There appears to have been some mistake on the part of the writer of the above note concerning the name of the fish landed at Girvan on December 9th. I am able to state conclusively that the fish landed were Tope or Topers, *Galeus canis*, and not *Scyllium catulus* as stated (see page 289). I have been unable hitherto to obtain any satisfactory information as to the occurrence of the Larger Spotted Dog-fish within the Clyde estuary.

especially during the herring fishing. Few of those taken by the "Garland" are of large size; they seldom reach beyond 2 feet in length; whereas the picked dog-fish is stated by Day to attain a length of at least 4 feet off our shores; but such large specimens are probably of exceptional occurrence.

Fam. Rhinidæ.

Rhina squatina (Linné). The Angel-fish or Monk-fish.

"In the Firth of Clyde it is by no means uncommon, and is frequently found there after gales; one was harpooned while asleep, but broke away."[*] This appears to be the only Clyde record for the angel-fish hitherto published. In the *Vertebrate Fauna of Argyle and the Inner Hebrides* it is stated that two specimens of the angel-fish in the Hunterian Museum, Glasgow, are from the West Coast, but no locality is given for them.

Mr. Wright, the chief Fishery Officer for the Barrow-in-Furness district of the Lancashire Sea Fisheries, states that he trawled in the Clyde for about eighteen winters previous to the closing of the estuary to that mode of fishing, and that during that time he occasionally captured angel-fish on the Ballantrae Bank. He knows the fish very well, and has brought specimens from Morecambe Bay to the Sea Fish Hatchery at Piel. The name he uses for the angel-fish is the "abbot."

Fam. Raiidæ.

Raia batis, Linné. The Skate, the Grey Skate or Blue Skate.

Moderately common and generally distributed, especially off-shore. Those captured by the "Garland" are seldom of large size, being usually mere pigmies when compared with the immense specimens sometimes landed at the fish market at Aberdeen. The largest taken by the "Garland" in the Clyde rarely exceed three feet in width.

Raia intermedia, Parnell. The Flapper Skate.

"A female with a disk 19 inches wide was obtained between Sanda Island and Ailsa Craig in 24 fathoms, March 6th, 1888" (Günther). The flapper skate has been taken by the "Garland" in the deep water to the east of Arran, and also in the vicinity of Ailsa Craig. (This is the *Raia macrorhynchus* of Day's *British Fishes*.)

Raia oxyrhynchus, Linné. Long or Sharp-nosed Skate.

Two specimens were captured by the "Garland" in the deep water to the east of Arran; both were comparatively small, one being only 21 inches and the other 29 inches across the pectoral fins.

Mr. Duthie, Fishery Officer, Girvan, informs me that on February 1st (1900) he saw a fine specimen of the long-nosed skate which had just been sent to Girvan from Maidens Village—a village not far from Girvan. The following measurements were taken by Mr. Duthie—Extreme length, 51 inches; extreme breadth, 34 inches; length from tip of snout to centre of closed mouth, 13 inches; length of tail, 18 inches; while from the tip of the tail to the vent was 21 inches.

Raia fullonica, Linné. Shagreen Ray, Fuller's Ray

"A female example, 24 inches across the disk, from Loch Fyne, off Skate Island, was obtained in 100 fathoms. An adult male, 19 inches across the disk, was caught in Kilbrennan Sound in 20 fathoms"

[*] A. Norman, *Zoologist*, Vol. XV. (1857), p. 5366.

(Günther). The shagreen ray is occasionally captured by the "Garland," but is apparently not very common in the Clyde.

Raia clavata, Linné.   Thornback Ray or Thorny.

"One adult female was obtained in 26 fathoms in Kilbrennan Sound, and an immature male 17 inches wide in 24 fathoms between Sanda Island and Ailsa Craig.   The contents of the stomach of the latter specimen consisted of small fish and crustaceans" (Günther).   The Thorny is one of the more common of the fishes captured in the Clyde by the Fishery steamer "Garland," but the specimens are mostly of small size, and very few of them extend beyond 25 or 30 inches across the disk.

Raia maculata, Montagu.   The Spotted Ray, or Homelyn Ray.

"An immature female was captured between Cumbrae and Wemyss Point in 30 to 40 fathoms.   Another from the Sound of Sanda from 20 fathoms.   One or two very young specimens from $2\frac{1}{2}$ to $4\frac{1}{2}$ inches across the disk were obtained off Ardrossan and off Whiting Bay" (Günther).   The spotted ray is occasionally captured by the "Garland," but the specimens are usually of small size.   One measuring 18 inches across the disk was taken in the vicinity of Ailsa Craig on April 27th, and one 10 inches across the disk at Station VI. on May 25th, 1897.   A small specimen a little over 7 inches across the disk is in the collection of fishes in the Laboratory at Bay of Nigg.

Raia circularis, Couch.   The Cuckoo Ray, Sandy Ray.

"An adult male, 14 inches across the disk, and a very young female, $3\frac{2}{3}$ inches broad, were obtained in the Sound of Sanda at a depth of 20 fathoms on March 10th; another adult male, also from the Sound of Sanda, was obtained at 49 fathoms on March 17th, 1888" (Günther).   The cuckoo ray has been taken by the "Garland" in Whiting Bay, in the vicinity of Sanda Island, in the vicinity of Ailsa Craig, and in various parts of Upper Loch Fyne.   It does not appear to be a very rare fish in the Clyde area.

SUB-CLASS CYCLOSTOMATA.

Fam. MYXINIDÆ.

Myxine glutinosa, Linné.   The Hag-fish.

The hag-fish has been taken at Rothesay in the Clyde area.   It is said to be numerous and most destructive to the line fishes off Girvan (H.B., p. 225).   I do not remember having seen *Myxine* in the Clyde.

SUB-CLASS LEPTOCARDII

Fam. CIRRHOSTOMI.

Branchiostoma lanceolata, Pallas.   The Lancelet.

"Amphioxus" is "taken plentifully in suitable ground in the Clyde area" (H.B., p. 225).   Mr. Gray, of the Millport Marine Station, states, in a letter to me, that "over a dozen specimens of this little fish were taken on a bank between Little Cumbrae and Hunterston, locally known as the *Dogger Bank*."   Dr. James Bryce, in his *Geology of Arran and Clydesdale* (1865), p. 250, remarks that the *Amphioxus* has been dredged at the north end of Holy Island, and has been obtained in Lamlash Bay, as well as near Millport.   Dr. Landsborough also speaks of having found it when dredging near Cumbrae.

■

XI.—NOTES ON SOME GATHERINGS OF CRUSTACEA COLLECTED FOR THE MOST PART ON BOARD THE FISHERY STEAMER "GARLAND" AND EXAMINED DURING THE PAST YEAR (1899).

By THOMAS SCOTT, F.L.S., Mem. Zool. Soc. de France.

(Plates XIII. and XIV.)

The following "Notes" are intended to supplement a somewhat similar series published in Part III. of the *Seventeenth Annual Report*. These notes deal entirely with the Crustacea, and refer to species that have been observed in various gatherings of tow-net and dredged materials examined during the past year, and collected for the most part on board the "Garland." The majority of the species referred to have been obtained in gatherings collected in Loch Fyne and in the seaward portion of the Firth of Clyde, but a few are also from other parts of the Scottish coasts; moreover, with the exception of one or two brackish-water forms those recorded are all marine species.

A few of the copepods mentioned in the sequel are apparently undescribed, but most of the other forms have already been recorded. Further information concerning these has, however, been obtained bearing on their structural details or on the distribution of the species, which it will be of interest to notice.

My son, Mr. Andrew Scott (assisted by Mrs. Scott), has prepared drawings to illustrate where necessary the various objects described. Several forms other than those mentioned in the sequel have had to stand over, but these will be described later.

COPEPODA.

Eucalanus crassus, Giesbrecht.

> 1888. *Eucalanus crassus*, Giesb., Atti Acc. Lincei, Rend. (4), vol. iv,, sem. 2, p. 333.

A considerable number of specimens of this species were obtained in a bottom tow-net[*] gathering collected in Dornoch Firth, November 19th, 1898. The depth at which the tow-net was worked varied from 8 to 16 fathoms. A large proportion of the specimens obtained were more or less immature. This *Eucalanus* has been obtained in the Moray Firth district on several occasions during the past few years, but it was apparently more frequent in the present gathering than in any of those previously examined. It is somewhat difficult to distinguish the immature forms of the species, but the adults are comparatively easily distinguished. When it is remembered that the distribution of *Eucalanus crassus* extends south as far at least as the Gulf of Guinea, its presence in the Moray Firth from time to time is of more than usual interest.

---

[*] What is here called the "bottom tow-net" is the tow-net that is fastened to the trawl-head in such a way that when the trawl is working the tow-net just clears the sea-bottom.

Eucalanus elongatus (Dana).

1849.   *Calanus elongatus,* Dana, Proc. Amer. Acad., vol. ii, p. 18.

A single specimen—a female—was obtained in the same gathering with the last. This is a larger species than *Eucalanus crassus*; the specimen referred to here measured fully 6mm. in length (about a quarter of an inch). In this species the last thoracic segment has the sides produced posteriorly into short pointed processes. The presence of *Eucalanus elongatus* in the Moray Firth is also of interest as bearing on the question of the distribution of species. These copepods are probably carried into the estuary by regular tidal currents or by temporary currents set up by the wind, should it happen to blow from one particular direction for a more or less lengthened period.

Neither of the two species named have yet been observed in the Firth of Forth or the Firth of Clyde. These two estuaries, though extending for a considerable distance inland, are comparatively narrow, and the entrance to each is obstructed to some extent by an island; the Moray Firth, on the other hand, presents an extensive opening to the North Sea and narrows very gradually westwards. Such a difference in the general contour of these inlets has probably a more or less distinct influence on the distribution of their local faunas.

Paracalanus parvus (Claus).

1863.   *Calanus parvus,* Claus, Die freileb. Copep., p. 173, Pl. XXVI., figs. 10–14; Pl. XXVII, figs. 1–4.

This small species has been observed in gatherings collected during the past year both in the Clyde and Loch Fyne, and usually in bottom tow-net gatherings. It was, for example, observed in a bottom-gathering from Station V. (Whiting Bay, Arran), collected September 8th, and in another from Station XVII. (near the head of Loch Fyne), collected September 29th.

Stephus fultoni, T. and A. Scott.

1898.   *Stephus fultoni,* T. and A. Scott, Ann. and Mag. Nat. Hist. (7). vol. i., p. 185, Pl. X., figs. 1–8; Pl. II., figs. 1–4.

Stephus fultoni has apparently not yet been observed outside the Clyde area. It was first observed in 1896 in some material dredged in Kilbrennan Sound, and subsequently in a small gathering of crustacea collected near the Spit, Loch Fyne, in 1897. I have now to record its occurrence for the second time at this Loch Fyne Station; the material in which it was obtained on the present occasion was collected during the early part of this year (1899), and a few specimens only were observed. *Stephus fultoni,* which seems to be a rare species, is comparatively easily distinguished from the other members of the genus by the somewhat larger size of the female and by the male having the fifth thoracic feet conspicuous and furnished with horn-coloured processes.

Bradyidius armatus (Vanhöffen).

*1897.   *Bradyanus armatus* (Vanhöffen in:) Chun, Arkt. antarkt. Plankton, p. 28.

It was pointed out in my " Notes" last year that *Bradyidius armatus* was one of the more widely diffused of the Clyde copepods, and that it was also occasionally observed on the East Coast. During the past year

* *Vide* Das Tierreich, p. 32.

it has again been observed in several of the gatherings sent to the Fishery Board's Laboratory from the Clyde. In 1897 a new genus was instituted for this copepod by Dr. W. Giesbrecht, of Naples, but it appears to be one of those species whose lot it is to undergo several removals ere it reaches its ultimate destination, as indicated by the following quotation from a work lately published by Herr. O. Nordgaard, entitled, "Report on Norwegian Marine Investigations, 1895–97." At page 21 of his report, Herr. Nordgaard refers to the species under consideration as follows :—

"In 'Undersögelser over Dyrelivet i Arktiske Fjorde' Herr Sparre Schneider has mentioned a copepod that is called *Undinopsis bradyi*, G. O. Sars. This species is said to have been found in Kvænangen and at Tromsö. In the summer of 1897 I showed Professor Sars a preparation of a copepod that I was unable to identify. He then declared it to be the very *Undinopsis bradyi*, and afterwards informed me in a letter that the said copepod had been described by Mr. Brady (Monograph of the Copepoda of the British Islands, i., p. 46, Pl. IV., figs. 1–11).

The species was, however, there wrongly identified with *Pseudocalanus armatus*, Boeck, which, according to Mr. Sars, is another species. As Mr. Sars in his gigantic revision of Norwegian Crustacea will soon come to the Copepoda, I shall do nothing but here note the occurrence of *Undinopsis bradyi* at the following places :—

March 14th, 1896, Vestfjord (67° 32·5′ N. ; 130° 24·5′ E.) in Plankton 0·200m.

March 5th, 1897 ; Ostnes-fjord in Lofoten.

March 7th, 1897, Irold-fjord in Lofoten, Plankton 0·65m.

Besides, I have this year (1899) taken several specimens of the species in fjords near Bergen."

This note by Herr. Nordgaard is of interest, if for nothing else than the information he gives concerning the distribution of the species, but it also indicates that Professor Sars' designation is likely to take precedence over that of *Bradyidius armatus* of Drs. Giesbrecht and Vanhöfen. It is doubtful, however, if this copepod will be allowed to rest under *Undinopsis bradyi*, for it is by no means a rare species, and, as Herr. Nordgaard and Professor Sars have shown, it has a wide distribution. It is probable, therefore, that it has not escaped the notice of some of the earlier naturalists, and may be described and named in their published works by a designation different from any of those referred to.

Isias clavipes, Boeck.

1864. *Isias clavipes*, Boeck, Forh. Vid. Selsk., Christiania, p. 18.

This, which is a moderately rare species, has again been observed in several of the tow-net gatherings forwarded from the Clyde during recent months. The specimens obtained were found for the most part in gatherings collected in the tow-net fixed to the head of the trawl, and appeared to be most frequent in the gatherings collected in September. The species was taken in Kilbrennan Sound at Stations III. and IV., near Sanda Island, and in the vicinity of Ailsa Craig. I have no records of *Isias* from Loch Fyne this year. It may be of interest to mention that it was also during September last year that *Isias* was chiefly observed.

Eurytemora velox (Lilljeborg).

1853. *Temora velox*, Lillj., De Crustac. ex ordin. tribus; Cladoc. Ostrac. et Copep. in Scania occurr., p. 177, Pl. XX., figs. 2–7.

This species was found in a small pond near the New Zoological

Station at Millport, Cumbrae, and also in shore pools at the south-west corner of the island on May 6th, 1899. Professor G. S. Brady obtained the species from Cumbrae many years ago, and it is interesting to find it in the same localities in which it was then observed.

Eurytemora affinis (Poppe).

> 1880. *Temora affinis*, Poppe, Abhandl. d. Naturw. Ver. Bremen, Bd. vii., pp. 55–60, Pl. III.

I find this species in rock pools near low-water mark at Bay of Nigg, Aberdeen (just in front of the New Laboratory of the Fishery Board for Scotland), where it is not uncommon. It also occurred in a large pool left by the ebbing tide near the bridge where the railway crosses the River Dee. These species of *Eurytemora* require careful examination in order to distinguish the one from the other.

Metridia lucens, Boeck.

> 1864. *Metridia lucens*, Boeck, Forh. Vid. Selsk., Christiania, p. 14.

This species was moderately frequent in a bottom tow-net gathering of Crustacea collected during the past year in Aberdeen Bay. In the Firth of Clyde, *Metridia lucens* appears to be one of the resident copepods, as it may be obtained there all the year round, though usually in small numbers. The more recent Clyde gatherings in which the species occurred, and of which I have notes, were collected in the bottom tow-net near the seaward part of the estuary.

Paramesophria cluthæ, Th. Scott.

> 1897. *Paramesophria cluthæ*, Th. Scott, *Fifteenth Rep. Fish. Board Scotl.*, Part III., p. 147, Pl. II., figs. 3–8; Pl. III., figs. 13–16.

This moderately large and distinct species was described from specimens dredged off Largabruach, Upper Loch Fyne, and I have now to record its occurrence at Tarbert Bank (off East Tarbert), Lower Loch Fyne; it was obtained amongst some material dredged from about 17 to 20 fathoms on October 21st, 1899. In one of the more recently published works on the Copepoda (Das Tierreich, Lief. 6), *Paramesophria* takes its place amongst the Centropagidæ.

Labidocera wollastoni (Lubbock).

> 1857. *Pontella wollastoni*, Lubbock, Ann. and Mag. Nat. Hist. (2), vol. xx., p. 406, Pls. X.–XI.

Last year I recorded this fine species from two Clyde stations, both of which are near the seaward limits of the estuary. This year I have to record its occurrence in a bottom tow-net gathering (12 to 26½ fathoms), collected in the vicinity of Sanda Island, near the mouth of the Clyde, on September 5th, 1899; it was also obtained in a second gathering collected two days later in about 55 fathoms and somewhat further seaward. In a gathering collected in November a number both of males and females were found. This gathering was also from the mouth of the estuary. I also take this opportunity to record *Labidocera wollastoni* from the Firth of Forth. During the past summer I overhauled some tow-net gatherings that had been collected a few years ago, and found that one or two of them had not before been examined; in one collected to the east of

Inchkeith on June 8th, 1891, I found amongst other things a single male specimen of the *Labidocera* referred to. I do not think that this species has been before recorded from the Firth of Forth.

Cyclopina gracilis, Claus.

1863. *Cyclopina gracilis,* Claus, Die frei-leb. Copep., p. 104, Pl. X., figs. 8–15.

Specimens of this small but distinct species were obtained in a gathering of material dredged in the vicinity of Otter Spit, Upper Loch Fyne, during the early part of the year. This gathering contained a large number of comparatively rare copepods, several of which are referred to in the present " Notes."

Notodelphys prasina, Thorell.

1859. *Notodelphys prasina,* Thorell, Bidr. till Känned. om Krustac. i Ascid., p. 46, Pl. V., fig. 7.

This copepod has occurred frequently in Ascidians dredged from Tarbert Bank, Loch Fyne, during recent months. *N. prasina* has very short caudal furca, and by this character alone it may be readily distinguished from the other species of *Notodelphys* hitherto recorded from Scotland. *Notodelphys allmani,* Thorell, is also occasionally found in Clyde Ascidians, but does not seem to be just so common as the species first named.

Doropygus pulex, Thorell.

1859. *Doropygus pulex,* Thorell, op. cit., p. 46, Pl. VI., fig. 8.

A few specimens of this curious species were obtained in Ascidians dredged at Tarbert Bank, Loch Fyne (17-20 fathoms), in October 1899. In this species the caudal furca are "slightly elongate" and "becoming gradually thin, or tapering, towards the distal end " (Canu, Copep. du Boulonn., p. 195). *Doropygus pulex* appears to be somewhat rare in the Clyde area.

Doropygus (?) *gibber,* Thorell.

1859. *Doropygus gibber,* Thorell, op. cit., p. 52, Pl. VIII., fig. 11.

One or two specimens of a *Doropygus,* apparently belonging to this species, were found in some dredged material from Tarbert Bank, Loch Fyne. Dr. Giesbrecht * considers this to be more nearly related to *Notopterophorus* than to *Doropygus.*

Doropygus porcicauda, G. S. Brady.

1878. *Doropygus porcicauda,* Brady, Mon. Brit. Copep., vol. i., p. 138, Pl. XXVII., figs. 1-9 ; Pl. XXXIII., figs. 14-16.

Several specimens of this large and apparently distinct species were obtained in the same dredged material as the last. In this species the caudal rami are of considerable length, somewhat divergent and more or less curved. It does not seem to be very uncommon in Loch Fyne Ascidians.

Enterocola (?) *fulgens,* Van Beneden. (Pl. XIII., figs. 21-27.)

1860. *Enterocola fulgens,* Van Ben., Bull. Acad. Belg. (2), vol. ix., p. 151, Pl. I.

I have on one or two different occasions found in gatherings of dredged material from the Clyde odd specimens of an *Enterocola* usually of a

* *Mittheil. Zool. Stat. Neapel,* vol. iii., p. 328 (1882).

larger size than either *Enterocola fulgens*, Van Beneden, or *Enterocola betencourti*, Canu, and which to some extent differs also in some of its structural details from both these forms. I prefer, however, in the meantime to regard our specimens as a "form" or variety of Van Beneden's *Enterocola fulgens* rather than institute a new species for their reception.

These Clyde specimens are found in the intestine—not the branchial cavity—of a small Ascidian. Usually only one Copepod is observed in each specimen of the Ascidian in which the parasites occur, and it also usually so fills up the part of the intestine in which it is lodged that it is with difficulty detached with its ovisacs *in situ*, one or both frequently breaking away while removing the Copepod from its environment.

All the specimens of the Copepods obtained as described, and which I have examined, appear to belong to the one species, but they vary greatly in size. The specimen represented by the drawing (fig. 21), measures little more than two millimetres (2·2mm.) in length, whilst another that I have measured extends to at least four millimetres. The ovisacs are of a distinctly reddish colour, so that when examining the Ascidians in which the Copepods occur one can see at a glance and without dissection whether a parasite is present by the red colour of the ovisacs showing itself through the thin wall of the intestine of the host.

The mouth-organs of the *Enterocola* are difficult to make out; they are all simple, and do not show much structure. Figures 22 and 23 represent what appear to be one of the antennules and one of the antennæ. The mouth takes the form of a short and somewhat cone-shaped process. There appears to be no mandibles properly so called, and, according to Dr. Canu, the absence of mandibles is one of the distinctive characteristics of the genus *Enterocola*. The (?) maxillæ (fig. 24) are broad foliaceous appendages, bearing on their distal margin a number of stout ciliated spines. The posterior foot-jaws are very stout, and terminate in very short but strong claw-like processes (fig. 25).

The first four pairs of thoracic feet are all very much alike. They are two-branched; the outer branches appear to be one-jointed, and are moderately stout, and taper towards the distal end; they are also furnished with two moderately long and plumose terminal setæ; the inner branches consist of short, stout, tapering appendages of a simple and almost rudimentary character (figs. 26 and 27). The fifth pair are conspicuous and broadly dilated appendages. The caudal furca are about twice as long as broad. The female carries two ovisacs, which in well grown specimens are about as long as the body of the copepod.

Gunenotophorus (?) *globularis*, Costa. (Pl. XIII., figs. 28–34; Pl. XIV., figs. 37 and 38).

> 1852.   *Gunenotophorus globularis*, Costa, Fauna del Regno di Napoli, Entom. (1840).

A somewhat curious copepod, agreeing in almost every detail of structure with the species described by O. G. Costa under the name of *Gunenotophorus globularis*, was obtained in some dredged material from the vicinity of Sanda Island, Firth of Clyde, in December 1898. The species is said to occur in the branchial cavity of Ascidians, but this Clyde specimen (only one was obtained) occurred free amongst the dredged material, having probably come from a dredged Ascidian.

The entire length of the specimen was about five millimetres (about one-fifth of an inch). The body was considerably dilated, but the abdomen was more slender, and was quite distinct; the whole animal was strongly incurved, as shown by the figure (fig. 28, Pl. XIII). The specimen, which had a somewhat macerated appearance, did not exhibit much segmentation of the cephalon or thorax, but the abdomen, which was moderately elongated and cylindrical, was divided into four distinct segments, the first three of which were smaller than the last one; the last—or anal segment—was about as long as the combined length of the two that preceded it. The caudal furca were short and very divergent, extending outwards at almost right angles to the abdominal segment (fig. 34, Pl. XIII.). The antennules, which showed very little jointing, were very short and stout except the end joint, which was a small one (fig. 29, Pl. XIII.) The antennæ (fig. 30, Pl. XIII.) were also short and stout, and armed with a moderately strong but short and slightly clawed terminal spine. The mandibles with their palps resembled very closely the figures of these appendages given by Dr. Canu in his interesting work on the marine Copepoda (*Les Copepodes du Boulonnais*, Pl. XI., figs. 3 and 4). The biting part of the mandible is armed with five large teeth, arranged widely apart, and several minute, close-set, and slightly elongated spinules, while the palp ends in two short setiferous branches (fig. 31, Pl. XIII.). The maxillæ are broadly foliaceous, the masticatory lobe is armed with a series of spiniform setæ along the margin, and a number of stout, elongated plumose setæ adorn the margins of the maxilla-palp. The anterior foot-jaws are stout but simple one-jointed appendages furnished with several stout plumose terminal setæ (fig. 37, Pl. XIV.). The posterior foot-jaws (fig. 32, Pl. XIII.) are moderately stout, and armed with a short but comparatively strong terminal claw. There are also a number of setæ on the inner margin of these appendages.

The first pair of thoracic feet (fig. 38, Pl. XIV.) are stout and moderately short; both branches are three-jointed and of nearly the same length, and they are both provided with elongated and densely plumose setæ on the inner margins. The next three pairs, which are somewhat similar to each other, have the inner branches short and slender and apparently three-jointed, while the outer branches, which are also three-jointed, are moderately stout and elongate; neither the inner nor outer branches were observed to carry setæ, their only armature appeared to consist of one or two minute spines. The fifth feet appeared to be obsolete, but this appearance may have been due to the specimen being somewhat imperfect.

When it is remembered that the specimen here described was found free amongst a quantity of mixed dredgings, and not *in situ* in any Ascidian, and that, moreover, from the habitat of the animal its whole structure is more or less flaccid and more liable to injury than the stronger free-swimming forms, it need not be surprising that it should differ to a small extent from the more perfect and better preserved specimens. After a careful study of the characters of this Clyde specimen, I have little doubt that it belongs to the species to which it has been ascribed.

There does not seem to be any previous British record of *Gunenotophorus globularis*, and its occurrence in the Clyde estuary is therefore of interest.

Botryllophilus (?) *ruber*, Hesse.

1864. *Botryllophilus ruber*, Hesse, Ann. Sci. Nat. Zool. (5), t. i., Pl. XII., figs. 1 and 2.

I have noticed two, or perhaps three, specimens of a *Botryllophilus* in some material that was dredged at Tarbert Bank, Lower Loch Fyne, in the vicinity of East Tarbert. The specimens were not found *in situ* within any Ascidian, but were mixed up amongst the debris; their host had probably been damaged by the lip of the dredge so that they escaped.

According to Dr. Canu, the characteristics of the genus *Botryllophilus* are shown particularly—First, in the structure and position of the thoracic feet; second, in the almost constant existence of a single ovigerous sac of a strictly spherical form sheltered between the fifth feet.

The peculiarity in the fifth thoracic feet in *Botryllophilus* consists mainly in their position on the last thoracic segment. Instead of occupying a position more or less on the ventral aspect of the segment, as is usual amongst the copepoda, the position of the fifth feet is more or less round towards the dorsal aspect, and as they each consist of a single-jointed, elongated, and somewhat curved spine-like appendage which projects more or less out from the body, they impart to the copepod a rather curious appearance.

After the above note on *Botryllophilus* had been sent to the printer, several specimens of the copepod were obtained *in situ* in a specimen of *Botryllus* sp. collected at Station X. in the Moray Firth on the 16th of June 1898. Some of the specimens carried a globular ovisac on the dorsal aspect between the fifth feet as stated above, but the ovisacs appear to be easily detached ; they were of a pale cream colour, due, probably, to the long immersion in spirit of the *Botryllus.*

Canuella perplexa, T. and A. Scott.

     1893.   *Canuella perplexa,* T. and A. Scott, Ann. Scot. Nat Hist., vol. ii., p. 92, Pl. II., figs. 1–3.

This copepod was obtained in shore gatherings of Crustacea collected at Cumbrae, Firth of Clyde, between tide-marks May 6th, 1899, and in shore pools at Inverkip on the 13th of the same month. The species appeared to be moderately rare at both places. *Canuella* is widely distributed, but is apparently more frequent amongst weed and where the bottom is of a sandy nature.

Ectinosoma gracile, T. and A. Scott.

     1896.   *Ectinosoma gracile,* T. and A. Scott, Trans. Linn. Soc. (2. z.) vol. vi., p. 429. Pls. XXXVI., XXXVII., XXXVIII.

A few specimens of this very small species were obtained in shore pools at Inverkip, Firth of Clyde, May 13th, 1899. *Ectinosoma gracile* has already been recorded from near Sanda Island, Firth of Clyde ; it appears to be widely distributed, but being very small is easily overlooked.

Tachidius brevicornis (Müller).

     1776.   *Cyclops brevicornis,* Müller, Zool. Dan., Prodr. (2414).

Though *Tachidius brevicornis* appears to be generally distributed, there are apparently few or no records of it as a member of the Clyde fauna. Being a brackish-water species, it need not be sought for except where such conditions exist, and as there are few shores around the British Islands where brackish-water pools are not to be found, the distribution of the species is correspondingly extensive. There are two species of *Tachidius* recorded for Britain, but the one referred to is readily distinguished by the structure of the fifth thoracic feet, which are of the form of two comparatively broadly, roundish plates, the free margins of which are fringed with setæ. The species was found in shore pools at Cumbrae in May 1899. The second species *Tachidius littoralis,* Poppe, has already been recorded from Hunterston, Firth of Clyde.*

Amymone nigrans, T. and A. Scott.

     1894.   *Amymone nigrans,* T. and A. Scott, Ann. and Mag. Nat. Hist. (6), vol. xiii., Pl. VIII., figs. 1–7.

This curious copepod is rather less than half a millimetre across the

* *Proc. Nat. Hist. Soc. Glasg.,* vol. V. (N.S.), p. 351, 1900.

longest diameter, and is of a blackish colour; it is therefore easily overlooked. The species have only hitherto been observed in Cromarty Firth, where it is not uncommon. I now record it for Loch Fyne, some specimens having been obtained in a gathering of dredged material collected near Otter Spit, Upper Loch Fyne, on January 12th, 1899.

Jonesiella fusiformis (Brady and Robertson).

1875. *Zosime fusiformis*, B. and R., Brit. Assoc. Rep., p. 196.

This seems to be the most frequent representative of the genus in Loch Fyne; in the Firth of Forth it is *Jonesiella spinulosa*, B. and R., that is the more frequent species. *Jonesiella fusiformis* occurs not rarely in material dredged on Tarbert Bank, Lower Loch Fyne; it is collected here in almost every dredging that is taken. The latest record I have of *Jonesiella fusiformis* from this locality is December 12th, 1899.

Stenhelia blanchardi, T. and A. Scott.

1895. *Stenhelia blanchardi,* T. and A. Scott, Ann. and Mag. Nat. Hist. (6), vol. xvi. p. 353, Pl. XV., figs. 1–10.

This distinct but apparently rare copepod was dredged off Arisaig, Argyleshire, in 1892, though not described till 1895. No further specimens were observed till the present year (1899), when the species was again found; this time in some material dredged in the "Fluke Hole," off St. Monans, Firth of Forth, in 1896, and the examination of which had been delayed for want of time. This species is readily distinguished from others of the same genus by the form of the secondary branches of the fifth pair of thoracic feet, which terminate in hook-like processes. The occurrence of the species in the Firth of Forth tends to indicate that, though it seems to be rare, it may at the same time be widely distributed.

Canthocamptus inconspicuus, sp. n.   (Pl. XIV., figs. 1–8.)

Description of the Female.—In general appearance this species is somewhat similar to *Canthocamptus parvus,* T. and A. Scott. The length of the specimen figured measures from the forehead to the end of the caudal furca 54mm. (about $\frac{1}{48}$ of an inch). The antennules, as in the species mentioned, are short and six-jointed; the third joint is longer than any of the others, being about equal to the entire length of the two joints preceding it as well as of the two that follow it. The formula shows approximately the proportional lengths of the different joints—

Proportional lengths of the joints,   13 · 12 · 32 · 15 · 13 · 20
Number of the joints,   1 · 2 · 3 · 4 · 5 · 6.

The antennæ are each furnished with a one-jointed secondary branch. The mandibles are small, and they are provided with a small one-branched but moderately elongated palp (fig. 3). The first pair of thoracic feet have both branches three-jointed; the inner branches are considerably longer than the outer, and the first joint exceeds in length that of the second and third combined, as shown in the figure (fig. 5). The inner branches of the second, third, and fourth pairs, which are shorter than the outer branches, are only two-jointed, but the outer branches are three-jointed, moderately elongated, and sparingly setiferous (fig. 6). The fifth pair are small and foliaceous; the basal joint is broadly sub-triangular, while the secondary one is small and ovate; both are provided with a few setæ, arranged as shown in the drawing (fig. 7). The caudal segments are slender and about as long as the anal segment (fig. 8). The female carries one ovisac.

Remarks.—This copepod has a superficial resemblance to a small form, with six-jointed antennules, described by T. and A. Scott under the name of (?) *Canthocamptus parvus.** Like that form, the copepod now described has the antennules six-jointed, and the inner branches of the second, third, and fourth pairs of thoracic feet two-jointed, but the proportional lengths of the joints of the antennules are different, and the caudal furca, which in *Canthocamptus parvus* are very short, are in the species now described as long as the anal segment. No males have been observed.

Habitat.—Moray Firth ; obtained amongst some dredged material.

Mesochra spinicauda, T. and A. Scott.

   1895. *Mesochra spinicauda*, T. and A. Scott, Ann. and Mag.
    Nat. Hist. (6), vol. xv., p. 52, Pl. V., figs. 12–25.

This was one of several curious species that were found in shore pools at Musselburgh, Firth of Forth ; the pools occurred between tide marks, but nearer low water, and were surrounded on all sides by beds of mussels. I am now able to record the occurrence of the species in shore pools near Millport, Cumbrae; it was obtained in some gatherings collected by hand-net on May 6th, 1899.

Tetragoniceps (?) *malleolata*, Brady. (Pl. XIV., figs. 9–17.)

   1880. *Tetragoniceps malleolata*, Brady, Mon. Brit. Copep., vol.
    ii., p. 66, Pl. LXXVIII., figs. 1–11.

In Part III. of the *Tenth Annual Report of the Fishery Board for Scotland* (1892), p. 252,† I recorded the occurrence of a species of copepod which had been obtained in the Firth of Forth off St. Monans, and which I had ascribed to *Tetragoniceps malleolata*, G. S. Brady. I then pointed out, however, that this copepod, while agreeing in most points with the genus and species named, differed in so far as it possessed nine-jointed instead of eight-jointed antennules, and in the fifth pair of thoracic feet being two-jointed instead of being composed of only one joint. At the time the record was published, I was quite aware that the second of these two differences was, in view of the definition of the genus, a somewhat important one, but considered that, as the copepod referred to resembled the species named in almost all the other details of structure, it was better to ascribe it to that species rather than to institute a new genus or species for its reception.

During the past year I have obtained, in some dredged material from the Firth of Forth collected in 1896, but only recently examined, a few more specimens of the copepod referred to above, as well as of another and somewhat closely allied form that appears to be undescribed.

When the supposed *Tetragoniceps malleolata* was recorded in the *Tenth Annual Report* no detailed description of the form was given ; a reference to the two principal points of difference was considered to be at that time all that was necessary for the identification of the form. The occurrence, however, of the closely allied and apparently undescribed species which I have alluded to makes it desirable that a description of both forms should now be given, so that the differences that have been observed between them may be more clearly indicated.

It may be considered doubtful whether the two forms to be described should be retained in the genus *Tetragoniceps*, but for the present, at

---

 * *Ann. and Mag. Nat. Hist.* (6), vol. xviii., p. 6, Pl. II., figs. 14–22 (1896).
 † Additions to the Fauna of the Firth of Forth, Part IV.

least, I prefer to leave them there. The earlier recorded form will be described first.

Description of (?) *Tetragoniceps malleolata*, Brady.

The body of this copepod is elongated and slender, tapering more or less gradually from the head to the extremity of the abdomen; the rostrum is short, the cephalic and thoracic appendages are moderately elongate, and the entire length of the specimen figured is ·89mm. (the $\frac{1}{28}$ of an inch). The antennules in the female are nine-jointed; the first joint is long, and the inner distal angle is produced into a stout and somewhat triangular tooth-like process; the next three joints are considerably shorter than the first; the last joint is about as long as the fourth, but the four joints that precede the last one are small; a moderately long sensory filament or *asthetask* springs from the end of the fourth joint as shown in the drawing (fig. 10). The antennæ, mandibles, and maxillæ are nearly similar to those in *Tetragoniceps bradyi.** The posterior foot-jaws are three-jointed, but the end joint is very small; there are two terminal setæ—one moderately elongate, the other smaller and slightly plumose. The outer branches of the first to the fourth pairs of thoracic feet are all three-jointed, but all the inner branches are two-jointed. In the first pair, which are comparatively slender, the inner branches are elongate, the first joint being rather longer than the entire length of the outer branches; the second joint, which is scarcely half the length of the first, carries two stout terminal setæ—the inner one being the longer; there is also a small seta on the lower half of the inner edge of the first joint. The inner branches of the second, third, and fourth pairs are considerably shorter than the outer branches; those of the second and third pairs extend slightly beyond the second joint of the outer branches; but in the fourth pair the inner scarcely reach the middle of the second joint of the outer branches, and this difference is owing, in part at least, to the outer branches of the fourth pair being proportionally more elongated than the outer branches of the two preceding pairs. The fifth pair of feet are two-jointed; the basal joint is foliaceous and somewhat triangular in outline; it is provided with three small setæ on the lower half of the inner margin, and with a small apical seta. The second joint is elongated and narrow, and it tapers gradually till it becomes somewhat attenuated at the extremity; this joint is provided with a few small setæ on the outer edge and one on the inner, and also with a slender terminal hair. The caudal segments are slender, and about as long as the last abdominal segment.

The female carries one ovisac, which contains a few moderately large ova arranged in a single series.

The male differs little from the female, except that the antennules are hinged, and otherwise modified for grasping; the fifth pair of feet are also less fully developed; the basal joint is sub-quadrate, and the inner portion slightly produced distally and furnished with two moderately stout, spiniform apical setæ; the secondary joint is sub-cylindrical, and is armed with a moderately stout and elongated spine near the distal end of the inner margin. The first abdominal segment in the male bears slightly produced lateral appendages provided with three moderately long setæ; these appendages are situated immediately posterior to the fifth thoracic feet.

Habitat.—Firth of Forth, off St. Monans; rather rare.

Tetragoniceps brevicauda, sp. n.   (Pl. XIV., figs. 18–22.)

As already stated, this copepod does not differ very greatly from

* *Vide* Part III. *Tenth Ann. Report Fishery Board for Scotl.*, p. 253, Pl. IX. (1892).

(?) *Tetragoniceps malleolata*, but, for the reasons stated below, I prefer to describe it under a distinct name rather than as a "variety" of the species referred to; for, after all, the question as to whether a thing is a "species" or a "variety" is very much a matter of opinion.

Description of the Female.—In general appearance the female of *Tetragoniceps brevicauda* is not unlike the form just described, but is somewhat smaller. The specimen figured (fig. 18) is only about ·7 mm. (about $\frac{1}{36}$ of an inch in length). The antennules have a structure somewhat similar to those of (?) *Tetragoniceps malleolata*, and there is the same hook-like process on the distal extremity of the first joint; the proportional lengths of the nine joints are, however, somewhat different. The mouth-organs and swimming-feet resemble those of the species named, except that the first feet appear to be rather more slender, and the fifth pair are proportionally somewhat smaller, but the secondary joint of the fifth pair is distinctly more elongated proportionally than that of the fifth pair in (?) *Tetragoniceps malleolata* (fig. 21). The caudal segments (fig. 22) are distinctly shorter than those of the species named, and they are also proportionally stouter; the size and form of the caudal furca of the species under description are so different from those of the closely allied form previously recorded as not only to have suggested the name that has been applied to it, but were the chief characters that first attracted my attention when examining the material in which it was found.

No males of this form and only very few females have been observed.

Habitat.—Firth of Forth, off St. Monans.

Pseudolaophonte spinosa (I. C. Thompson).

> 1893.   *Laophonte spinosa*, I. C. Thompson, Revised Report on the Copepoda of Liverpool Bay, Trans. L'pool. Biol. Soc., vol. vii., p. 24, Pl. XXX., figs. 1–13.

> 1896.   *Pseudolaophonte aculeata*, A. Scott, Report Lancashire Sea Fisheries (1895), p. 11, Pl. III., figs. 7–23.

This rare copepod species occurred in a gathering of dredged material collected near Otter Spit, Loch Fyne; a male and a female specimen were obtained. The antennules in this species are each furnished with a prominent and strong spine on the lower (exterior) aspect of the second joint; both the male and female possess these spines; the female antennules appear to be only four-jointed. The species has a close general resemblance to *Laophonte*, so much so that, like Mr. I. C. Thompson, I was at first inclined to regard it as a member of that genus, but a close examination of the thoracic appendages, and especially of the swimming-feet, bring to light structural differences that must exclude it from the genus *Laophonte*. The principal differences, as pointed out by Mr. A. Scott, are observed in the structure of the second and third pairs of swimming-feet. In the second pair each foot consists of a single one-jointed branch, and in the third pair, though each foot is two-branched, both branches are only two-jointed. This interesting and somewhat anomalous copepod has not before been recorded from the Clyde district.

Leptopsyllus minor, T. and A. Scott.

> 1895.   *Leptopsyllus minor*, T. and A. Scott, Ann. Scot. Nat. Hist. (Jan. 1895), p. 31, Pl. II., figs. 15–22.

This species belongs to a group of peculiarly slender copepods, the first of which was added to the British fauna in 1894.[*] Hitherto all the

---

[*] Part III. of *Twelfth Ann. Report of the Fish. Board for Scot.* (1894), p. 254.

described species have been found either in shore pools or in comparatively shallow water. The females do not appear to be very prolific; they usually carry but one ovisac, which contains only a few—frequently not more than three or four—ova. It is interesting to note, however, that though the creatures are small their ova are comparatively of large size.

Leptopsyllus minor, which has not before been recorded from the Clyde, was obtained in shore pools near Millport, Cumbrae, and also at Inverkip during the past summer.

Leptopsyllus herdmani, I. C. Thompson and A. Scott.

> 1900. *Leptopsyllus herdmani*, I. C. Thomp. and A. Scott, Trans. L'pool Biol. Soc., vol. xiv., p. 141, Pl. VIII.

A few specimens of this minute species were obtained, along with the species just recorded, in the shore pools at Millport, Cumbrae, in May 1899. One of the principal differences between this species and *Leptopsyllus minor* is in the comparative lengths of the inner and outer branches of the first thoracic feet ; in the present form the inner branches are considerably longer than the outer ones, while in *Leptopsyllus minor* the inner are scarcely longer than the outer branches. There are some other differences, but they are less obvious than the one referred to.

Nannopus palustris, G. S. Brady.

> 1878. *Nannopus palustris*, G. S. Brady, Mon. Brit. Copep., vol. ii., p. 01, Pl. LXXVII., figs. 18–20.

This curious copepod was obtained in brackish-water pools at Inverkip, Firth of Clyde, on May 13th, 1899, but it did not appear to be very common: It has a superficial resemblance to *Platychelipus*, and may have sometimes been passed over as such. There are very few Clyde records for *Nannopus*.

Cylindropsyllus minor, T. Scott. (Pl. XIV., figs. 23–32.)

> 1892. *Cylindropsyllus minor*, T. Scott, Part III., *Tenth Ann. Report Fish. Board for Scot.*, p. 260, Pl. XI., figs. 17–24.

The copepod described under this name was discovered in 1891 off St. Monans, Firth of Forth. At that time no males had been observed, and therefore, though the characters of the female, so far as they could be made out, agreed very well with the definition of the genus *Cylindropsyllus*, there was still the probability that the species might not after all be a true member of that genus.

During the past year the examination of a gathering of entomostraca collected in the same locality where the species was first discovered yielded several additional specimens to those already observed, and this time both males and females were obtained. The occurrence of these specimens has enabled me not only to revise the previous description of the female, but to add to that a description also of the male, and to show conclusively that the species is a true *Cylindropsyllus*.

Description of the Female.—Body cylindrical (fig. 23) ; length of the specimen figured, 97mm. (fully $\frac{1}{25}$ of an inch). The antennules, which are comparatively short, are nine-jointed ; the second joint is considerably longer than any of the other joints. Their proportional lengths are shown approximately by the formula :—

Proportional lengths of the joints,	7	43	14	10	10	8	16
Numbers of the joints,	1	2	3	4	5	6	7

The antennæ closely resemble those of *Cylindropsyllus lævis,* but the end joints are proportionally rather longer ; the secondary branches (protopodites) appear also to be slightly more elongated.   The mandible-palp, which is moderately slender, is of greater length than the same appendage in *Cylindropsyllus lævis,* and is composed of two joints, but the last joint is small (fig. 24).   The maxillæ appear to be similar to those of the species named.   The anterior foot-jaws (first maxillipedes) are small and apparently one-jointed ; the single joint is somewhat dilated, and bears two elongate processes at the distal end of the inner margin ; the terminal claw is also moderately large and stout (fig. 25). The posterior foot-jaws (second maxillipedes) are slender and two-jointed, and armed with a moderately long, slender, almost setiform, terminal claw (fig. 26).   All the thoracic feet ·are as previously described.   The caudal segments are nearly as long as the anal segment, and they are each furnished with a broad, sabre-like terminal spine nearly of the same length as the furcal segment, and each spine bears a secondary setiform process on the outer margin (fig. 29). The segments are also provided with one or two minute hairs.

Description of the Male.—The antennules of the male are modified for grasping.   The cephalo-thoracic and other appendages are similar to those of the female, except in the following particulars :—(1) The outer branches of the second pair of thoracic feet, which are moderately stout and elongate, are each armed with a stout elongated falciform terminal process, bent inwardly at nearly right angles to the joint from which it springs.   These processes are somewhat similar to those on the outer branches of the second feet of the male of *Cylindropsyllus lævis,* but the apex is somewhat differently modified, as shown by the drawing (fig. 30). (2) The short inner branches of the third pair differ from those of the female in having the first joint produced interiorly into a stout tapering spine, which is slightly sinuate, and extends beyond the end of the second joint ; the second joint is dilated—both margins being convex (fig. 31). (3) The caudal segments are provided with terminal spines that are stout and tapering (fig. 32).   It may also be noted that the exterior spine with which each of the fifth feet in the female is armed is wanting in those of the male.

Cylindropsyllus minor, though apparently not very common, is a widely distributed species ; it has been obtained not only off St. Monans, Firth of Forth, but also at Ballantrae Bank, Firth of Clyde.   I have not, how-ever, observed male specimens other than those referred to in the pre-ceding description, which are from the Forth estuary ; probably the males are scarcer than the other sex.

Huntemannia jadensis, S. A. Poppe.

1884.   *Huntemannia jadensis,* S. A. Poppe, Abhandl. d. Nat. Ver. Bremen, Bd. IX., p. 59.

In previous years I have recorded this curious species from the head of West Loch Tarbert (Cantyre), and from the Cromarty Firth, which hitherto appeared to be the only two Scottish localities where this copepod was known to occur.   I have now to report two additional stations for *Huntemannia,* both of which are in the Clyde district.   It was taken with the hand-net in shore-pools a little below high-water mark at the south-west corner of the Greater Cumbrae on May 6th, 1899, and in shore-pools at Inverkip on the 13th of the same month.   At the latter place I obtained for the first time one or two females with ovisacs.   I find that the females of this species carry two ovisacs of average size, which contain a considerable number of small ova.   What

AA

we know of the distribution of this species tends to show that it is more or less restricted to brackish water.

Ilyopsyllus coriaceus, Brady and Robertson.

> 1873. *Ilyopsyllus coriaceus,* B. & R., Ann. and Mag. Nat. Hist. (4), vol. xii., p. 132, Pl. IX., figs. 1–5.

I have to record the occurrence of this small but interesting species from the Cromarty Firth. It was obtained in a brackish-water pool at the mouth of the River Alness in the summer of 1893, and only one specimen was observed. It was not recorded at that time, as it was expected that other specimens might be found, when a description with drawings of the species would have been prepared. No more specimens have, however, been discovered, and I now therefore place on record the solitary specimen obtained, which appears to be a female.

Quite recently the Rev. A. M. Norman very kindly presented me with a few specimens of this species collected by himself at Birterbuy Bay, Ireland, in 1874. These at first sight looked as if they belonged to another species, for, instead of the broad spathulate furcal setæ referred to in Prof. G. S. Brady's description and figures, the principal furcal setæ were long and slender; but this, it was afterwards found, was merely a sexual difference, the specimens I had received from Dr. Norman being males, whereas the Cromarty Firth specimen, like that described by Prof. Brady, was a female. Moreover, it was observed that a form of *Ilyopsyllus,* which, in my report on some Entomostraca from the Gulf of Guinea, I had described under the name of *Ilyopsyllus affinis,* resembled so closely these Birterbuy males that it is probably only a southern form of *Ilyopsyllus coriaceus.* The Gulf of Guinea specimens were obtained in a shore-gathering collected at the Island of Sao Thome. These Copepods are strongly gibbous on the dorsal aspect, and the peculiar spathulate furcal setæ of the female of *Ilyopsyllus coriaceus* serve to distinguish it readily from its congeners.

Scutellidium tisboides, Claus.

> 1866. *Scutellidium tisboides,* Claus, Die Copep.-Fauna v. Nizza, p. 21, t. iv., figs. 8–15.

This somewhat rare copepod has been obtained at various times in shore-pools between tide-marks at Bay of Nigg, Aberdeen. The species does not appear to be very rare in some of the gatherings obtained here. On the other hand, I have as yet failed to obtain it in the Firth of Clyde, and neither Mr. Robertson nor Prof. Brady appear to have observed it there; neither do I remember of its having been observed by us in the Firth of Forth—probably its distribution is local rather than rare. The colour of the Bay of Nigg specimens was generally not very pronounced; some were colourless, but usually they were tinged more or less with a light brownish pigment.

Clausia cluthæ, T. and A. Scott.

> 1896. *Clausia cluthæ,* T. and A. Scott, Ann. and Mag. Nat. Hist. (6), vol. xviii., p. 1., Pl. I., figs. 1–12.

Several specimens of this curious copepod have been obtained in dredged material from Tarbert Bank, Lower Loch Fyne. Though this is apparently the first time that *Clausia cluthæ* has been recorded from Loch Fyne, it is not the first time for the Clyde generally; the specimens from which the species was described were discovered in Ayr Bay in

1896. In this species the fifth thoracic feet project outward from each side of the body, and are more or less conspicuous. The genus *Clausia* was established by Claperède in 1863. Boeck, not knowing this, established a genus of free-swimming Copepods under a similar name in 1864, but in 1872 Boeck changed his "*Clausia*" to "*Pseudocalanus.*"

Corycæus anglicus, Lubbock.   (Pl. XIII., figs. 1–14.)

> 1857.   *Corycæus anglicus*, Lubbock, Ann. and Mag. Nat. Hist. (2), vol. XX., Pl. XI., figs. 14–17.

This pretty species was added to the British fauna by Sir John Lubbock in 1857 from specimens which had been obtained at Weymouth. For a considerable number of years afterwards our knowledge of the British distribution of the species was almost entirely limited to the information contained in the description which Sir John had published.

In 1880 Prof. G. S. Brady, by the publication of the third volume of his monograph of British Copepoda, was able to considerably extend the known distribution of our *Corycæus*. But though our knowledge of its British distribution continued to increase from year to year, there has apparently been no record of it from the Scottish seas till 1896, when a report of its occurrence in the Firth of Forth was published in Part III. of the *Fourteenth Annual Report of the Fishery Board for Scotland*. So far as I know, no further captures of *Corycæus* have been made in Scottish waters till the past summer, when it was taken in the Firth of Clyde. It occurred in a surface tow-net gathering collected in the vicinity of Ailsa Craig on the 29th of May.

The presence of *Corycæus anglicus* in our Scottish estuaries may be owing to changes in the trend of oceanic currents induced by the prevalence of certain winds,* or it may be that the methods of research being now more perfect than formerly, the presence of such organisms is more readily detected. Several specimens of *Corycæus* were obtained in the Clyde gathering collected on the 29th of May, and some of the colouring of the species still remained when they first came under my observation. Both males and females were obtained.

The female represented by the drawing on Plate XIII. measured slightly over one millimetre in length, while the length of the male, which is represented by one or two detailed figures on the same plate, was slightly less than that of the female. The female antennules are short and six-jointed. The proportional lengths of the joints are shown approximately by the formula :—

$$\text{Proportional lengths of the joints,} \quad \frac{8 \cdot 8 \cdot 8 \cdot 10 \cdot 7 \cdot 5}{1 \cdot 2 \cdot 3 \cdot 4 \cdot 5 \cdot 6}$$
$$\text{Number of the joints,}$$

The antennæ (fig. 3) are stout; each is armed with a moderately strong and slightly-hooked terminal claw; an elongated spine springs from the inner distal angle of the first joint, while one or two smaller spines occur on the other joints. In the male the terminal claws of the antennæ are long and sickle-shaped (fig. 12). The biting part of the mandible is armed with a few moderately long teeth, and one or two spine-like lateral appendages. The palp is very small, and composed of a single one-jointed branch (fig. 4). The maxillæ are simple, one-jointed and moderately stout, and armed with a few short, stout, apical, and sub-apical spines (fig. 5). The anterior foot-jaws (fig. 6) are short and very stout, their structure is somewhat rudimentary, and their armature consists of several

---

* My son, Mr. Andrew Scott, in a letter to me on July 28th, incidentally mentioned that Mr. I. C. Thompson "had been getting *Corycæus anglicus* in abundance off Port Erin, Isle of Man, a week or two ago." That would be nearly about the time it was observed in the Clyde.

apical and sub-apical setæ. The posterior foot-jaws of the female (fig. 7) have moderately short and slender terminal claws, but those of the male (fig. 13) are armed with terminal claws of considerable length   The first three pairs of swimming-feet have both branches three-jointed. The inner branches, which are considerably shorter than the outer, are provided with a number of setæ on the inner margin and apex, but have apparently no terminal spines ; the outer branches are also furnished with several setæ on the inner edge. Moreover, the first and second joints bear a short but moderately stout spine on the exterior distal angle, while the third joint carries two marginal and two apical spines, the inner one of the two apical spines being longer and stouter than the other (figs. 8 and 9).   In the fourth pair (fig. 10) the inner branches are reduced to a single minute joint ; the outer branches are also comparatively small, and they want, to a large extent, the spiniform armature of the outer branches of the preceding pairs.   The fifth pair are small, and each consists of a single one-jointed branch, which is furnished with two apical setæ (fig. 11).

In the female the lateral processes of the fourth body segment extend backward to about the middle of the penultimate segment of the abdomen. This abdominal segment appears to be larger in the male than in the female, as shown by the figure (fig. 14). The caudal segments in both male and female are moderately elongated, being about one and a half times the length of the anal segment.

This species when living is one of the more brilliantly coloured of the British Copepods, but spirit extracts the colour very quickly.

Monstrilla (?) *danæ*, Claparède.   (Pl. XIII., figs. 15–20.)

      1863.   *Monstrilla danæ*, Clap., Beobacht. üb. Anat. u. Enwickl.-wirbellos Thiere an der Küste v. Normandie angestellt., p. 95.

Representatives of this curious genus of copepods have, as in previous years, been occasionally observed in tow-net gatherings of entomostraca from the Clyde.   Two or three species of the Monstrillidæ have been recorded from North Britain, but the only one that has hitherto been observed in the Clyde estuary is the species now referred to, and which I have for the present ascribed to Claparède's *Monstrilla danæ*.   The genus *Monstrilla* was added to the British fauna in 1857 by Sir John Lubbock, when he described the *Monstrilla anglica*.   For nearly thirty years afterwards little or nothing further appears to have been known concerning these organisms, so far at least as regards their distribution in the British seas, and in view of this it is somewhat remarkable that now not a year passes without a lesser or greater number of specimens representing sometimes two or three different species, being observed.

In 1890 Gilbert C. Bourne published in the *Quarterly Journal of Microscopical Science* a little paper on the genus *Monstrilla*, and gave a short summary of the characters of the different forms.   He divided them into two groups, distinguished by the number of furcal hairs.   In the one group the number of setæ on each furcal member is said not to exceed three, while in the other there are said to be six setæ on each of the caudal furca.   *Monstrilla danæ* was placed by Mr. Bourne in the first of these groups.

The Clyde specimens which I record here, and which I am inclined to ascribe to *Monstrilla danæ*, do not fit in with either of Mr. Bourne's groups as regards the number of furcal hairs.   In the more perfect of the specimens there are five hairs on each of the caudal rami, four of which are prominent and one very small ; of the four large hairs, three spring from the end of the furca and one from a notch on the outer margin ; the small hair is also marginal.

The male and female have each three abdominal segments. Those of the male are nearly of equal size, but in the female the genital segment is about as long as the combined length of the next two, and it is provided on the ventral aspect with two moderately long setæ. In the specimen figured (fig. 15) several minute ova were observed attached to these genital setæ, as shown in the drawing. The fifth thoracic feet in the female (fig. 18) are sub-cylindrical, and rather longer than broad, and carry two apical setæ.

Monstrilla danæ appears to be more frequent in Upper Loch Fyne than in the seaward part of the Clyde. In a tow-net gathering collected on the 28th November last (1899) near the head of the loch as many as twenty-seven specimens of *Monstrilla* were obtained, apparently all belonging to this species. But a much larger number of specimens was obtained in a gathering collected, also near the head of the loch, in the month of September immediately preceding. This gathering, which was collected with the surface tow-net on the 29th of the month referred to, was a small gathering, and contained a considerable quantity of fibrous matter. It was not examined until the following month of March, when over eighty specimens were obtained! The specimens comprised both males and females, but whether they all belong to the one species I am not yet in a position to say. The fact that such a large number of specimens was found in a single small gathering is of no little interest in its bearing on the distribution of these curious animals.

Pseudanthessius thorellii (Brady and Robertson).

> 1875.   *Lichomolgus thorellii*, B. and R., Brit. Assoc. Report, p. 197.

This species, which is one of the Lichomolgidæ distinguished by the possession of elongated caudal furca, has been obtained in dredged material from various parts of the Clyde area. It is quite easily distinguished from *Lichomolgus forficula*, which also has long furca, not only by the structure of the inner branches of the fourth pair of swimming-feet and the difference in the proportional lengths of the abdominal segments, but also by the difference in *habitat*. *Lichomolgus forficula* lives in the branchial cavity of large Ascidians, while *Pseudanthessius thorellii* appears to live free amongst weed or zoophytes, and perhaps also amongst *Filograna*. I have not on any occasion found it naturally inside an Ascidian, and neither does Professor Brady in his description of the species refer to it as a commensal.

Hermanella arenicola (G. S. Brady).

> 1872.   *Boeckia arenicola*, G. S. Brady, Nat. Hist. Trans., Northumberland and Durham, vol. iv., p. 430.

A specimen of this fine species was obtained in a gathering of entomostraca from Loch Gilp (near Ardrishaig, Loch Fyne), which is a new Clyde Station for this species. The vicinity of Otter Spit is the only other locality within the Clyde area that I know of where *Hermanella arenicola* has been obtained.

Asterocheres (?) *echinicola* (Norman).   (Pl. XIV., figs. 33–36.)

An *Asterocheres* is obtained in the water passages of a sponge (*Suberites* sp.), common both in the Clyde and Loch Fyne, which is closely allied to *Asterocheres echinicola* (Norman), and which may probably be only a variety of that species. It differs from the typical *Asterocheres echinicola* in having the caudal segments rather shorter than the anal segment, while in typical specimens these are slightly longer than that segment. The general outline of the cephalothorax, and especially the outline of the posterior margin of the third segment, seem also to be slightly different, as shown by the drawing (fig. 33). Moreover, the outer lobe of the maxillæ is also apparently somewhat shorter than that of the same appendages in *Asterocheres echinicola*, but whether such differences are constant seems somewhat doubtful.

A species of *Asterocheres* described by Dr. Giesbrecht, of Naples, under the name of *Asterocheres suberites*, seems to have a *habitat* similar to this Loch Fyne form, and it also agrees with the same form in having the caudal segments shorter than the anal segment ; indeed the difference in the length of the caudal segments appears to constitute one of the principal points of distinction between *Asterocheres echinicola* (Norman) and *Asterocheres suberites* (Giesbrecht). Notwithstanding the points of agreement observed between this Clyde copepod and Giesbrecht's *A. suberites*, I prefer for the present at least to ascribe it to the *A. echinicola*, Norman. It may also be noted that *Asterocheres boecki* (G. S. Brady) was occasionally obtained in the water-passages of *Suberites* in company with the *A.* (?) *echinicola*.

Rhynchomyzon purpurocinctum (T. Scott).

> 1893. *Cyclopicera purpurocinctum*, T. Scott, *Eleventh Ann. Report Fish. Board of Scot.* (III.), p. 209, Pl. III., figs. 29-40.

This species, which has already been recorded from one or two places on the East Coast of Scotland, has only recently, and for the first time, been observed in the Clyde district. It occurred in the washings of dredged materials (weed, gravel, sand, etc.) collected in the vicinity of Otter Spit, Upper Loch Fyne, on January 12th, 1899, at a depth of about 8 to 15 fathoms. The occurrence of the species here would seem to indicate that though it may be, and probably is, a scarce one, it may be more or less generally distributed around our shores. The species is readily distinguished not only by its general form, but by the fact that it is adorned by a dark purple-coloured band extending across the thorax, and it is worth noting in regard to this band that a lengthened immersion in methylated spirits appears to have had little effect on it. Usually the spirit extracts the colouring matter very quickly from these microforms, but it would appear that in this case the purple pigment is of a more permanent character than that generally observed in the colours of copepoda.

Artotrogus orbicularis, Boeck.

> 1859. *Artotrogus orbicularis*, Boeck, Forh. Vid.-Selsk., Christiania, p. 2. Pl. I.

One specimen, about one millimetre in length and 0·78 millimetre in breadth, was obtained amongst some washings of material dredged on Tarbert Bank, Loch Fyne, on the 28th October last (1899). This is the second specimen of this rare species from the same locality.* I. C. Thompson has recorded this species from the Liverpool Bay district,† but Tarbert Bank, Lower Loch Fyne, is the only Scottish locality I know of where *Artotrogus orbicularis* has been obtained.

---

* *Sixteenth Ann. Report Fish. Board of Scot.* (III.), p. 272, Pl. XIV., figs. 12–21 (1898).
† *Trans. Lit. and Phil. Soc., Liverpool*, vol. viii., p. 87.

Parartotrogus richardi, T. and A. Scott.

> 1893. *Parartotrogus richardi*, T. and A. Scott, Ann. and Mag. Nat. Hist. (6), vol. xl., p. 210, Pl. VII.

This curious little species has not before been observed in the Clyde area. It is a form that is readily missed, and may therefore be more widely distributed than at present it appears to be. In Scotland it has only hitherto been observed in the Firth of Forth, but it has also been found in the vicinity of Naples by Dr. W. Giesbrecht.

AMPHIPODA.

A few of the amphipods observed in the tow-net and other gatherings sent from the Fishery steamer "Garland" may now be noticed. Only the rarer forms are recorded here.

HYPERIIDÆ.

Hyperia galba, Parathemisto (?) *oblivia*, and *Hyperoche tauriformis* have been occasionally observed in the tow-net gatherings sent from the Clyde and Loch Fyne. These may still be reckoned as comparatively rare amphipods in the Clyde area. Their scarcity here is in somewhat marked contrast to the frequency of the species on the East Coast.

PONTOPOREIIDÆ.

Urothoë marina has been obtained at Tarbert Bank, Loch Fyne, while *Argissa hamatipes* (Norman) has been observed in tow-net gatherings collected both in Loch Fyne and in the seaward portion of the Clyde estuary. *Argissa* is sometimes frequent in under surface tow-net gatherings from the Clyde. The somewhat remarkable difference in the dorsal aspect of the urosome in the male and female is an interesting feature of this species. *Argissa* was also obtained in a tow-net gathering collected in Aberdeen Bay in May 1898.

AMPHILOCHIDÆ.

A number of species belonging to this group have been observed in the gathering of tow-netted and dredged material sent from the "Garland" during the past year. One of these appears to be identical with a form discovered a few years ago in the Moray Firth. This form was described in the "Annals and Magazine of Natural History"* under the name of (?) *Cyproidia brevirostris*, T. and A. Scott; it is a very small amphipod, scarcely reaching to two millimetres in length ; the Clyde specimens, which are of a somewhat chocolate-brown colour are easily overlooked. *Cyproidia brevirostris* comes very near *Cyproidia danmoniensis*, Stebbing. *Stegoplax longirostris*, G. O. Sars, is also another closely allied form. A few specimens of *Cyproidia brevirostris* were obtained in some dredged material from Tarbert Bank, Loch Fyne—a rocky bank which rises to within 15 or 17 fathoms of the surface, while all around the water is deep. This amphipod has not before been recorded from the Clyde area.

EPIMERIDÆ.

Epimeria corniyera (Fabricius), var. In Part III. of the *Fifteenth Annual Report of the Fishery Board for Scotland* (1897), p. 169, I recorded from the Clyde a specimen of what appeared to be *Epimeria tuberculata*, G. O. Sars ; and since then a few more specimens of the

---

* Ser. 6, vol. xii., p. 244, Pl. XIII. (1893).

same form have been obtained. They are usually found in bottom tow-net gatherings from the deeper parts of the Clyde. It now appears that these Clyde specimens do not belong to Sars' *Epimeria tuberculata*, but are a deep-sea variety of *Epimeria cornigera.* They seem to form a connecting link between the two species named, and to belong nearly as much to the one as to the other. It is a form which is in some respects as handsome as typical specimens of either species, both in size and coloration.

Eusiridæ.

Specimens of *Eusirus longipes*, Boeck, have been occasionally observed during the year in bottom tow-net gatherings both from the Clyde and Loch Fyne, but few of the specimens appeared to be mature.

Gammaridæ.

Mœra othonis (M.-Edw), *Cheirocrates intermedius*, G. O. Sars, and *Lilljeborgia kinahani* (Spence Bate) have all been obtained in material dredged at Tarbert Bank, Loch Fyne, at a depth of 15 to 17 fathoms. *Megaluropus agilis* was captured in a tow-net gathering from the vicinity of Sanda Island, Clyde, collected September 5th, 1899.

Photidæ.

Leptochirus pilosus.—A few specimens of this species were recently obtained in material from Tarbert Bank. In most of the Clyde specimens I have seen, some of which were females with ova, the secondary branches of the antennules are only two-jointed, the end joint being quite small. Another species, *Microprotopus maculatus*, Norman, was obtained in a bottom tow-net gathering, recently examined, collected in Aberdeen Bay in May 1898.

Podoceridæ.

Of species belonging to this group the following may be mentioned :— *Ischyrocerus minutus*, which was obtained in the same gathering as that in which the *Microprotopus* referred to above, was observed. *Erichthonius abditus* occurred in a bottom tow-net gathering from 28 fathoms collected in the vicinity of Ailsa Craig, Firth of Clyde, October 10th, 1899.

Caprellidæ.

Protella phasma (Mont.) is not very rare at Tarbert Bank, Loch Fyne. There are few hauls with the dredge taken here in which it does not occur. *Caprella linearis* (Linn.) was obtained at Inverneil Bay, Loch Fyne, by Mr. F. G. Pearcey in November 1899. Males and females of the same pecies have also been captured by the fishery steamer in the Morays Firth.

Caprella septentrionalis, Kröyer.—One specimen of this somewhat rare species was captured in the Cromarty Firth on June 6th, and another on November 23rd, 1898. I am not aware of any previous record of this species for the East Coast of Scotland.

Isopoda.

There are few isopods to record. A considerable number of the chelifera have been observed during the year, but most of the species to which they belong have already been recorded. A few, however, require further study.

Several parasitic forms have been observed, among which are the following :—*Phryxus abdominalis*, attached to the under side of the abdomen of *Spirontocaris securifrons*, captured in the shrimp-trawl net of the fishery steamer "Garland" near the seaward limit of the Clyde estuary. *Pseudione affinis*, attached under the carapace of *Pandalus montagui*, also from near the mouth of the Clyde. *Pleurocrypta marginata*, attached under the thoracic shield of *Galathea dispersa*, taken at Station XIII. (Upper Loch Fyne), October 10th, 1899. *Aspidophryxus peltatus*, attached to the back of *Erythrops serrata* and *Erythrops elegans*,[*] from deep water to the east of Arran, Firth of Clyde, July 18th, also obtained at Station XIII. (Upper Loch Fyne) on 29th December 1899.

CUMACEA.

Several species of cumacea have been obtained in recent gatherings of tow-net and dredged material, most of which have already been recorded, but the following may be mentioned :—*Campylaspis rubicunda* has again occurred in bottom-gatherings from the deep water of Upper Loch Fyne. *Cumella pygmæa* was obtained in dredged material from Tarbert Bank, Lower Loch Fyne. *Nannastacus unguiculatus* and species of *Diastylis* have also been observed in Clyde tow-net gatherings. *Cuma edwardsii* was obtained in a bottom tow-net gathering collected in Aberdeen Bay; while *Cumopsis edwardsii* (sp. Bate) (= *C. goodsiri*, Van Ben.[†]) was taken between tide-marks on the shore near Millport, Cumbrae, Firth of Clyde, on May 6th, 1899, where it had previously been found by Dohrn thirty years before.[‡] *Cumopsis longipes* (Dohrn) (= *C. lævis*, G. O. S.) has also been recorded from the Clyde. These two species are somewhat like each other in size and general appearance, but in *Cumopsis edwardsii* the cephalo-thoracic shield is adorned on both sides with two oblique and arcuate lateral folds ; while in that of the other species the lateral folds are altogether wanting. Moreover, the natatory branches of the first pair of feet in *Cumopsis edwardsii* are composed of ten joints, but of only eight joints in *Cumopsis longipes*.[§] *Cumopsis edwardsii* did not appear to be very rare between tide-marks at Cumbrae. A considerable number of adult and young specimens were included in the gathering I collected at Cumbrae in May last.

SCHIZOPODA.

Thysanoessa neglecta (Kröyer).—I have again to report the occurrence of this Euphausid from the Firth of Clyde. In my Notes published in the *Seventeenth Annual Report*, *Thysanoessa neglecta* is recorded for the Clyde for apparently the first time. The specimens referred to in that Note had been obtained in a bottom tow-net gathering from Station X., near the seaward limit of the estuary, collected on January 16th, 1899, at a depth of 26 fathoms. The specimens referred to on the present occasion were obtained in two separate bottom tow-net gatherings from Station XII., between Arran and Turnberry Head. These gatherings were collected, the one on the 18th and the other on the 24th of July 1899. In the first one five specimens were obtained, but only two were observed in the other. This species, though apparently rare in the Clyde district, is one of the more common schizopods on the East Coast As pointed out by Dr. Norman in his useful "Synopsis of the British

---

[*] A specimen of the curious *Aspidæcia normani* occurred also on the back of a specimen of *Erythrops elegans*, from the same part of the Clyde estuary. (See also record of *Aspidæcia* in Part III. of the *Sixteenth Ann. Report of the Fishery Board for Scotland*.

[†] Cf. Scott, *S. F. B. Rept.*, 1888, p. 253.

[‡] Cf. *Jen. Zeitschr.*, vol. v., 1869 ; *Unters. üb. Bau u. Entw. d. Arthropoden*, 1870, p. 23.

[§] *Vide* Prof. G. O. Sars' description of the two species in his work *Middelhavets Cumaceer*.

Schizopoda," this *Thysanoessa* is readily distinguished from its near ally by having a spine over the base of the telson.

Mysidopsis gibbosa and *Mysidopsis angusta* have both been obtained in tow-net gatherings forwarded from the Clyde, as also have *Leptomysis gracilis* and *Praunus inermis.* *Neomysis vulgaris*, a Mysid which appears to be rare in the Clyde district, formed part of the contents of the stomach of a fifteen-spined stickleback (*Gasterosteus spinachia*) captured near the head of Loch Fyne on the 16th of April last (1899). The same Mysid has been obtained in the Dhu Loch, near Inveraray.

Siriella armata (M. Edw.) and *Siriella clausii*, G. O. Sars, were obtained in a gathering of crustacea from Loch Gilp by Mr. F. G. Pearcey, the naturalist on board the "Garland." The first—*Siriella armata*—has been occasionally captured in the Clyde during recent years, but the second—*Siriella clausii*—has, so far as I remember, only once before been recorded from the Clyde district, viz., in 1886, when one or two specimens were taken in East Loch Tarbert. The specimen now recorded from Loch Gilp is a male, apparently full-grown. It was captured in three to five fathoms on 31st October 1899.

Siriella armata, besides having a strongly produced rostrum, has usually some of the cephalic and caudal appendages ornamented with chocolate-coloured blotches. The body, especially on the ventral aspect, is also occasionally coloured. *Siriella clausii* appears to be colourless.

Erythrops serrata, G. O. Sars, and *Erythrops elegans*, G. O. Sars, have both been obtained in tow-net gatherings of the "Garland" forwarded from the Clyde. *Erythrops serrata*, which appears to be of more frequent occurrence than the other, has been observed for the most part in gatherings from the seaward portion of the estuary; while *Erythrops elegans* is taken occasionally in Loch Fyne, as well as further to seaward. Both are sometimes infested with parasites.

Anchialus agilis, G. O. Sars.—Two specimens of this rare Schizopod were obtained in a tow-net gathering collected at Station VI., Firth of Clyde (a little to the east and north of Sanda Island). They occurred in a bottom tow-net gathering from a depth of 20 to 27 fathoms collected on December 15th, 1898. Sars obtained the *Anchialus* in the Bay of Naples at a depth of six to eight fathoms, and he also obtained one near Messina at a depth of 20 fathoms.

Anchialus is one of the many interesting species which Dr. A. M. Norman has added to the British fauna. The single female specimen recorded by him was obtained at Plymouth in 1890. There does not appear to be any previous record of *Anchialus* from the Clyde estuary. Dr. Norman has seen my specimens.

DECAPODA.

Xantho hydrophilus (Herbst).—A single specimen of this species—a male—was captured with the shrimp-trawl of the "Garland" at the mouth of the Clyde, at a depth of 60 fathoms, on June 15th, 1899, and forwarded by Mr. Pearcey to the Laboratory, Bay of Nigg, and is now in the collection there. One of the characters that seems to distinguish this form from *Xantho incisus*, Leach, is that the claws have the movable finger grooved on the upper aspect; the grooves extend nearly the whole length of the fingers. All the joints of the feet in *Xantho hydrophilus* are also ciliated on the upper edge, while in *Xantho incisus* the third only is ciliated.

Corystes cassivelaunus (Pennant).—A small male specimen of this species was captured in the same gathering as the last, and is now in the collection at Bay of Nigg. It measures about 20mm. from the extremity of the rostrum to the base of the dorsal shield.

Jaxea nocturna, Nardo.

In my paper on Clyde tow-net and other gatherings published in Part III. of the *Seventeenth Annual Report* (1899) I reported the occurrence of an interesting lucifer-like crustacean in the Firth of Clyde. I stated further that this crustacean had been identified with a form, also from the Clyde, which had been described in the Proceedings of the Royal Society, Edinburgh, vol. xv., p. 420, figs. 1 and 2 in the text (1889), by the late George Brook under the name of *Trachelifer.* In some additional remarks which immediately follow what had been stated in regard to Brook's description of *Trachelifer,* it is clearly shown that this "*Trachelifer*" was really the young of *Calliaxis adriatica,* Heller. Nothing further transpired concerning these Clyde organisms till last summer, when I received from Mr. F. G. Pearcey, the naturalist on board the "Garland," a number of fragments of a small *Nephrops*-like crustacean which he had found in the stomachs of some gurnards captured in the vicinity of Ailsa Craig, near the mouth of the Clyde estuary. It was at once evident that these fragments did not belong to *Nephrops norvegicus,* though in some respects they had a more or less close resemblance to that crustacean. The species, however, could not be made out for a considerable time. At first it was thought that the fragments might represent one or other of the described species of *Nephropsis,* but with none of these would they fit in satisfactorily. Failing, for various reasons, to arrive at a satisfactory solution of the difficulty, I applied to the Rev. T. R. R. Stebbing, who has not unfrequently proved in such matters to be a "friend indeed;" and he, after some investigation, found that the fragments which had given us so much trouble belonged to a species which Nardo in 1847 had described under the name of *Jaxea nocturna.* He, moreover, pointed out (as he does also in his *History of Crustacea,* p 187) that *Jaxea nocturna* is identical with *Calliaxis adriatica,* Heller, described in 1856 ; and as *Trachelifer* is the young of *Calliaxis,* so also, as a matter of course, is it the young of *Jaxea.* The position of the species may therefore be stated thus :—

Jaxea nocturna, Nardo (1847).

= 1856.   *Calliaxis adriatica,* Heller.

= 1889.   *Trachelifer,* sp. (jun.), Brook.

Another point of interest that may now be considered is the *habitat* of *Jaxea.* Can we claim it as a member of the Clyde fauna ? In regard to this point I am inclined, after a careful consideration of all the circumstances, to consider that we may fairly make this claim. We find these juvenile forms occurring at more or less frequent intervals in various parts of the Clyde area,* and occasionally in considerable numbers, two or three different stages of development being represented, and latterly, as pointed out, fragments of several adult specimens have been found in the stomachs of gurnards caught in the vicinity of Ailsa Craig. From the state of preservation in which these fragments were found it is scarcely likely that the time that had elapsed between the capture by the gurnards of the specimens to which the fragments belonged and the capture of the gurnards themselves in the "Garland's" trawl-net could have been very great. All this seems to indicate that the adult *Jaxea* are not very far off from the places where these larvæ and fragments were obtained. It

---

* *Trachelifer* was obtained in a bottom tow-net gathering collected at Station V. (Whiting Bay)—a station well within the limits of the Clyde estuary—on October 11th, 1899.

might be thought that if *Jaxea* were present in the Clyde, specimens occasionally ought to be taken in the trawl or dredge, yet none have ever been observed. This, however, does not militate against the supposition that this crustacean occurs within the Clyde estuary, for its *habitat* may be about rocky ground, where neither trawl nor dredge could be used, but which would offer no obstruction to gurnards in their search for food. Moreover, it was shown in my paper in Part III. of the *Seventeenth Annual Report* (1899) that at Naples, though the larval forms of *Calliaxis* (*Jaxea*) are met with amongst the surface fauna, the adult has only been found once in 25 years. But whatever be the opinion concerning the *habitat* of this apparently rare species—that is, "rare" as regards its adult form—the fact that fragments of adults were found in the stomachs of gurnards caught in the vicinity of Ailsa Craig is in itself of much interest to students of the British crustacea. The fragments referred to above are now in the Laboratory at Bay of Nigg.

After the preceding notes had been sent to the printer, Mr. Pearcey kindly forwarded the posterior portion of another specimen of *Jaxea*, which he had obtained in the stomach of a witch sole (*Pleuronectes cynoglossus, L.*) captured at Station VIII., Firth of Clyde, on the 20th of November last (1899). Station VIII. is about five miles west by south of Ailsa Craig. (This specimen is in our collection at Bay of Nigg with the others previously referred to.)

EXPLANATION OF THE PLATES.

PLATE XIII.

Corycæus anglicus, Lubbock.

Fig. 1.	Female, dorsal view	× 70.
Fig. 2.	Antennule	× 190.
Fig. 3.	Antenna .	× 190.
Fig. 4.	Mandible	× 380.
Fig. 5.	Maxilla .	× 380.
Fig. 6.	Anterior foot-jaw	× 380.
Fig. 7.	Posterior foot-jaw	× 190.
Fig. 8.	Foot of first pair of swimming-feet	× 190.
Fig. 9.	Foot of third pair ,, ,,	× 190.
Fig. 10.	Foot of fourth pair ,, ,,	× 190.
Fig. 11.	Foot of fifth pair	× 760.
Fig. 12.	Antenna, male .	× 190.
Fig. 13.	Posterior foot-jaw, male	× 190.
Fig. 14.	Abdomen and caudal stylets, male	× 160.

Monstrilla (?) *danæ,* Claparède.

Fig. 15.	Female, dorsal view	× 20.
Fig. 16.	Antennule, male	× 40.
Fig. 17.	Foot of first pair of swimming-feet, female	× 80.
Fig. 18.	Fifth thoracic feet, female	× 40.
Fig. 19.	Last thoracic segment, abdomen and caudal stylets, male	× 80.
Fig. 20.	Fifth thoracic feet, male	× 40.

Enterocola (?) *fulgens,* Van Beneden.

Fig. 21.	Female, dorsal view	× 40.
Fig. 22.	Antennule	× 160.
Fig. 23.	Antenna .	× 160.
Fig. 24.	Maxilla .	× 190.
Fig. 25.	Posterior foot-jaw	× 253.
Fig. 26.	Foot of first pair of swimming-feet	× 126.
Fig. 27.	Foot of second pair ,, ,,	× 126.

Gunenotophorus (?) *globularis*, Costa.

Fig. 28.	Female, lateral view	. . . .	× 17.
Fig. 29.	Antennule	. . . .	× 190.
Fig. 30.	Antenna .	. . . .	× 19.
Fig. 31.	Mandible and palp	. . . . . .	× 84.
Fig. 32.	Posterior foot-jaw	. . . . .	× 190.
Fig. 33.	Foot of fourth pair of swimming-feet	. . . .	× 80.
Fig. 34.	Last segments of the abdomen and caudal stylets	. .	× 40.

PLATE XIV.

Canthocamptus inconspicuus, sp. n.

Fig. 1.	Female, side view	. . . . . .	× 80.
Fig. 2.	Antennule	. . . . . .	× 380.
Fig. 3.	Mandible	. . . . . .	× 380.
Fig. 4.	Posterior foot-jaw	. . . . .	× 380.
Fig. 5.	Foot of first pair of swimming-feet	. . .	× 380.
Fig. 6.	Foot of fourth pair ,, ,,	. . .	× 380.
Fig. 7.	Foot of fifth pair	. . . .	× 380.
Fig. 8.	Last segments of the abdomen and caudal stylets	. .	× 80.

Tetragoniceps (?) *malleolata*, Brady.

Fig. 9.	Female, side view	. . . . . .	× 80.
Fig. 10.	Antennule	. . . . . .	× 253.
Fig. 11.	Posterior foot-jaw	. . . . .	× 380.
Fig. 12.	Foot of first pair of swimming-feet	. . .	× 253.
Fig. 13.	Foot of third pair ,, ,,	. . .	× 150.
Fig. 14.	Foot of fourth pair ,, ,,	. . .	× 150.
Fig. 15.	Foot of fifth pair	. . . .	× 166.
Fig. 16.	Last abdominal segment and caudal stylets	. .	× 126.
Fig. 17.	Foot of fifth pair and appendage of first abdominal segment	.	× 253.

Tetragoniceps brevicauda, sp. n.

Fig. 18.	Female, side view	. . . . . .	× 80.
Fig. 19.	Antennule	. . . . .	× 253.
Fig. 20.	Foot of first pair of swimming-feet	. . .	× 380.
Fig. 21.	Foot of fifth pair	. . . .	× 380.
Fig. 22.	Last abdominal segment and caudal stylets	. .	× 126.

Cylindropsyllus minor, T. Scott.

Fig. 23.	Female, back view	. . . . .	× 80.
Fig. 24.	Mandible and palp	. . . . .	× 380.
Fig. 25.	Anterior foot-jaw	. . . . .	× 380.
Fig. 26.	Posterior foot-jaw	. . . . .	× 380.
Fig. 27.	Foot of first pair of swimming-feet	. . .	× 253.
Fig. 28.	Foot of fifth pair ,,	. . .	× 380.
Fig. 29.	Last abdominal segment and caudal stylets	. .	× 190.
Fig. 30.	Foot of second pair of swimming-feet, male	.	× 380.
Fig. 31.	Foot of third pair of swimming-feet, male	.	× 380.
Fig. 32.	Last abdominal segment and caudal stylets, male	.	× 190.

Asterocheres echinicola, Norman.

Fig 33.	Female, dorsal view	. . . . .	× 80.
Fig. 34.	Antenna	. . . . .	× 190.
Fig. 35.	Mandible and palp	. . . .	× 190.
Fig. 36.	Maxilla .	. . . .	× 126.

Gunenotophorus (?) *globularis*, Costa.

Fig. 37.	Anterior foot-jaw	. .	× 126.
Fig. 38.	Foot of first pair of swimming-feet	.	× 95.

PLATE XIII.

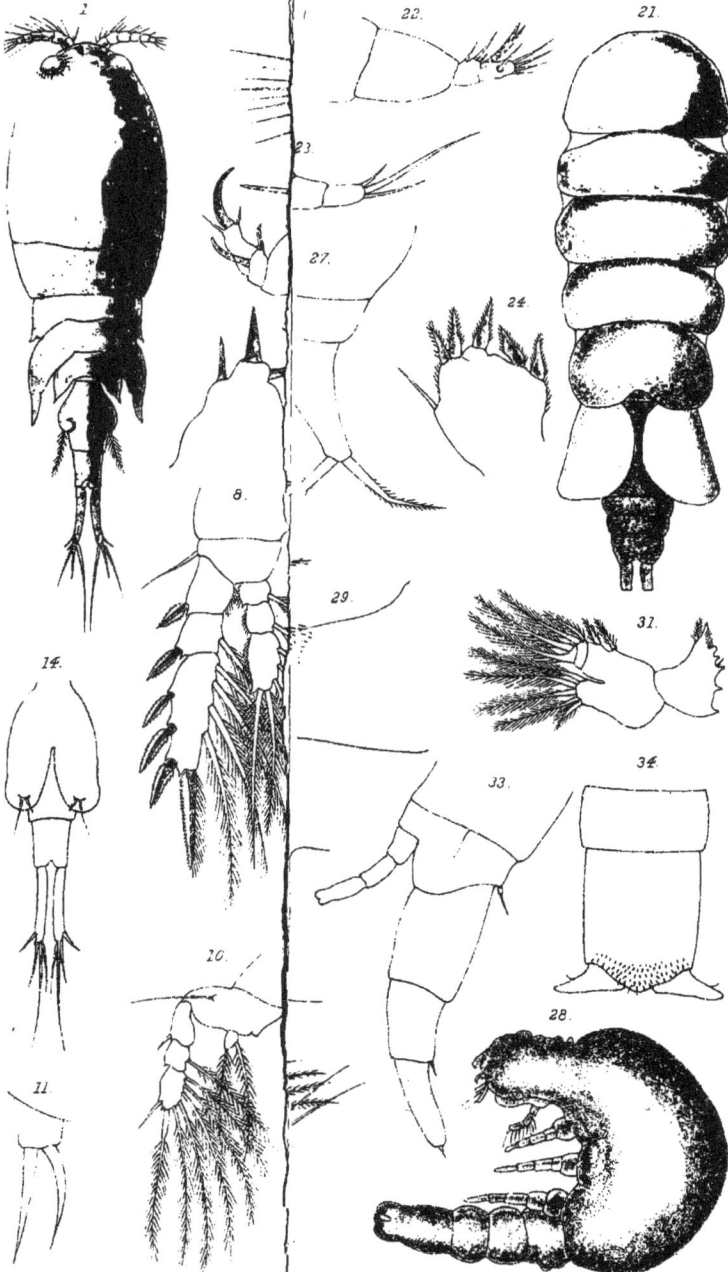

Fig. 18 (?) globularis (Costa).

.

PLATE XIII.

PLATE XIV.

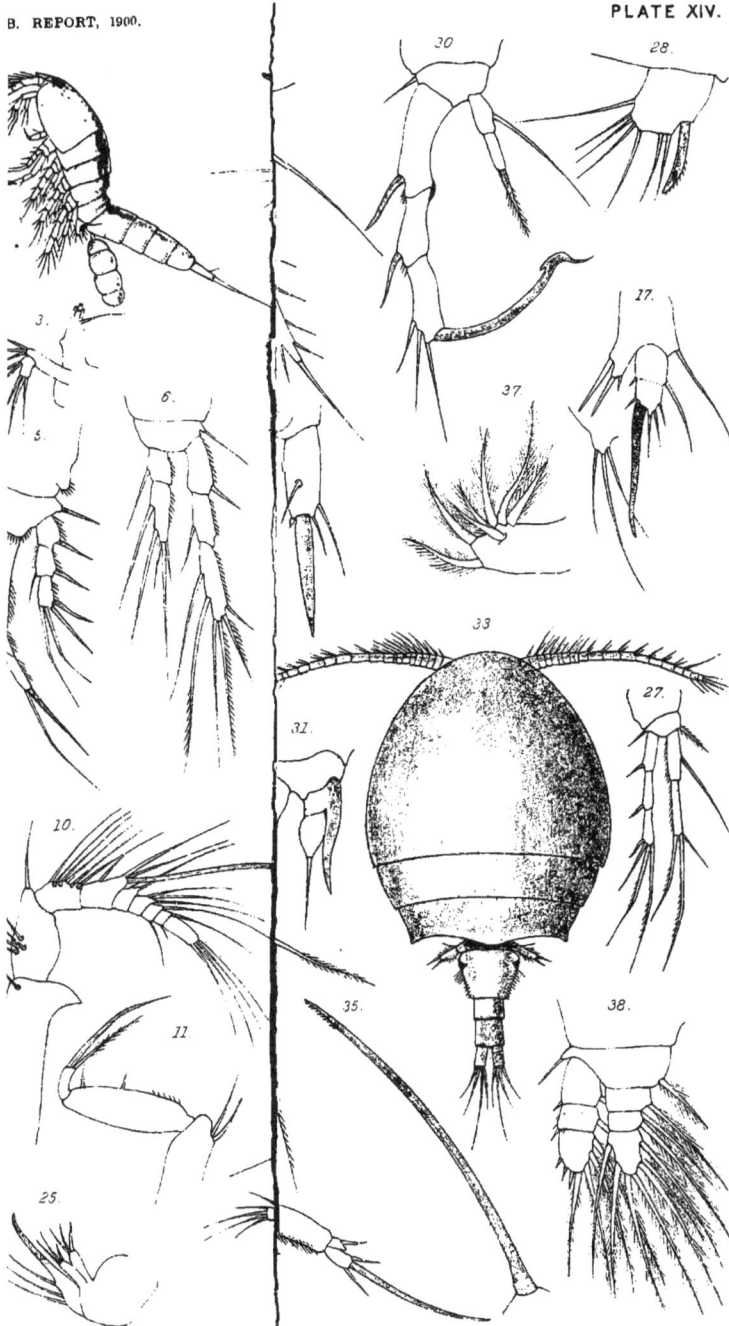

, sp. n.
pilaris (Costa).

PLATE XIV.

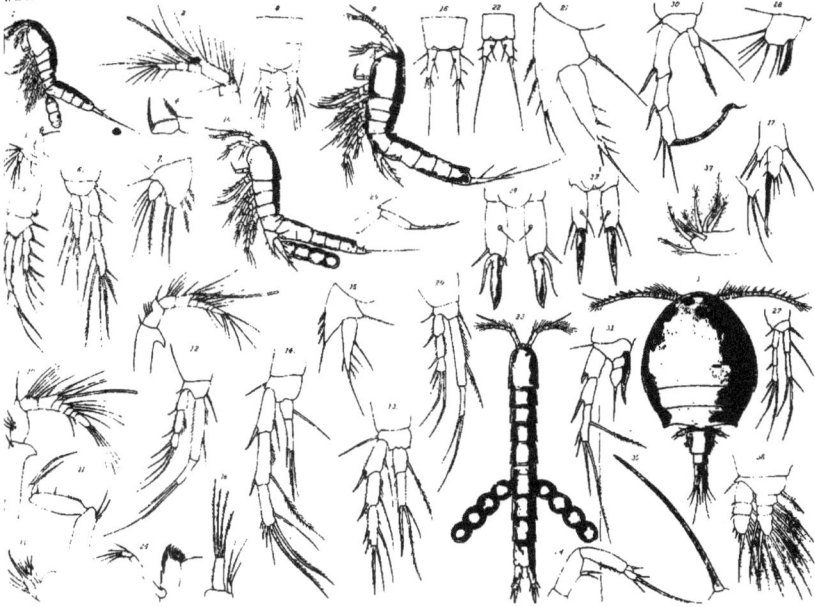

NOTES ON GATHERINGS OF CRUSTACEA, COLLECTED
FOR THE MOST PART BY THE FISHERY STEAMER
"GARLAND" AND THE STEAM TRAWLER "ST.
ANDREW" OF ABERDEEN, AND EXAMINED DURING
THE YEAR 1900.

By Thomas Scott, F.L.S., Mem. Soc. Zool. de France.

(Plates XVII. and XVIII.)

Contents.

INTRODUCTORY REMARKS.

These Notes form a continuation of the series of papers which have from time to time been published in Part III. of the Annual Reports of the Fishery Board for Scotland. The purpose of these notes is to record, in a more or less accessible form, some, at least, of the interesting organisms that have been, and still are being, obtained in connection with the investigations carried on under the authority of the Fishery Board. It is now generally acknowledged that the Crustacea constitute a considerable portion of the food of our food-fishes, and, as will be shown in the sequel, several of the Crustaceans to be recorded here have only so far been obtained in the stomachs of fishes; a knowledge, therefore, of the kinds and distribution of these organisms is necessarily of more than merely zoological importance.

During the past year a considerable amount of information concerning the distribution of the Crustacea has been obtained, and a number of species have been added to the marine fauna not only of Scotland but also of the British Islands, while a few are apparently new to science. A considerable number of rare species have been obtained in gatherings collected by the "Garland" and forwarded to the Laboratory at Bay of Nigg. Several rich gatherings of Crustacea were also collected by the steam trawler "St. Andrew" of Aberdeen, which carried on some experiments to the east of the Orkney and Shetland Islands during the months of September and October last. One or two of the richest gatherings examined during the past year were collected by the "St. Andrew" off the Fair Island in October.   The following are a few of the Crustacean species observed in the "St. Andrew" gatherings :—

Eucalanus elongatus (Dana).
Eucalanus crassus, Giesbrecht.
Rhincalanus (?) *gigas*, Brady.
**Cyclopina longifurcata*, sp. n.
**Eucanuella spinifera*, gen. et sp. n.
Macrocypris minna (Baird).
Euonyx chelatus, Norman.
Byblis gaimardi (Kröyer).
Aceros phyllonyx (M. Sars).
Lœmatophilus tuberculatus, Bruzel.
Dulichia monacantha, Metz.

Arcturella delatata, G. O. Sars.
Paramunna bilobata, G. O. Sars.
Macrostylis spinifera, G. O. Sars.
Echinopleura aculeata, G. O. Sars.
Diastylis cornuta (Boeck).
Diastylopsis resima (Kröyer).
Leptostylis villosa, G. O. Sars.
Pseudocuma similis, G. O. Sars.
Campylaspis rubicunda (Lillj.).
Campylaspis costata, G. O. Sars.
Erythrops serrata, G. O. Sars.

Several organisms other than Crustacea were also observed in the gatherings collected by the "St. Andrew," the following of which may be mentioned :—

> *Clio pyramidata*, Browne, obtained fifty miles S.E. of Fair Island (living when captured), October 19th, 1900.
> *Natica grœnlandica*, Beck, dead, but fresh shell.
> *Siphonodentalium lofotense*, Sars, a few specimens ; and the following Foraminifera :—*Saccamina sphœra, Psammosphœra fusca, Astrorhiza arenaria, Placopsilina bulla.*

Some interesting collections were also sent from the "Garland," and especially those from the deep water off Aberdeen.   Some moderately rare species belonging to the Amphipoda, Isopoda, Sympoda, and Macrura were obtained off Aberdeen, comprising such forms as the beautiful *Epimeria cornigera*, the curious *Lepidepecrium longicorne*, and *Euonyx chelatus*, and also the following rare forms :—*Amphilochoides intermedius,* T. Scott, *Metoponia robusta* (G. O. Sars), *Cuma pulchella*, G. O. Sars, *Petalosarsia declivis*, G. O. Sars, *Pleurogonium inerme*, G. O. Sars, and a number of other minute species.

Copepoda were not very plentiful, neither in these gatherings nor in those collected by the "St. Andrew," but nevertheless interesting species will be found recorded in the sequel.   A curious and somewhat abnormal form which was dredged by the "Garland" at Tarbert Bank, Loch Fyne, I have been unable to identify with any known species or genus, and have therefore recorded it under the name of *Cancerina confusa* ; it resembles in some measure the *Cancerilla tubulosa* discovered by Sir John Graham Dalyell as the parasite of a Starfish, but it differs from that Copepod in several important points.

In concluding these introductory remarks, I desire to acknowledge my indebtedness to Mr. F. G. Pearcey of the "Garland" and to Mr. Dannevig for the valuable gatherings of Crustacea which they have collected, and which have yielded so many interesting species.   I have also to thank Mr. H. C. Williamson for a few rare forms, and especially for *Ischyrocerus anguipes*, Kröyer, which he discovered in the Bay of Nigg, and of which there is no previous authentic British record.

My son, Mr. Andrew Scott, assisted by Mrs. Scott, has prepared the drawings necessary to illustrate several of the species described in the sequel.

---

* These were obtained in the stomach of a small Haddock captured about sixty-five miles south-east by east of Sumburgh Head, Shetland, on September 4th, 1900.

CRUSTACEA.

SUB-CLASS ENTOMOSTRACA.

Order I. COPEPODA.

CALANIDÆ.

Eucalanus elongatus (Dana) ; and *Eucalanus crassus*, Giesbrecht.

I have at different times during the past few years recorded these two species of *Eucalanus* from the north-east and north of Scotland, and this year I have again to note their occurrence in Scottish waters, but from a locality different from those previously mentioned—viz., at about fifty miles south-east of the Fair Island, between Orkney and Shetland. They were obtained in at least two tow-net gatherings collected on October 19th, 1900, by the steam trawler "St. Andrew." A number of specimens of both species were observed in this gathering.

Rhincalanus (?) *gigas*, Brady. (Pl. XVII., figs. 1–4.)

> *Rhincalanus gigas*, Brady, Challenger Reports, vol. viii., Copepoda, p. 42, Pl. VIII., figs 1–11.

Several specimens of a *Rhincalanus*, apparently belonging to *R. gigas*, Brady, were obtained in one of the tow-net gatherings collected to the eastward of the Fair Island, in which the two species of *Eucalanus* mentioned above were observed. Most of the specimens were females, and were apparently adult ; about half a dozen males were also observed, but they were considerably smaller than the females, and appeared to be scarcely mature.

The only apparent difference between these female specimens and *R. gigas*, Brady, is in their size. The specimens of *Rhincalanus gigas* described by Dr. Brady measured from 8·5 to 10 millimeters, while those collected to the eastward of the Fair Island were 5 mm. in length. In these specimens the fifth pair of feet in the female are identical in form and armature with those of *R. gigas*. Figures 10 and 11 of Plate VIII. in Dr. Brady's " Challenger " Report probably represent the fifth feet and abdomen of a male specimen ; they resemble closely figures 2 and 4 on Plate XVII. of this paper, which represent similar parts of one of the immature males in the collections from the Fair Island. Figure 3 (Plate XVII.) represents the last thoracic segment and abdomen of one of the female specimens. I am inclined to ascribe the Fair Island specimens to *R. gigas*, though they are somewhat smaller than those recorded by Dr. Brady, rather than to *R. nasutus*, Giesbrecht. It may also be noted in passing that *R. nasutus* is in some respects, and especially in the form of the female fifth feet, not very unlike *R. gigas*, Brady.

Stephus gyrans (Giesbrecht).

> 1892. *Möbianus gyrans*, Giesbrecht, Pelagischen Copep. d. Golfes v. Neapel, p. 205, Pl. 5, 9, 35.
> 1897. *Stephus gyrans*, T Scott, Fifteenth Ann. Rept., Part III., p. 146, Pl. II., fig. 9 ; Pl. III, figs. 17, 18.

This distinct species was recorded for Loch Fyne in 1897, and up till last year this was the only Scottish locality where *Stephus gyrans* had been observed ; now, however, I am enabled to record it from a second locality, also in the West of Scotland, viz., in the Sound of Mull, where it was collected by the "Garland" in 72 fathoms on March 31st, 1900. This record shows that, though the species may be rare, its distribution in Scottish waters is probably fairly extensive.

Ætidius armatus, G. S. Brady.

A few specimens of this curious species were obtained in tow-net gatherings collected to the eastward of the Fair Island in October, 1900, but they were all females, and in connection with this it is interesting to note that the specimens in the collections made by H.M.S. "Research" in the Shetland-Faröe Channel in July and August, 1896, were also all females.[*] In *Ætidius armatus* the bifid rostrum with its stout spine-like branches, which projects from the front of the head at nearly right angles to the body, appears, when seen in profile, not unlike the beak of a bird. It was this character of the rostrum which suggested to Dr. Brady the name of the genus.[†]

Bradyidius armatus, Vanhöffen.

This species was collected with the tow-net in October, 1900, off the Fair Island.

Scolecithrix hibernica, A. Scott.

This species, which was discovered in the Irish Sea by my son, and which has been observed in the Firth of Clyde and the Moray Firth by myself, has now to be recorded from a new station—viz., about fifty miles south-south-east of the Fair Island, between Orkney and Shetland. It occurred in a gathering collected with a tow-net fixed to the head of one of the trawls of the steam trawler "St. Andrew," on October 16th, 1900, while that vessel was engaged in carrying on the investigations referred to in .the preliminary remarks. This is the farthest north of any of the localities where *S. hibernica* has yet been observed.

CENTROPAGIDÆ.

Centropages typicus, Kröyer.

Specimens of this species were obtained in Lerwick Bay, Shetland, by the steam trawler "St. Andrew" while at anchor there in October last. It was also observed to be moderately frequent off the Fair Island on the 19th of the same month.

Centropages hamatus (Lilljeborg).

This *Centropages* occurred in a tow-net gathering collected in Lerwick Harbour about the same time as the specimens of *C. typicus* were collected.

Eurytemora affinis (Poppe).

This Copepod occurred in great abundance in some of the hatching boxes at the Sea-Fish Hatchery at Bay of Nigg, at the end of the hatching season last year (1900).

---

[*] Notes on the Animal Plankton from H.M.S. "Research." *Fifteenth Annual Report of the Fishery Board for Scotland*, Part III., p. 313 (1897).
[†] Report on the Copepoda collected by H.M.S. "Challenger," p. 75.

Metridia lucens, Boeck.

This species was taken with the surface tow-net in Lerwick Bay on October 15th, 1900.   It was also obtained in a tow-net gathering collected about fifty miles south-east of the Fair Island on the 19th of the same month.   *M. lucens* is not only smaller than *Metridia longa* (Lubbock), but the structure of the fifth feet is different.

Paramisophria cluthæ, T. Scott.

P. cluthæ, which was described in Part III. of the Fifteenth Annual Report of the Fishery Board for Scotland, has only hitherto been recorded for Loch Fyne, Firth of Clyde ; I have now to record the species from the Cromarty Firth. It occurred in some material collected there on January 17th, 1900, and forwarded from the Fishery steamer "Garland" to the Laboratory at Bay of Nigg.

CANDACIIDÆ.

Candacia pectinata, G. S. Brady.

This species was observed during the past year not only in tow-net gatherings from the Clyde and the Moray Firth, but also in one or two gatherings collected in October to the eastward of the Fair Island, between Orkney and Shetland, and also in Lerwick Bay.   The distribution of *C. pectinata* would thus seem to be co-extensive with the British Islands.

PONTELLIDÆ.

Anomalocera patersonii, Templeton.

A number of specimens of this fine species occurred in one of the tow-net gatherings collected to the eastward of the Fair Island on October 19th.   These specimens appeared to be even more highly coloured than those observed in the Clyde and the Moray Firth.

Acartia discaudata, Giesbrecht.

The only Scottish locality where, till recently, I have observed *Acartia discaudata* is the Firth of Forth, but I have now to record it from other two stations, from both of which specimens have been sent to the Laboratory at Bay of Nigg from the "Garland." At one of these stations—viz., Loch Eil, at the head of Loch Linnhe—specimens of *A. discaudata* were dredged in 10 to 30 fathoms on March 4th, 1900 ; while at the other—viz., Dornoch Firth, near Muckle Ferry—the species was moderately common in gatherings collected with the surface tow-net between the 16th and 17th of May.

Acartia bifilosa, Giesbrecht.

A second species of *Acartia—A. bifilosa*—was obtained in a tow-net gathering collected in Lerwick Harbour on October 15th, 1900, by the steam trawler "St. Andrew." It may be mentioned that another species of *Acartia—A. clausii,* Giesbrecht—was obtained during the investigations in which the "St. Andrew" was engaged. It was collected about seven miles off shore between Lerwick and Sumburgh Head on October the 16th, by passing the water from the donkey-pump through a tow-net. *Acartia clausii* occurred also in tow-net gatherings collected to the eastward of the Fair Island on the 19th of the same month.

CYCLOPIDÆ.

Pterinopsyllus insignis, G. S. Brady.

A few specimens of this somewhat rare species were obtained in some material collected in the Moray Firth about thirteen or fourteen miles north-east of Buckie, in about 50 to 55 fathoms, on November 3rd, 1900 ; one or two of them were females, and carried small roundish ovi-sacs, —these are the first specimens of *Pterinopsyllus* which I have seen carrying eggs.

Cyclopina gracilis, Claus.

Specimens of this small but distinct species were obtained in the store pond of the Sea-Fish Hatchery at Bay of Nigg on October 1st, 1900.

Cyclopina longifurcata, T. Scott (sp. n.).   (Pl. XVII., figs. 5–14.)

Description of the female.—In this species the body is moderately stout, the forehead is narrowly but evenly rounded, and the abdomen is slender and elongated ; the caudal furca are very long and somewhat lamelliform, being rather more than half the length of the abdomen ; the entire length of the animal from the forehead to the end of the furca is about 1·56 mm. (about $\frac{1}{16}$ of an inch) ; (fig. 5).

The antennules (fig. 6) are slender and about as long as the cephalic segment ; they appear to be composed of twenty-six joints, and are sparingly setiferous ; all the joints are short, but the ultimate and penultimate joints are rather longer than the others ; the formula shows the number and approximate proportional lengths of all the joints :—

Number of the joints,	1	2	3	4	5	6	7	8	9	10	11	12
Proportional lengths of the joints,	11	4	4	3	3	3	4	5	6	4	5	4

No. of joints,	13	14	15	16	17	18	19	20	21	22	23	24	25	26
Pro. len. of jts.,	5	5	7	9	9	8	9	10	8	7	8	6	12	13

The antennæ are four-jointed, the joints are sub-equal and of moderate length, the third and fourth joints carry one or two long, slender, plain setæ at their distal extremities, the first joint is furnished with a few plumose hairs, three of which spring from a small tubercle situated near the distal end of the outer margin, as shown in the figure (fig. 7). This tubercle, with its fascicle of three plumose setæ, is probably a rudimentary secondary branch.

The mandibles are moderately stout, and provided with well-developed two-branched palps, the branches of which arise from a moderately stout basal joint, the upper branch is two- but the lower is four-jointed, both branches carry several long plumose setæ (fig. 8). The mandibles of this species have a general resemblance to those of *Cyclopina littoralis*, G. S. Brady.

The maxillæ, which are small, possess a distinctly two-branched palp, both branches of which are furnished with long plumose setæ, the manducatory lobe bears on its inner distal aspect one or two moderately stout and several small setæ, all of which are plumose (fig. 9).

The first maxillipedes resemble those of *Cyclopina littoralis*, they are moderately stout, apparently six-jointed, but though the basal joints are dilated, the end ones are small ; the marginal lobe-like processes of the second and third joints are each armed with a moderately stout and long claw-like spine and a few setæ, while a number of stout plain setæ spring from the margins and apex of the three end-joints (fig. 10).

The second maxillipedes are also somewhat similar to those of *Cyclopina littoralis*. The first and second joints are elongated and moderately stout ;

in the first the inner margin is slightly lobate towards the distal end, and one or two moderately stout setæ spring from the rounded apex of each lobe ; the gently rounded inner margin of the second joint is fringed with minute hairs, and carries one or two setæ near its distal end ; the remaining five joints are small, the end one is provided with two short and two long terminal setæ, a long seta also springs from the inner aspect of the penultimate joint, while the three preceding joints are furnished with a few setæ as shown by the drawing (fig 11).

The first four pair of swimming feet are, each of them, two-branched and both branches are three-jointed.   The outer branches of the first pair have their exterior margins armed with elongated spines arranged as follows :—One near the distal end of the first joint, one on the second joint, three on the margin and one (larger than the others) at the end of the third joint ; a single long plumose seta springs also from the inner margins of the first and second joints, and four from the inner edge of the third joint.   The first and second joints of the inner branches have no setæ or spines on the outer margin, but the first joint bears one and the second two setæ on the inner edge, while the third joint bears one seta near the middle of the outer margin, three on the inner margin, and two at the apex.   The armature of the other three pairs of swimming feet is somewhat similar to that of the first, except that the third joint of the outer branches of the fourth pair carries five setæ on the inner margin ; moreover, in the fourth pair the third joint in both branches is proportionally rather more elongated, being about equal to the entire length of the first and second joints combined (figs. 12, 13).   The spines on the outer margins of the outer branches in all the four pairs are of a sabre-like form, with the edges minutely serrated.

The fifth pair of thoracic feet are each composed of a single two-jointed branch, the first joint is somewhat dilated, being about as broad as long, but the second is narrow and moderately elongated, and furnished with five setæ, one being situated near the middle of the outer margin, and four around the distal end (fig. 14).

Habitat.—In the stomach of a small Haddock, *Gadus æglefinus*, L., 88mm. (about 3½ inches) in length, captured 65 miles east by south of Sumburgh Head, Shetland, September 4th, 1900, by the steam trawler "St. Andrew."

Remarks.—The species now described agrees in some of its more important characters with the genus *Cyclopina* of Claus ; but in that genus the antennules are described as being shorter than the cephalic segment, while the number of joints in those of the species hitherto recorded is not more than twenty-two or twenty-three ; in *C. longifurcata*, on the other hand, the antennules are not only as long as the cephalic segment, but they are also apparently twenty-six-jointed ; then again the antennæ of *Cyclopina* are described as four-jointed, but the third joint is very short, whereas in our specimen the third is as long as the fourth joint.   Notwithstanding these and one or two other minor differences, the specimen seems to be a true *Cyclopina*.   Only the one specimen was observed ; and as it was apparently quite uninjured, it had probably only been a short time in the Haddock's stomach before the Haddock itself was captured.

ASCIDICOLIDÆ.

(?) *Enteropsis vararensis*, T. Scott (sp. n.).   (Pl. XVII., figs. 28–34.)

Description of the female.—Length 1·2 mm. (about 1/20 of an inch) Body robust, cylindrical, slightly recurved.   The cephalon is very small.

and the cephalic appendages are therefore somewhat crowded together near the anterior end and on the ventral aspect of the body. The thorax exhibits a slightly articulated structure, but there is no apparent segmentation of the abdomen, this part of the body is not differentiated from the thorax ; the caudal appendages are very small (fig. 28).

The antennules are very short and moderately stout ; they taper towards the distal end, and are each composed of three joints of nearly equal length, and bear three short terminal spines, but are otherwise unprovided with spines or setæ (fig. 29).

The antennæ are simple, and are each composed of two dilated joints and armed with a strong terminal claw ; there are no secondary branches (fig. 30).

The mandibles are of moderate length, and they each consist of a stout and slightly curved claw-like appendage ; there does not appear to be a mandible-palp (fig. 31).

The maxillæ, which are of a simple structure, appear to be two-jointed, the basal joint is stout, but the end one consists of a somewhat narrow and elongated plate, rounded and slightly crenulate at the extremity (fig. 32).

Only one pair of maxillipedes could be made out, and these are very rudimentary ; they each consist of a slightly elevated and broadly rounded prominence armed with a small apical spine (fig. 33).

The thoracic feet comprise four pairs, they are very small but comparatively stout, and appear to be all more or less uniform in structure. Each foot is composed of a single two-jointed branch, both joints are somewhat dilated, but the end one is much smaller than the other, and is armed with two short and apparently movable terminal spines (fig. 34). The position of the first pair is somewhat abnormal, they are not in line with the other three pairs, as is more or less usually the case with free-living species, but are situated round towards the lateral aspect ; this position of the first pair does not appear to be accidental, but has been observed in the other specimens examined.

It will also be observed that the genital opening is situated on the dorsal aspect of the abdominal part of the body (see fig. 28).

Habitat.—In the branchial chambers of a compound Ascidian (*Botryllus* sp.), associated with *Botryllophilus* (?) *ruber* ; not very common. The *Botryllus* was dredged in the Moray Firth in 1896.

Remarks.—This species, if not a true *Enteropsis*, is very closely allied to that genus. The most important point of difference seems to be the apparent absence of a mandible-palp. If the presence of this appendage is clearly established in the other members of the species *Enteropsis*, the one now described may have to be removed to some other genus or a new one instituted for its reception.

Botryllophilus (?) *ruber*, Hesse.   (Pl. XVII., figs. 15–27.)

1864.   *Botryllophilus ruber*, Hesse, Ann. Sci. Nat., Zool. (5), t. i., Pl. XII., figs. 1–7.

1900.   *Botryllophilus* (?) *ruber*, T. Scott, 18th Ann. Rept. Fishery Board for Scot., Part III., p. 388.

In my paper "Notes on some gatherings of Crustacea," published in Paper III. of the Eighteenth Annual Report, I recorded the occurrence of a *Botryllophilus*, in the Clyde and the Moray Firth, under the name of *Botryllophilus* (?) *ruber*, Hesse, but only the females had at that time been observed. In consequence of further research I am able to give a more detailed description of the species than was done last year, and to include in it an account of the male as well as of the female. In my previous record of the species I stated that one of the peculiarities of *Botryllophilus* was "the almost constant existence of a single ovigerous sac of a strictly spherical form sheltered between the fifth feet." I now find that this sac in the specimens observed, both in the Clyde and the Moray Firths, is really composed of two sacs, partly adherent to each other and so closely joined as to appear as if they formed but a single globular mass.

Description of the female.—Length of the specimen figured, 1·16 mm. (about $\frac{1}{18}$ of an inch). The cephalothorax, which terminates anteriorly in a short bluntly rounded rostrum, is somewhat narrow in front, but becomes gradually dilated towards the posterior end and especially on the dorsal aspect. Scarcely any trace of segmentation can be made out in the cephalothoracic region, but this is probably due to the fact that the *Botryllus* in which the specimen was obtained had been for several years (since 1896) preserved in alcohol. The abdomen is narrow and elongated, but is rather shorter than the cephalothorax (fig. 15).

The antennules are very short and composed of four joints, the first is considerably dilated, the second is moderately stout but much smaller than the first; the third and fourth are small; the antennules are furnished with a number of long and moderately stout setæ, which appear to be devoid of feathering (fig. 17).

The antennæ, which are two-jointed, are comparatively long and slender and of nearly equal thickness throughout, they are each composed of two (or (?) three) joints; the end joint is provided with three stout spines on its inner margin, and a few spines and setæ at the apex, the other joints appear to be unarmed (fig. 18). The antennæ are not provided with secondary branches.

The mandibles and maxillæ appear to be obsolete or very rudimentary.

The first maxillipedes are small and feeble, they are each composed of three joints, the second and third joints are sub-equal and rather smaller than the first; the first joint is unarmed, the second bears two or three spine-like setæ on its inner aspect, and the third is provided with a few terminal setæ of unequal lengths, the outermost being considerably longer than any of the others (fig. 19).

The second maxillipedes are short but somewhat robust, the basal joint is moderately distended, and larger than the second one, the third joint is very small, the end-joint is narrow and fully twice the length of the third and bears a short but strong terminal claw (fig. 20).

The thoracic feet.—All the thoracic feet, with the exception of the fifth pair, are short and robust and composed of two branches. In the first and second pairs, the inner branches are very short and two-jointed; about eight plumose setæ of moderate length spring from the rounded apex of the second joint, and one from the inner margin of the first joint. The outer branches, which appear to be uniarticulate, are rather longer than the inner ones and are armed with a few small marginal and apical spines, as shown in the drawing (fig. 21). In the third and fourth pairs the outer as well as the inner branches are two-jointed; the inner branches, which are somewhat alike in both pairs, are rather longer than those of the first pair, but their armature is somewhat similar, except that the end-joints bear six instead of eight plumose setæ; the outer branches of the third pair are somewhat similar to those of the first except that there is an articulation between the two proximal marginal spines, dividing the branches into two joints (fig. 22); the outer branches of the fourth pair are less robust than the inner branches, but are nearly equal to them in length; a single small seta springs from near the end of the outer margin of the first joint, while the second is furnished with four apical and sub-apical setæ, the two outer sub-apical setæ are small, but the other two are moderately stout and unequal in length, the inner one being the longest, both are feathered on the distal half (fig. 23).   Each foot of the fifth pair consists of a single uniarticulated branch, elongated and somewhat slender and spine-like ; this pair, instead of being situated on the ventral aspect, which is the usual position in the majority of Copepods, are placed well round towards the back, and are probably utilised for supporting between them the egg-sacs which combine to form a globular mass on the dorsal aspect of the genital segment of the abdomen.   This globular mass is easily detached, and would be even more so were it not supported and kept in position by the fifth feet (fig. 15).

The first segment of the abdomen is of moderate size, but the other segments are small ; the caudal furca are very short and somewhat divaricate.

Description of the male.—The male is about half the size of the female, being only ·825 mm. (about the $\frac{1}{30}$ of an inch) in length.   The cephalothorax is distinctly segmented, and of a somewhat oval outline, the cephalic segment is moderately large, being equal to about two-fifths the entire length of the cephalothorax, but the thoracic segments are small. The genital segment of the abdomen is somewhat dilated, and equal to about half the entire length of the remaining segments and the caudal furca combined, the posterior portion of the abdomen is composed of what appear to be seven or eight distinct articulations, but with the exception of the ultimate segment they are for the most part very small ; the caudal furca, which are scarcely equal in length to the last abdominal segment, are each provided with a small seta near the base of the outer margin in addition to a few short apical setæ (fig. 16).

The antennules are very short and four-jointed, the first joint is moderately large and somewhat inflated, numerous delicate hairs spring from its rounded upper surface, as shown in the drawing (fig. 24).   The remaining joints are small, the second one and the last are furnished with several delicate hairs similar to those on the first joint.

The thoracic feet.—The first four pairs of thoracic feet are all two-branched, and the outer branches are all three-jointed and armed on their exterior margins with strong sabre-like spines, the first and second joints being each furnished with one of these spines, and the third joints with three, except that the third joints of the first pair carry four spines ; in all the outer branches the terminal spines are considerably larger than the others.   Moreover, in the outer branches of all the four pairs, the inner margins of the first joints appear to be devoid of spines or setæ, but the second joints are each provided with one plumose seta on the inner margin ; on the other hand, the third joints of the outer branches of the first pair have only four plumose setæ on their inner margin, while on the inner margin of the third joints of the next three pairs there are five plumose setæ, so that though the outer branches of the first pair have one spine more than the other three pairs, the number of setæ is correspondingly fewer.

But although all the four pairs of thoracic feet are somewhat similar as regards the structure and armature of the outer branches, a considerable divergence is observable when the inner branches are compared.   In the first pair of feet the inner branches are composed of a single small but somewhat tumid joint bearing one or two claw-like spines (fig. 25).   In the second and third pairs the inner branches are three-jointed, but considerably shorter than the outer branches ; in each of the inner branches the first joint bears one and the second and third two moderately elongated plumose setæ, in addition to two small sabre-like terminal spines (fig. 26). In the fourth pair the inner branches are two-jointed, and not much more than half the length of the outer branches; a moderately long plumose seta springs from the inner margin of the first joint, and three similar setæ from the inner margin of the second joint; the inner branches are also provided with two small apical spines (fig. 27).

Remarks.—It is probable that the species described above may be new to science, as it does not agree very satisfactorily with any described species known to me; but in the meantime I prefer to place it under the *Botryllophilus ruber* of Hesse, with which it agrees in some of its more characteristic features.

HARPACTICIDÆ.

Longipedia minor, T. and A. Scott.

Specimens of this Copepod were obtained in a gathering collected at 72 fathoms in the Sound of Mull, on March 31st, 1900. There can be no reasonable doubt as to there being two distinct forms of *Longipedia,* a larger (*L. coronata,* Claus), and a smaller (*L. minor,* T. and A. Scott), for ova-bearing females of both are occasionally obtained. Whether the smaller form should be regarded as a species or a variety is a question that is comparatively unimportant; the two terms are in not a few instances convertible, as their use depends to a large extent merely on one's opinion as to the value that should be placed on certain observed differences either in structural details or otherwise.

Canuella perplexa, T. and A. Scott.

This fine species was obtained in tow-net gatherings collected in Dornoch Firth, near Muckle Ferry, on May 19th, 1900. This is a new station for *Canuella perplexa.*

Eucanuella, T. Scott (gen. nov.).

Eucanuella is, in some respects, not unlike *Longipedia,* Claus, but is somewhat intermediate in structure between that genus and *Canuella.* *Eucanuella* differs from both of the genera named in the structure of the antennæ, of the maxillæ and of the second maxillipedes, and to some extent in that of the first thoracic feet; it agrees with *Canuella* in the structure of the second, third, and fourth pairs of feet, and to some extent in that of the first pair, but differs in that of the fifth pair; it agrees with *Longipedia* in the structure of the third, fourth, and fifth pairs of thoracic feet, and to some extent in that of the first pair, but differs in that of the second pair. As there is but the one species of *Eucanuella,* a detailed description of the various points referred to will be found in the specific definition.

Eucanuella spinifera, T. Scott (sp. n.). (Pl. XVIII., figs. 1–10.)

Description of the female.—Body elongated, moderately stout, gradually tapering towards the extremity of the abdomen; length, 1·33 mm. (about $\frac{1}{19}$ of an inch). The forehead terminates in a broadly conical rostrum, the apex of which is rounded. The first cephalothoracic segment is about as long as broad, and equal to fully the entire length of the remaining segments of the thorax, which are all moderately short. The posterio-lateral angles of the fourth joint of the thorax and the first joint of the abdomen are produced backward into distinct tooth-like processes. The caudal furca are moderately stout and rather longer than the last abdominal segment (fig. 1).

The antennules are short and seven-jointed, and ¯also moderately setiferous ; the first four joints are moderately large and tumid, but the last three are very small ; the second joint is armed with a short conical tooth on its exterior aspect, and a stout and elongated plumose seta springs from the interior distal angle of the fourth joint (fig. 2). The formula shows the proportional lengths of the joints of the antennules :—

Numbers of the joints,	.	.	.	1	.	2	.	3	.	4	.	5	.	6	.	7
Proportional lengths of the joints,	.			13	.	7	.	16	.	9	.	4	.	4	.	9

The antennæ.—The primary branches of the antennæ are three-jointed, the joints, which are sub-equal in length, are moderately elongated, a secondary four-jointed branch is articulated to the distal end of the first and is nearly equal to it in length ; the last three joints of the secondary branches are small and their entire length scarcely exceeds that of the preceding joint. Both branches of the antennæ are provided with a number of plumose setæ of moderate length (fig. 3). In *Longipedia* and *Canuella* the antennæ differ from those just described, not only in the character of the basal joints and in the manner in which the secondary branches are hinged to them, but also in the secondary branches, in both being six- instead of four-jointed.

The mandibles are moderately stout, sub-cylindrical, and truncate at the apex, each mandible is armed with several coarse apical teeth, and provided with a well-developed two-branched palp ; the branches of the palp are supported on a slightly tumid basal joint ; the terminal branch appears to be uniarticulate, but the posterior branch is composed of about four joints, the three end ones of which are minute, both branches are moderately setiferous (fig. 4).

The maxillæ are each furnished with a distinct manducatory lobe armed with several moderately long and awl-shaped apical spines ; the palp is obscurely two-branched and turned toward the same side as the manducatory process, it is also furnished with a number of slender and moderately elongated setæ (fig. 5).

The first maxillipedes are robust and somewhat similar in structure to those of *Canuella* (fig. 6).

The second maxillipedes are smaller than the first pair, and are apparently four-jointed ; the first joint is larger than the other three combined, but the third and fourth are very small. These maxillipedes are armed with several stout and spiniform marginal setæ and two or three elongated spines, two spines at the end and on the inner aspect of the first joint have a fringe of minute hairs along the interior edge, while another at the end of the second joint is lancet-shaped, and has both margins of the distal half minutely serrated (fig. 7).

The thoracic feet.—The first four pairs of thoracic feet are each composed of two branches, and both branches are three-jointed. The outer branches of the first pair are considerably more elongated than the inner branches, and they are also more robust, they are armed with moderately elongated, stout, and spiniform marginal and terminal setæ ; a single and somewhat slender plumose seta also springs from the inner margin of each of the three joints of the outer branches, as shown in the figure (fig. 8). The inner branches are not much more than half the length of the outer ones, they are composed of three subequal and sparingly setiferous joints ; the inner distal angles of the first and second joints are continued into tooth-like projections. A strong spine with ciliated edges extends downwards from the interior distal angle of the second basal joint of the first pair. The second, third, and fourth pairs are very similar to the same appendages in *Canuella*, in each pair the length of the inner and outer branches is subequal and both are slender and elongated (fig. 9). The fifth pair are somewhat like those of *Longipedia*, in each foot the basal joint is almost rudimentary, and consists of a narrow plate which extends laterally outwards into a small cylindrical lobe, bearing at its apex an elongated and slender plumose seta; the secondary branch is elongate-narrow and sub-cylindrical, about four times longer than broad, and furnished with one long and plumose terminal seta and a sub-terminal and shorter one; a small seta also springs from near the middle of the outer margin. Interiorly the basal joint is not produced, nor does it appear to carry either spines or setæ (fig. 10).

Habitat.—Found in the stomach of a small Haddock captured about sixty-five miles south-eastward of Sumburgh Head, Shetland, September 4th, 1900. Obtained also in some bottom material collected with a tow-net about fifty miles south-eastward of the Fair Island, between Orkney and Shetland, October 19th, 1900. The tow-net had touched the bottom, and when hauled up was found to contain a quantity of sand, amongst which were several rare Crustaceans. The vessel on which these collections were made was the steam trawler "St. Andrew," of Aberdeen.

Remarks.—The species described above, while agreeing in some respects both with *Longipedia*, Claus, and *Canuella*, T. and A. Scott, presents too many points of difference to permit of its being ascribed to either of these two genera. I have already drawn attention to the fact that the secondary branches of the antennæ in *Eucanuella* are only four-jointed, whereas in *Canuella* and *Longipedia* the secondary branches are composed of six joints; and it has also been shown that the second maxillipedes in *Eucanuella* are more fully developed than they are in either of these two genera. It has also to be noted that whereas in *Eucanuella* in the branches of the second pair of thoracic feet the inner and outer branches are about equal in length, the inner branches of the same pair in *Longipedia* are remarkably elongated; and, further, in *Eucanuella* the fifth pair of feet have the secondary joint moderately developed and somewhat similar in form and armature to the fifth pair in *Longipedia*, but in *Canuella* the fifth pair are quite rudimentary. There are other points of difference, but these are quite sufficient to distinguish *Eucanuella* from either *Canuella* or *Longipedia*, to which it is no doubt closely allied.

Bradya typica, Boeck.

This species was obtained in material dredged in the Moray Firth, and examined during last year.

Bradya hirsuta, T. and A. Scott.

One or two specimens of this distinct species were obtained in the same gathering as the last; it is a moderately large species with strongly divergent caudal furca.

Bradya elegans, T. and A. Scott.

This species has not only been observed in the Moray Firth along with the others just referred to, but it has also been obtained in material dredged by the "Garland" in Loch Eil, at the head of Loch Linnhe, in 10 to 15 fathoms, on the 3rd of April last year.

R

Brayda similis, T. and A. Scott.

Bradya similis has been obtained during the past year in material dredged by the "Garland," both in the Moray Firth and in 45 fathoms off Aberdeen ; this species is considerably smaller than *Bradya hirsuta*, but the caudal furca are, as in that species, distinctly divergent.

Tachidius littoralis, Poppe (1881).   [=*T. crassicornis*, T. Scott (1892).]*

This small species, distinguished at first sight from *T. brevicornis* by its short, stout antennules, which terminate so abruptly that they appear as if the ends had been broken off, was found in the store-pond of the Sea-Fish Hatchery at Bay of Nigg.   It is a littoral species, and its presence in the store-pond seems to indicate that it may also occur in the Bay.   It was observed in the pond on November 29th, 1900.

Zosima typica, Boeck.

This species was obtained in some material dredged in Loch Eil, at the head of Loch Linnhe, on April 3rd, 1900.   It has also been obtained during the past year in a gathering of Crustacea from the Moray Firth forwarded from the "Garland."

Stenhelia hispida, G. S. Brady.

This fine species occurred in the same gathering from Loch Eil in which the *Zosima* was obtained ; although the species appears to have a fairly extensive distribution, and is occasionally obtained in moderate numbers, it seems nevertheless to be somewhat local.

Stenhelia intermedia, T. Scott.

This species, which was described in Part III. of the Fifteenth Annual Report from specimens obtained in Kilbrennan Sound, Firth of Clyde, has been obtained also in the Moray Firth, having occurred very sparingly in some dredged material sent from the "Garland."

Ameira longipes, Boeck.

This does not appear to be a very common form, and I have only occasionally met with it.   Its distribution, however, seems to be moderately extensive ; it was obtained in a gathering of dredged material collected at 72 fathoms in the Sound of Mull, March 31st, 1900, and forwarded from the "Garland" to the Laboratory at Bay of Nigg.

Laophonte intermedia, T. Scott.

A number of *Laophontes* have been observed during the past year. *L. intermedia* is a distinct and moderately rare species, and I have now to record its occurrence in the Bay of Nigg, near Aberdeen, where it was obtained on October 1st, 1900.

Laophonte denticornis, T. Scott.

This species, which is also moderately rare, was dredged from 72 fathoms in the Sound of Mull, March 31st, 1900.   The specimens from which the species was described were dredged off St. Monans, Firth of Forth, in 1893.

* *Tenth Ann. Rept. Fishery Board for Scotland*, Part III., p. 250, Pl. VIII., figs. 14-27 (1892).

Cletodes linearis, Claus.

This was contained in a gathering of Clyde Crustacea dredged between Sanda and Bennan Head on November 3rd, 1896, but only recently examined. The species seems to be rare in the British seas.

Cletodes monensis, I. C. Thompson.

This somewhat curious species was first made known to science by Mr. I. C. Thompson, of Liverpool, in 1893.* Some time afterwards, but also in the same year,† I recorded its occurrence in the Moray Firth, where, along with some other rare Crustaceans, it had been obtained by washing lumps of *Filograna implexa*, which had been brought up in the net of the shrimp-trawl from a depth of about 40 fathoms, about five to seven miles to the northward of Rosehearty. On the present occasion I have again to record *Cletodes monensis* from the Moray Firth, but this time at about thirteen or fourteen miles north-east of Buckie ; it occurred along with some other interesting Crustaceans in a gathering collected by tow-net which had touched bottom at 50 to 55 fathoms, November 3rd, 1900. (This gathering was from the steam trawler " St. Andrew.") I have also to record the same interesting species from the head of Loch Eil (at the upper end of of Loch Linnhe), where it was dredged by the "Garland" in 8 to 10 fathoms on April 3rd last year. This is the first time *Cletodes monensis* has been recorded from the West of Scotland.

Cletodes hirsutipes, T. Scott.

This species was described in Part III. of the Fifteenth Annual Report (1897), from Clyde specimens. I have now to note its occurrence in the Moray Firth in some dredged material collected off Nairn in 1898, but only recently examined. One peculiarity by which this species may be distinguished is the character of the fifth thoracic feet, the secondary joints of which are lamelliform, and of a narrow-oblong outline ; their outer edges are fringed with a dense border of short hairs, but this fringe is frequently so coated with mud that it becomes necessary to clean it ere it can be clearly made out.

Cletodes perplexa, T. Scott.

This is a small species, but of so marked a character as to be readily distinguished without dissection ; it was described in 1899 in Part III. of the Seventeenth Annual Report from specimens dredged by the Fishery steamer " Garland " at Smith Bank, Moray Firth. I have now to record the species from the head of Loch Eil ; it occurred in the same gathering in which *Cletodes monensis* was obtained. It was collected on April 3rd, 1900.

Cletodes irrasa, T. and A. Scott.

This appears to be a somewhat rare species in the Copepod fauna of Scotland, and I have obtained it in only one or two localities ; it was described from specimens obtained near the Bass Rock, Firth of Forth, in 1893. It has more recently been obtained in the Clyde, and I have now

* Revised Report on the Copepoda of Liverpool Bay (*Trans. L'pool. Biol. Soc.*, vol. vii., 1893), p. 26, Pl. XXXIV.
† *Ann. and Mag. Nat. Hist.*, (6), vol. xii., p. 243 (Oct. 1893). (*Cletodes monensis* was here by a slip recorded as *Laophonte monensis*, but the error was rectified in a paper published in the *Ann. and Mag. Nat. Hist.* in the following February.)

to add another station for it—viz., Loch Eil (off Loch Linnhe)—where it was dredged along with *Cletodes monensis* and *Cletodes perplexa* on April 3rd, 1900.

Pseudothalestris major (T. and A. Scott).

This somewhat rare species was obtained in the store-pond of the Sea-Fish Hatchery at Bay of Nigg in August, 1900.

Cervinia bradyi, Norman.

This interesting and well-marked species was obtained in the same gathering with *Cletodes monensis* referred to above, collected 13 or 14 miles north-east of Buckie, Moray Firth, in 50 to 55 fathoms, November 3rd, 1900. This is the second time that *Cervinia* has been taken in the Moray Firth ; it was taken the first time about five to seven miles off Rosehearty, in 1893, and at that time it was also associated with *Cletodes monensis*.

Zaus goodsiri, G. S. Brady.

This fine species was collected in 45 fathoms off Aberdeen, on November 7th last year, by the Fishery steamer "Garland." The gathering in which the *Zaus goodsiri* occurred contained a remarkable assemblage of rare and interesting Crustacea ; they comprised nearly fifty species, and included many genera, and nearly all the more important groups were represented—viz., the Copepoda, Amphiphoda, Isopoda, Sympoda, Schizopoda, and the Macrura.

Peltidium purpureum, Philippi.

This species was added to the British fauna in 1886 ; it was obtained at Tarbert, Loch Fyne, during some investigations carried on there under the direction of the Fishery Board for Scotland. Though *Peltidium purpureum* has since that time been observed in one or two other parts of Loch Fyne and the Clyde estuary, it does not appear to have been recorded from any other district of Scotland. In the present paper, however, I am able to add two new stations for this *Peltidium*, both of which are on the West Coast—viz., Loch Etive, off Abbot's Island, in a gathering dredged at 9 fathoms on March 30th, 1900, and in a gathering collected a few days later at from 8 to 10 fathoms near the head of Loch Eil (off Loch Linnhe). Both these gatherings were collected by the "Garland," and forwarded to the Laboratory at Bay of Nigg.

Alteutha purpurocincta, Norman.

This fine species was collected in Lerwick Harbour, Shetland, by the steam trawler "St. Andrew," October 14th, 1900. Rev. A. M. Norman collected this species at Hillswick, Shetland, in 1868.*

Idya cluthæ, T. Scott.

This species, which appears to be a deep-water form, was described in 1899 from specimens which had been obtained in Loch Fyne at from 50 to 70 fathoms, and in the Clyde at over 40 fathoms.† The same species of *Idya* was recently observed in a tow-net gathering from the Moray Firth

---

* *Report Brit. Assoc.*, 1868, p. 298.
† *Seventeenth Annual Report of the Fishery Board for Scotland* (1899), Part III., p. 260, Pl. XII., figs. 2–6

collected about thirteen to fourteen miles north-east of Buckie, in 50 to 55 fathoms, November 3rd, 1900. *Idya cluthæ* is quite distinct from any other species of *Idya*. The difference in the armature of the first thoracic feet and of the form of the fifth pair is marked, and are of themselves sufficient'for distinguishing the species.

Idya minor, T. and A. Scott.

A few specimens of this *Idya* were obtained in a gathering of Copepods collected in the store-pond of the Sea-Fish Hatchery at Bay of Nigg, in August, 1900. They occurred amongst crowds of *Idya furcata,* and were readily distinguished by their small size. The same species was obtained in a gathering collected by hand-net in the Bay of Nigg on October 1st.

Idya longicornis, T. and A. Scott.

This species was obtained in Lerwick Harbour, Shetland, in the same gathering with *Alteutha purpurocincta* already referred to. The occurrence of *Idya longicornis* here extends the distribution of the species considerably.

Synatiphilus luteus, Canu and Cuénot.

> 1892. *Synatiphilus luteus,* Canu and Cuénot, Commens. paras. Echinod., *Rev. Biol. du Nord de la France,* Oct. 1872, p. 19, Pl. T., figs. 6, 7.

> 1893. *Remigulus tridens,* T. and A. Scott, Ann. and Mag. Nat. Hist. (6), vol. xii., p. 242, Pl. XI., figs. 15–20 ; Pl. XII., figs. 1–3.

This species occurred in some dredged material collected near the head of Loch Eil (off Loch Linnhe), in 10 to 15 fathoms, April 3rd, 1900. *S. luteus* was added to the British fauna in 1893 as a new species under the name of *Remigulus tridens,* from specimens collected near the mouth of Loch Spelve, Island of Mull, in 1892 (but not recorded till the following year). As the species, however, had already been described by Canu and Cuénot, as indicated above, the name *Remigulus tridens* necessarily becomes a synonym. So far as I know, *Synatiphilus luteus* has not been obtained anywhere else in Scotland than in the district of Loch Linnhe. It seems to be a rare species.

CORYCÆIDÆ.

Corycæus anglicus, Lubbock. (Pl. XVIII., fig. 11.)

In my paper on tow-net gatherings published in the Eighteenth Annual Report, Part III., I recorded *Corycæus anglicus* from the Firth of Clyde, and submitted a description of the species, which was illustrated by a number of figures ; these figures were prepared from a female specimen, as no males had been observed. One of the gatherings which were collected during the investigations carried out on board the steam trawler "St. Andrew" was obtained while the vessel was proceeding along the south-east coast of Shetland between Lerwick and Sumburgh Head, and about seven miles off shore; the gathering was collected by passing the water from the donkey-pump through a fine tow-net. In this gathering, collected on October 16th, three specimens of a *Corycæus* were found, which I at first thought might be *C. obtusus,* Dana, as the genital segment of the abdomen possessed a small but distinct ventral hook at its proximal end, similar to the hook observed in the male of that species (fig. 11); a further examination of the specimens showed, however, that they were the males of *C. anglicus*, Lubbock (fig. 11). *Corycæus anglicus* has also been obtained in a gathering from the Moray Firth, collected thirteen to fourteen miles north-east from Buckie, in 50 to 55 fathoms, on November 3rd, and in a gathering collected off Aberdeen in 45 fathoms on the 7th of the same month, but the specimens were females.

LICHOMOLGIDÆ.

Lichomolgus fucicolus (G. S. Brady).

One or two specimens of *L. fucicolus* occurred in a gathering collected in 72 fathoms in the Sound of Mull, March 31st, 1900.

ASTEROCHERIDÆ.

Collocheres gracilicauda (G. S. Brady).

This species was obtained in some dredged material collected near the head of Loch Eil in 10 to 15 fathoms, April 3rd, 1900. *C. gracilicauda* is a moderately rare species, but it has apparently an extensive distribution. It has been recorded off the Yorkshire coast by Dr. Brady, from the Liverpool Bay district by Mr. I. C. Thompson, and also from the Firth of Forth; Dr. Canu has obtained it in the neighbourhood of Boulogne, and Dr. Giesbrecht in the Bay of Naples.

INCERTA SEDIS.

Cancerina, T. Scott (gen. nov.).

Description of the genus.—Antennules short, six-jointed. Antennæ small, and simple in structure, not fitted for grasping. Mandibles narrow and of moderate length, toothed on the distal half of the inner margins, and somewhat like the mandibles of *Lerneopoda*. First maxillipedes somewhat rudimentary, unprovided with terminal spines or claws. Second maxillipedes large and strong, armed with stout but very short terminal claws. Thoracic feet, two pairs, they both are somewhat alike, and each foot consists of a single biarticulated branch. As there is but the one species a detailed description will be found in the specific definition.

Cancerina confusa, T. Scott (sp. n.).   (Pl. XVIII., figs. 12–20.)

Description of the female.—The length of the specimen represented by the figure is fully 1 mm., from the forehead to the end of the caudal furca. The cephalothorax, seen from above, is sub-rhomboidal in outline. The cephalic segment is small, but moderately distinct; it is very short, being scarcely half as long as it is broad. The thorax, which appears to be unsegmented, is widest at about one-fourth of its length from the anterior end; from the widest part, the thorax, on each side, tapers towards both ends, the front slope is short and terminates at the cephalic segment, the posterior slope, which is longer and slightly sinuate, extends to the genital segment of the very short abdomen; this segment is moderately broad, rather more so than the posterior part of the thorax to which it is adjacent; the remaining segment of the abdomen is very small, and the caudal furca, which are short, are placed somewhat widely apart. The ovisacs are very large, and the point at which they are attached to each side of the genital segment is nearer their posterior than their anterior ends; they thus occupy a rather peculiar position in relation to the body of the Copepod; the ovisacs are of an elongate-oval form, and from the peculiar manner in which they are attached to the genital segment, they extend forward along each side of the body of the Copepod instead of backwards, as in the majority of Copepods which carry two ovisacs. It was the peculiar position of the ovisacs that attracted my attention to this specimen, (fig. 12).

The antennules are short and six-jointed, the second joint is considerably longer than any of the others; the formula shows the proportional lengths of all the joints :—

Numbers of the joints,	1	.	2	.	3	.	4	.	5	.	6
Proportional lengths of the joints	14	.	24	.	17	.	12	.	13	.	10

The joints are sparingly setiferous, and taper very gradually from the proximal to the distal end (fig. 14).

The antennæ are very small, they each consist of a single two-jointed branch, armed with about six terminal setæ; the three middle setæ are moderately elongated, but the others are short; the joints of the antennæ are sub-equal and nearly three times longer than broad (fig. 15).

The mandibles have a remarkable resemblance to those of the Lernæopodidæ; they are moderately elongated, narrow, and somewhat cylindrical; they taper slightly from about the middle of the distal extremity, the inner margin is denticulate from about the middle to the apex of the mandible (fig. 16).

The maxillæ are very rudimentary; they are somewhat papilliform, and furnished with about four moderately stout spines (fig. 17).

The first maxillipedes are moderately stout, but of a somewhat rudimentary structure; they appear to be each composed of two joints, the first is comparatively tumid, the second is smaller and terminates in a boldly rounded apex finely serrated on the margin (fig. 18).

The second maxillipedes are large and strong; they are each composed of two joints, the basal joint is considerably larger than the other, the second is armed with a short but stout claw, the margin of which is convex and fits into a corresponding hollow in the joint to which it is articulated (fig. 19).

Thoracic feet.—There appear to be only two pairs of thoracic feet, which are similar to each other in size and structure. They appear to be each composed of a single two-jointed branch of moderate length, furnished with two somewhat elongated and two or three very small setæ (fig. 20). The position of the thoracic feet is somewhat abnormal; one pair is almost in line with, and outside of the second maxillipedes, the other pair occupy a somewhat intermediate position and somewhat further forward, as shown in figure 13.

Habitat.—Dredged at Tarbert Bank, Lower Loch Fyne, in 17 to 20 fathoms, October 28th, 1899. One specimen only—a female—was observed.

Remarks.—This somewhat curious Copepod, which I have now tried to describe, is in some respects not very unlike *Cancerilla tubulata,* Dalyell. It has a somewhat similar squat form, large ovisacs, and short six-jointed antennules; while, as regards the form of the mandibles and maxillæ, and especially of the latter, the disparity is not very great; it is only in the structure of the antennæ, the maxillipedes, and the thoracic feet that the most marked differences are observed. The antennæ are simple in structure and do not form grasping organs as in *Cancerilla;* the maxillipedes are more rudimentary ; and the thoracic feet, of which there appear to be only two pairs, are each composed of two moderately stout basal joints and a single biarticulated branch ; in *Cancerilla* the first and second pairs of feet, though two-branched, are more rudimentary than those of *Cancerina*, and the third and fifth pairs, though present, are also rudimentary.

In consequence of these differences I have considered it necessary to institute a new genus for this Copepod, and, from its general resemblance to the *Cancerilla* of Dalyell, have given to it the name of *Cancerina.*

NICOTHOIDÆ.

Nicothoë astaci, Aud. and M.-Edw.   (Pl. XVII., figs. 35–39 ; Pl. XVIII., figs. 21–26.)

> 1826. *Nicothoë astaci*, Aud. and M.-Edw., Ann. Sci. Nat., (1), vol. ix., p. 345, Pl. 49, figs. 1–9.
> 1850. *Nicothoë astaci*, Baird, Entom., p. 307, Pl. XXXIII., fig. 11.

Description of the female.—The length of the specimen represented by the drawing is 1·5 mm. (about $\frac{1}{17}$ of an inch), exclusive of the caudal setæ. The body is cyclopoid in its general outline, but the two great wing-like lateral expansions of the posterior part of the thorax destroy the symmetrical appearance of the animal. These expansions appear to be the result of an extraordinary development of the fourth segment of the thorax. The three segments intermediate between this one and the cephalic segment are very narrow ; they are represented on the dorsal aspect by more or less distinct articulations as shown in the drawing, while in front of the cephalic segment two distinct eyes can be observed. The abdomen is composed of four segments, the genital segment is somewhat dilated and rather longer than the combined lengths of the next two segments. The caudal furca are very short, and each is furnished with an elongated and moderately stout terminal seta and a few minute hairs. The two ovisacs are very large. (Pl. XVII., fig. 39, and Pl. XVIII., fig. 21.)

The antennules are of moderate length and sparingly setiferous, they are composed of eleven joints, but, with the exception of the second and last, all the joints are small (Pl. XVIII., fig. 35). The formula shows approximately the proportional lengths of all the joints :—

Numbers of the joints,	1 . 2 . 3 . 4 . 5 . 6 . 7 . 8 . 9 . 10 . 11
Proportional lengths of the joints,	11 . 33 . 13 . 10 . 10 . 10 . 10 . 10 . 10 . 10 . 20

The antennæ are small but moderately stout, and appear to be composed of four joints, the basal joint is considerably larger than the others ; each antenna is armed with two moderately strong apical spines, and as the end-joint to which they are attached is hinged at nearly a right angle to the preceding joint, the antennæ become, with the assistance of the apical spines, effective grasping organs ; the third joint is provided with a curved and somewhat lamelliform plate, which may be a sensory appendage (Pl. XVIII., fig. 36).

The mandibles are long and slender, slightly incurved, and tapering to a point (Pl. XVII., fig. 22).

The maxillæ are each composed of a moderately stout basal joint, and a small and apparently two-jointed secondary branch, which is articulated to the inner edge of the basal joint; the basal joint is furnished with three strong terminal setæ of moderate length, plumose on the distal half; the secondary branch is armed with a few small terminal spines (Pl. XVII., fig. 37).

The first maxillipedes are small and apparently composed of two joints, of which the end-joints are the smallest, each maxillipede is armed with two moderately strong terminal claw-like spines, and as they are slightly curved inwards, they form, with the end-joint, a moderately powerful hook (Pl. XVII., fig. 38).

The second maxillipedes, which are larger than the first pair, are each composed of five joints, the first and second joints are of moderate size but the three end ones are small, the terminal claw-like spine is longer than the entire length of the three end joints (Pl. XVIII., fig. 23).

The thoracic feet.—The first four pairs of thoracic feet are all somewhat similar, they are each composed of a stout two-jointed basal part, which bears two sub-equal three-jointed branches at its distal end; the outer branches are armed on the exterior margin with several small, slender spines, while the inner margins of both branches are furnished with plumose setæ (Pl. XVIII., figs. 24 and 25). The fifth pair consist each of a single uniarticulated lamelliform branch, about three times longer than broad; each branch is provided with six small plumose setæ—one near the middle of the inner margin and five arranged round the apex (Pl. XVIII., fig. 26).

The ovisacs are very large and contain numerous ova. No males have yet been observed.

Habitat.—On the gills of a scarcely full-grown Lobster, captured by Mr. H. C. Williamson in Bay of Nigg, near Aberdeen, June 30th, 1900.

Remarks.—Though *Nicothoë astaci* appears to be moderately frequent on the gills of Lobsters captured on various parts of the English coast, I do not know of any previous authentic record of its occurrence in Scottish waters; Edward of Banff, who was so successful a collector of Crustacea, though he records the common Lobster in his list of Moray Firth species, does not appear to have observed the Copepod parasite which is so intimately associated with that Crustacean. *Nicothoë astaci* is no doubt frequently overlooked, and if a careful examination were made of the lobsters captured on our shores the distribution of the parasite in the Scottish seas might be found to be co-extensive with its host.

The *Nicothoë* seems to be a remarkably sluggish animal. Milne-Edwards —quoted by Dr. Baird[*]—states that "they allowed themselves to be torn to pieces without making the least movement or quitting their hold," and further, that though "taken carefully off the gills of the lobster with all possible precautions not to injure the animals, and placed in a glass of sea-water, though watched for several hours, and though they lived during that period, as might be seen by the peristaltic movement of the intestine, they made no attempt themselves at locomotion." I may add that my son has kept specimens of *Nicothoë* alive for five weeks in ordinary sea water, which was changed about once a week, and though they were carefully watched during that time he never saw them make any attempt to move about, and the only indication that they were alive was the persistence of their semi-transparent pinkish colour and the peristaltic movement of the alimentary canal. The parasite is usually brightly coloured, and, as it is of moderate size, is readily observed *in situ* when that part of the lobster's carapace which covers the gills is removed.

[*] British Entomostraca, p. 304.

CHONDRACANTHIDÆ.

Lamippe proteus, Claparède.

This curious little species is common on the *Alcyonium digitatum,* or "Dead-man's Fingers," so frequently brought up in the trawl-net or on fishermen's lines. A number of specimens were recently obtained on pieces of *Alcyonium* trawled by the "Garland" off Aberdeen. *Lamippe proteus* was first recorded from the Scottish seas in 1895 from specimens observed in the Firth of Forth.

Lamippe forbesi, T. Scott.

1896. *Lamippe* sp., T. Scott, Fourteenth Ann. Rept. of the Fishery Board for Scotland, Part III., p. 164, Pl. IV. figs. 9–12.

This distinct form was first observed in the Firth of Forth in the early part of 1896, and the same species was almost immediately afterwards found by my son in Liverpool Bay. At that time it was thought it might be a species which was already described, and it was therefore recorded as "*Lamippe* sp." *L. forbesi* seems to be a rare species, for, although the portions of *Alcyonium* recently examined yielded between two and three hundred specimens of *L. proteus,* only two specimens of *L. forbesi* were observed after a careful search. *L. forbesi* is a distinctly larger species than *L. proteus.*

Change of Name.

Genus *Eurynotopsyllus,* nov. nom. [= *Eurynotus,* T. and A. Scott (1898) preoccupied.]

In 1898 a new Copepod from the Clyde was described and figured by T. and A. Scott in the "Annals and Magazine of Natural History," s. 7, vol. i., p. 188, under the name of *Eurynotus insolens;* it was also recorded under the same name in Part III. of the Sixteenth Annual Report of the Fishery Board for Scotland (1898), p. 279. I have now ascertained that *Eurynotus* has already been twice used—viz., by Kirby in 1817 for a genus of Coleoptera, and by Agassiz in 1835 for a genus of fishes; as, therefore, the name cannot be retained for this Copepod I propose to change it for the altered form *Eurynotopsyllus* (*Eurynotus* plus *psyllus,* a flea).

Order 2, OSTRACODA.

BAIRDIIDÆ

Genus *Bairdia,* M'Coy (1849).

Bairdia inflata (Norman).

1862. *Cythere inflata,* Norman, Ann. and Mag. Nat. Hist., vol. xi., p. 49, Pl. iii., figs. 6-8.

1896. *Bairdia inflata,* Brady and Norman, "Mon. M. and F.-W. Ostrac. of the N. Atlantic and N.-W. Europe" (*Sci. Trans. Roy. Dublin Soc.,* vol. iv., s. ii., p. 112.

One whole specimen and a valve of this species occurred amongst some sand dredged in about 80 fathoms to the east of Fair Island in October last.

Genus *Macrocypris*, G. S. Brady (1868).

Macrocypris minna (Baird).

> 1850. *Cythere minna*, Baird, Brit. Entom., p. 171, t. xx., fig. 4, 4 a–d.
>
> 1868. *Macrocypris minna*, Brady, Mon. rec. Brit. Ostrac., p. 392, t. xxvii., figs. 5–8; t. xxxviii., fig. 4.
>
> 1889. *Macrocypris minna*, Brady and Norman, *op. cit.*, vol. iv., s. ii., p. 117.

This fine species was frequent in a tow-net gathering collected on the 19th October, 1900, in 80 to 85 fathoms, about fifty miles S.S.E. of Fair Island, between Orkney and Shetland. The tow-net had touched bottom, and when hauled up was found to contain a considerable quantity of mud, this was washed through the net and the material that remained in the net was preserved; mixed up in this material was a considerable number of *Macrocypris*. In the Monograph of the Marine and Fresh-water Ostracoda of the North Atlantic and North-Western Europe, the authors, referring to *Macrocypris minna*, say (*op. cit.*, p. 117) that "the only British locality for this species is Shetland, where a single specimen was dredged by MacAndrew forty years ago, and a second by A. M. N. on the outer Haaf in 1861."

CYTHERELLIDÆ.

Genus *Cytherella*, Jones (1849).

Cytherella abyssorum, G. O. Sars.

> 1865. *Cytherella abyssorum*, G. O. Sars, Overs. of Norg. marine Ostrac., p. 127.
>
> 1865. *Cytherella beyrichi*, G. S. Brady, "On New or Imperfectly Known Ostrac.," *Trans. Zool. Soc.*, vol. v., p. 362, Pl. lvii., figs. 3 a, b.
>
> 1866. *Cytherella scotica*, G. S. Brady, Brit. Assoc. Rept. (1866), p. 211.
>
> 1896. *Cytherella abyssorum*, Brady and Norman, *op. cit.*, vol. v., s. ii., p. 716, Pl. lxvi., figs. 1, 2, 5; Pl. lxvii., figs. 13, 14.

A single perfect specimen and a valve of this species were obtained in the same material with the last; *Cytherella abyssorum* seems to be a rare species in our seas, as the only Scottish record for it hitherto appears to be that of G. S. Brady, who, in his "Monograph of Recent British Ostracoda," states that two or three specimens were obtained by himself and the Rev. A. M. Norman amongst sand dredged by Mr. Jeffreys in 60 fathoms in the Minch.*

It may be mentioned in passing that the Foraminifer *Saccamina sphæra* was very common in this gathering; a few other Foraminifera, such as *Astrorhiza arenaria, Psammosphæra fusca,* and *Placopsilina bulla,* were also observed.

---

* *Trans. Linn. Soc.*, vol. xxvi., p. 473, Pl. xxxiv., figs. 18–21.

Sub-Class MALACOSTRACA.

Sub-Order AMPHIPODA.

LYSIANASSIDÆ.

Normanion quadrimanus (Bate and Westwood).

In the Seventeenth Annual Report, Part III., I recorded this rare species from near Davaar Island, Firth of Clyde. I have now to add another Clyde station for the species—viz., Tarbert Bank, Lower Loch Fyne, in 17 to 20 fathoms; it occurred in some material dredged on the Bank on December 22nd, 1899, but not examined till some time afterwards. Besides the two stations mentioned, *Normanion* has also been obtained off Cumbrae in 20 to 25 fathoms by the Rev. A. M. Norman.*

Acidostoma obesum (Bate).

I have to record this species from the Moray Firth, in a gathering collected in 1898 but only examined last year. Also in a gathering of Crustacea collected off Fair Island in October; and in another from 45 fathoms off Aberdeen, collected in November, 1900.

Perrierella audouiniana (Bate).

Dredged in 72 fathoms, Sound of Mull, March 31st, 1900; and in Loch Etive, off Abbot's Island, in 9 fathoms, on the 30th of the same month (Fishery steamer "Garland").

Orchomene humilis (A. Costa).

Orchomene humilis was obtained in Loch Etive in the same gathering with *Perrierella audouiniana*. This is the *Anonyx edwardsii* of Bate and Westwood's "Sessile-eyed Crustacea" and the *Orchomene batei* of G. O. Sars' "Crustacea of Norway."†

Tryphosella horingii (Boeck).

This species occurred in a tow-net gathering collected about fifty miles south-eastward of the Fair Island, in October, 1900, and in moderately deep water.

Anonyx nugax (Phipps).

Quite recently several specimens of *Anonyx nugax* (Phipps) were forwarded to the Laboratory at Bay of Nigg, from Mr. F. G. Pearcey, the Naturalist on board the Fishery steamer "Garland." They were collected in the Cromarty Firth in 7½ fathoms on January 10th, 1901. This is only the second time I have seen *Anonyx nugax* in Scottish waters, and the first time in the Moray Firth district. The previous occurrence of the species was in 1889, when it was obtained near May Island in the Firth of Forth in the month of February, but was not recorded till 1893.‡ It may be of interest to note that on the present occasion the specimens have been obtained a month or so earlier than they were observed in 1889.

Ann. and Mag. Nat. Hist., (7), vol. v., p. 141 (1900).
† Norman, *op. cit.*, p. 202.
‡ *Eleventh Ann. Rept. of the Fishery Board for Scotland*, Part III., p. 212, Pl. V., figs. 18–21 (1893).

Lepidepecreum longicorne (Bate).

This species was dredged by the "Garland" in 45 fathoms off Aberdeen on November 7th, 1900.

Euonyx chelatus, Norman.

This somewhat rare species was obtained in some material dredged by the "Garland" in 32½ fathoms on October 12th, 1900 ; and in a bottom tow-net gathering from the steam trawler "St. Andrew," collected about fifty miles south-eastward of the Fair Island, in 70 to 80 fathoms, on the 19th of the same month. Only one or two specimens were obtained in either gathering.

<h2 align="center">PONTOPORIIDÆ.</h2>

Bathyporeia norvegica, G. O. Sars.

Obtained in a bottom tow-net gathering collected about fifty miles south-eastward of Fair Island, October 16th, 1900.

Argissa hamatipes, Norman.

Specimens of *A. hamatipes* occurred in gatherings collected off Aberdeen in June, and off Fair Island, collected October 12th, 1900.

<h2 align="center">PHOXOCEPHALIDÆ.</h2>

Phoxocephalus fultoni, T. Scott.

A few specimens of this species were obtained in material dredged in Loch Etive off Abbot's Island, in 9 fathoms, March 30th, 1900. The same species has been dredged in Loch Ryan, near the mouth of the Clyde, as well as in the Clyde itself.

Phoxocephalus oculatus. G. O. Sars.

This form occurred with several other Crustaceans in some dredged material collected in the Sound of Mull, in 72 fathoms, March 31st, 1900. So far as the male of *Phoxocephalus oculatus* is concerned, there is little danger of confounding it with *Phoxocephalus holbölli* (Kröyer). The large and conspicuous black eyes are sufficient of themselves to attract the attention of the observer; it was collected by the Fishery steamer "Garland."

<h2 align="center">AMPELISCIDÆ.</h2>

Ampelisca spinipes, Boeck.

This fine species was obtained off Aberdeen in a gathering from 45 fathoms collected November 7th, 1900 (Fishery steamer "Garland.")

Ampelisca macrocephala, Lilljeborg.

This species occurred in one or two of the gatherings collected off Fair Island on the 12th and 19th October, 1900.

Ampelisca brevicornis (A. Costa).

According to Norman, the *A. lævigata* of Lilljeborg is identical with *A. brevicornis* (A. Costa), and, as this name is prior to the other, *A lævigata* must be set aside. The species known as *Ampelisca lævigata*, Lilljeborg, was dredged during the past year in Loch Etive, near Abbot's Island, in 9 fathoms, on March 30th.

Byllis gaimardi (Kröyer).

A single specimen of this rare species was obtained in a bottom tow-net gathering collected about fifty miles S.S.E. of Fair Island at a depth of about 80 fathoms on October 12th, 1901. There appear to be extremely few British records of *Byllis gaimardi*. Norman gives the following three:—Off Seaham, Co. Durham; off St. Abbs Head, 40 fathoms; and near May Island, Firth of Forth (1890).* The latter seems to be the most recent British record of the species previous to the present one.

AMPHILOCHIDÆ.

Amphilochoides odontonyx, Boeck.

This species was obtained in a tow-net gathering collected about sixty-five miles east by south off Sumburgh Head, Shetland, September 4th, and about fifty miles south-south-east of Fair Island, October 16th, 1900. This is the form described by G. O. Sars under the name of *Amphilochoides pusillus*.

Amphilochoides intermedius, T. Scott.

One or two specimens of *Amphilochoides intermedius* occurred in some material dredged off Aberdeen in forty-five fathoms, November 11th, 1900. This species was described in 1896 in the Fourteenth Annual Report of the Fishery Board for Scotland, Part III., p. 159, Pl. IV., figs. 1–3.

Stegoplax brevirostris (T. and A. Scott).

Several specimens of this small species were obtained in the same gathering in which *Amphilochoides intermedius* occurred, collected in 45 fathoms off Aberdeen. This species was ascribed by T. and A. Scott to the genus *Cyproidia*,† but the Rev. A. M. Norman refers it to *Stegoplax* because the upper antennæ have no secondary appendage.‡

METOPIDÆ, Stebbing (1899).

Metopa rubrovittata, G. O. Sars.

This small, brightly red-coloured species occurred in a gathering collected sixty-five miles east by south of Sumburgh Head, Shetland, and in another collected off Fair Island; the first was collected in September and the other in October, 1900.

Metopa borealis, G. O. Sars.

This species was obtained in a gathering collected in Aberdeen Bay on June 28th, 1900.

Metopa norvegica (Lilljeborg).

This is the *Metopa pollexiana* of Bate and Westwood; several specimens of this fine species were obtained in a few of the gatherings collected off

---

* British Amphipoda ; *Ann. and Mag. Nat. Hist.*, (7) vol. v., p. 344 (1900).
† *Ann. and Mag. Nat. Hist.*, (6), vol. xii., p. 244, Pl. XIII., figs. 1-11 (1893).
‡ British Amphipoda ; *Ann. and Mag. Nat. Hist.*, (7), vol. vi., p. 38 (1900).

Aberdeen, in October and November last year. The species occurred in a gathering from 36 to 49 fathoms collected on October the 23rd, in another from 60 fathoms collected on the 25th of the same month, and in a third gathering from 45 fathoms collected on November 7th. These gatherings were collected by the Fishery steamer "Garland."

Metopella nasuta (Boeck).

This small species was moderately common in a gathering collected off Fair Island, in 80 to 85 fathoms, on October 12th, 1900. It was also obtained in some bottom tow-net material, collected about sixty-five miles to the south-east of Sumburgh Head, on September 4th, by the steam trawler "St. Andrew."

Metopina robusta (G. O. Sars).

This somewhat rare species occurred in one of the gatherings from the Fishery steamer "Garland," collected off Aberdeen, in 45 fathoms, November 7th, 1900. The Rev. A. M. Norman has instituted the new genus—*Metopina*—for this form,* and one of its principal characters is the peculiar structure of the first gnathopods. There appear to be only three British records for *Metopina robusta* previous to this one, viz. :— The Firth of Forth, where it was obtained by Dr. Henderson in 1884, and the Moray Firth, where it was dredged by the "Garland" in 1893,† and again in 1899.‡ The gathering in which it occurred in 1893 was collected a few miles to the north of Rosehearty, but the gathering collected in 1899 was from Station XV., in the vicinity of Smith Bank.

LEUCOTHOIDÆ.

Leucothoë lilljeborgii, Boeck.

This was obtained in a tow-net gathering collected about fifty miles south-south-east of Fair Island, October 16th, 1900. There does not appear to be any previous record of *L. lilljeborgii* from this part of the North Sea.

ŒDICERIDÆ.

Aceros phyllonyx (M. Sars).

Although a number of species belonging to the Œdiceridæ have been observed in the various tow-net gatherings examined, and notes taken of them, they are rather more common than the one specially referred to. *Aceros phyllonyx* appears to be comparatively rare as a British species ; the only British record known to me is that of Metzger of the German North Sea Expedition, who obtained it "sixty miles north of Peterhead." During the Fishery Board's recent investigations eastward of Fair Island, *Aceros* was obtained on two occasions in bottom tow-net gatherings, collected on the 12th October about fifty miles south-south-east of Fair Island, in 80 to 85 fathoms. A number of specimens were obtained, but all were more or less damaged ; they were, however, sufficiently perfect for identification.

---

* British Amphipoda ; *Ann. and Mag. Nat. Hist.*, (7), vol. vi., p. 45 (1900). It has to be noted, however, that *Metopina* has already been used by Macquart for a genus of insects.

† *Ann. and Mag. Nat. Hist.*, (6), vol. xiii., p. 148 (1894).

‡ *Seventeenth Ann. Rept. of the Fishery Board for Scotland*, Part III., p. 265.

PLEUSTIDÆ.

Paramphithoë monocuspis, G. O. Sars.

This species and the closely-allied form *Paramphithoë bicuspis* were obtained in gatherings collected off Fair Island in October, and in one collected off Aberdeen in 45 fathoms, November 7th, 1900. In the gathering collected off Aberdeen *P. monocuspis* was much more frequent than the other.

Sympleustes latipes (M. Sars). (= *Amphithoë latipes*, M. Sars.)

A few specimens of this species were obtained in a bottom gathering of Crustacea collected by the "Garland" off Aberdeen in 60 fathoms, October 25th, 1900.

EPIMERIDÆ.

Epimeria cornigera (Fabricius).

This fine species was moderately frequent in some material collected by the "Garland" off Aberdeen in $32\frac{1}{2}$ fathoms, October 12th, 1900; one or two of the specimens in this gathering were of a nearly white colour. *Epimeria cornigera* was also moderately frequent in some of the gatherings collected eastward of Fair Island in the same month.

EUSIRIDÆ.

Eusirus longipes, Boeck.

This was dredged by the "Garland" in the Sound of Mull in 72 fathoms, on March 31st, 1900; it also occurred in some material collected by the "Garland" off Aberdeen in $32\frac{1}{2}$ fathoms on October the 12th, and again on October 23rd in 36 to 49 fathoms; in this last gathering the species was moderately frequent.

CALLIOPIIDÆ.

Apherusa borealis (Boeck).

This was moderately frequent in material collected in Aberdeen Bay on June 28th. Another species—*Apherusa bispinosa* (Bate)—was obtained in a gathering collected off Aberdeen in 60 fathoms, on October 25th; while a third species—*Apherusa furinii* (M.-Edw.)—was obtained in a gathering of Crustacea collected in Bay of Nigg with a hand tow-net and rowing-boat by Mr. Williamson, July 27th (1900).

ATYLIDÆ.

Paratylus uncinatus, G. O. Sars.

This is the species which has no projections on the dorsum, but which otherwise somewhat resembles *Paratylus falcatus* (Metzger). *P. uncinatus* occurred in some material collected in Aberdeen Bay on June 28th, and also in the same gathering from Bay of Nigg which Mr. Williamson collected on July 27th, 1900. The species was very sparingly represented.

DEXAMINIDÆ.

Dexamine thea, Boeck.

This comparatively small species was captured in Bay of Nigg, near Aberdeen, and also near Muckle Ferry, Dornoch Firth. The gathering from Muckle Ferry was collected in May, and that from Bay of Nigg in July, 1900.

Tritæta gibbosa (Bate).

This curious little species was dredged by the "Garland" in Loch Etive, near Abbot's Island, in 9 fathoms, on March 30th, 1900.

Guernia coalita (Norman).

This curious little species was obtained in a gathering collected by the " St. Andrew " about fifty miles south-south-east of Fair Island, in about 85 fathoms, on October 16th, 1900. The species was described by the Rev. A. M. Norman in 1868 from specimens collected in the surface-net at Lerwick, Shetland, by the late Dr. Robertson of Millport.[*] The same species has been obtained at considerable depths in the Firth of Forth,[†] in the Firth of Clyde,[‡] and in the Irish Sea,[**] and it is interesting to note that, though its distribution is thus apparently extensive, it is one of the few British species of Amphipoda that have not yet been recorded from Norway.

MELPHIDIPPIDÆ.

Melphidippella macera (Norman).

Melphidippa and *Melphidippella* have been removed by Stebbing from the *Gammaridæ* into a new family, *Melphidippidæ.* The brick-red colour of the *Melphidippella* makes it a fairly conspicuous object even in a crowded gathering. The species was found in the gatherings collected eastward of Fair Island, and also off Aberdeen in October, 1900, but although of frequent occurrence in the gatherings from both places, scarcely a single specimen was undamaged, owing to the appendages being so brittle.

GAMMARIDÆ.

Amathilla homari (Fabricius).

A moderately large female specimen was found in the stomach of a Cod captured in the Bay of Nigg in June, 1900 ; young specimens were also taken in the Bay in July by means of a tow-net worked from a rowing-boat.

Gammarus duebeni, Lilljeborg.

This *Gammarus* is moderately frequent in brackish-water pools at the mouth of the River Don near Aberdeen.

---

[*] *Ann. and Mag. Nat. Hist.*, (4), vol. ii., p. 418 (1868) ; *Brit. Assoc. Rept.* for 1868 (pub. 1869), p. 280.
[†] *Eighth Annual Report Fishery Board for Scotland,* Part III., p. 326.
[‡] *Seventeenth Ann. Report Fishery Board for Scotland,* Part III., p. 265 (1899).
[**] A. O. Walker, Revision of the Amphipoda of the L.M.B.C. District (*Trans. L'pool. Biol. Soc.*, vol. ix. (1895), p. 307.

s

Mæra othonis (M.-Edw.).

This species was obtained in a gathering of material dredged by the " Garland " in 72 fathoms in the Sound of Mull, March 31st, and in another gathering collected off Aberdeen in 45 fathoms, November 7th 1900.

Mæra lovéni (Bruzelius).   (Pl. XVIII., figs 27 and 28.)

A damaged specimen of this species was obtained in a gathering collected about fifty miles south-east of Fair Island in about 60 to 65 fathoms, October 19th, 1900.   Figures 27 and 28 represent the first and second gnathopods of this specimen.

Cheirocrates assimilis (Lillejeborg).

One or two specimens of this species were dredged in 72 fathoms in the Sound of Mull on March 31st, and in 40 to 49 fathoms, off Aberdeen, on October 23rd, 1900.

LILLJEBORGIIDÆ.*

Lilljeborgia kinahani (Bate).

One or two specimens of this species occurred in the same gathering with the *Cheirocrates* dredged in the Sound of Mull on March 31st, 1900.

PHOTIDÆ.

Photis longicaudata (Bate).

This species was collected about fifty miles south-south-east of Fair Island in October and off Aberdeen in 45 fathoms in November, 1900.

ISCHYROCERIDÆ.†

Ischyrocerus anguipes, Kröyer.   (Pl. XVIII., fig. 29.)

This species was obtained in a tow-net gathering collected in the Bay of Nigg, Aberdeen, on July 17th, by Mr. Williamson, while carrying out some investigations in the Bay by means of a rowing-boat.   There does not appear to be any previous authentic records of *Ischyrocerus anguipes* for Britain.   Figure 29, Pl. XVIII., represents one of the second gnathopods of our specimen, which appears to be an adult male.   This specimen though somewhat smaller than those from higher latitudes, is in other respects quite characteristic ; the Rev. A. M. Norman has examined the specimen and considers it to be a true *Ischyrocerus anguipes*.

Ischyrocerus minutus, Lilljeborg.

This species occurred in some material collected by the " Garland " at Muckle Ferry, Dornoch Firth, on May 19th, and it was also obtained in the same gathering from Bay of Nigg in which the *Ischyrocerus anguipes* occurred.

---

* See Revision of Amphipoda, by Rev. T. R. R. Stebbing, *Ann. and Mag. Nat. Hist* (7), vol. iv., p. 211 (1899).
† Stebbing, *op. cit.*, p. 211.

Jassa falcata (Montagu). (= · *Podocerus falcatus* (Mont.), in Sars' Crustacea of Norway, vol. iii., p. 594, 1895).*

A fine male specimen of *Jassa falcata* [*Podocerus falcatus* (Mont.) of Sars' *Crustacea of Norway*, vol. iii., p. 594] was found in the stomach of a Long-spined Cottus, *Cottus scorpius*, captured in the Bay of Nigg on June 6th. Another specimen, scarcely mature, was found adhering to a second *Cottus scorpius* captured in the Bay in July, while several others were obtained in a tow-net gathering collected also in the Bay of Nigg in July.

Jassa pusilla (G. O. Sars).

This was obtained in a tow-net gathering, collected by rowing-boat, in Bay of Nigg, in July ; and in another collected by the "Garland" in 45 fathoms, off Aberdeen, November 11th, 1900.

†*Microjassa cumbrensis* (Stebbing and Robertson).

Dredged in 72 fathoms in the Sound of Mull, on March 31st, 1900, and forwarded to the Laboratory at Bay of Nigg, from the "Garland."

†*Parajassa pelagica* (Leach). [*Janassa capillata* (Rathke), G. O. Sars.]

The densely setose character of both pairs of antennæ, gives to this species a striking appearance. A few specimens were obtained by Mr. Williamson in some tow-net gatherings which he collected in the Bay of Nigg in July and the beginning of August (1900). *Parajassa* was also obtained in the stomach of a Pollack caught in the salmon-nets in the Bay of Nigg on the 4th of May previous.

It may be noted in passing that Mr. A. O. Walker, in the Fourteenth Annual Report of the Liverpool Marine Biology Committee (*Trans. L.-pool. Biol. Soc.* (1900), p. 16), objects to the use of the name *Parajassa* for a genus of Amphipods, and states that he fails to see why *Jassa* of Leach, 1815, should be displaced by a genus of fossil fish founded in 1839. But the restoration of *Jassa* for those amphipods for which *Parajassa* has been proposed can only hold good if it has been proved that they do not after all differ generically from such other forms as *Jassa* (*Podocerus*) *falcata* (Mont.). If, on the other hand, it is found that those Amphipods really differ generically from *Jassa falcata*, and if, for the reason stated, the name *Parajassa* may not be used for them, it seems to me that some other name than *Jassa* will require to be substituted, as it can hardly be utilised for two genera, even though they belong to the Amphipoda and are closely allied to each other.

Erichthonius hunteri (Bate).

Specimens of *Erichthonius hunteri* have on one or two occasions been observed in gatherings collected off Aberdeen, but usually in moderately deep water. It occurred in a gathering collected by the "Garland" in 40 to 49 fathoms on October 23rd, and in another collected in 45 fathoms on November 7th (1900). There appear to be few authentic British records of this species.

---

* Stebbing, Ann. and Mag. Nat. Hist. (7), vol. iii., p. 237 (March 1899) ; *see* also A. O. Walker, *op. cit.*, vol. iii., p. 294 (May 1899).

†For changes in generic names see Stebbing, Revision of Amphipoda, *Ann. and Mag. Nat. Hist.*, (7), vol. iii., p. 240 (1899).

COROPHIIDÆ.

Siphonoecetes colletti, Boeck.

This species was obtained in a gathering collected about fifty miles south-south-east of Fair Island on October 16th, 1900. British specimens of this species have sometimes been recorded as *Siphonoecetes typicus*, Kröyer; but it seems that Kröyer's *S. typicus* is a truly Arctic species.

Corophium grossipes (L.).

A few specimens of this *Corophium* occurred in a surface tow-net gathering collected by the "Garland" off the Ferry Slip at Invergordon, Cromarty Firth, on June 16th, 1900. The usual habitat of this species is in brackish-water pools which have a soft muddy bottom, and it occurs in pools of this description at the mouth of the River Alness, a few miles to the west of Invergordon; these pools are occasionally covered by the sea at high water. In such pools the *Corophium* may frequently be observed rising quickly out of the mud, and after swimming a short distance drop down and as quickly disappear, being apparently unable to swim continuously for any lengthened period. The presence of the specimens in the surface tow-net, therefore, may probably be due to their having been carried down by the ebbing tide while clinging to some floating rubbish; that this explanation of their presence in the surface tow-net is probably the correct one is borne out by the fact that several beetles and other insects were also observed in the same gathering.

Corophium affine, Bruzelius.

Though I only record this species for the Moray Firth now, the one or two specimens obtained were dredged in 1895 on July 25th. The material in which they were obtained was dredged from 100 fathoms a few miles to the northward of Rosehearty. The marked difference in the form of the third pair of uropods, together with the slender antennules, makes this species easily distinguished from the other members of the genus.

Unciola planipes, Norman.

This species was dredged by the "Garland" off Aberdeen in 45 fathoms, November 7th, 1900. *Unciola planipes* was described from specimens which Dr. Norman dredged in deep water off the Northumberland coast.*

DULICHIIDÆ.

Lætmatophilus tuberculatus, Bruzelius.

This rare species was represented by a damaged specimen in a gathering collected by the steam trawler "St. Andrew" about fifty miles south-east of Fair Island, in about 60 to 65 fathoms, on October 19th, 1900. *Lætmatophilus armatus* (Norm.) was obtained by Rev. A. M. Norman in material dredged from 100 to 110 fathoms, N. by W. of Burrafirth Lighthouse in 1867, and I at first thought that the specimen obtained off Fair Island might belong to the same species, but a careful examination showed that it was the *Læmatophilus tuberculatus* of Bruzelius. G. O. Sars describes this species as rather abundant in several localities off the west coast of Norway in depths of 20 to 50 fathoms. *Lætmatophilus*

Nat. Hist. Trans. of Northumb. and Durham, vol. i. (1865), p. 14, Pl. VII., figs. 9–13.

has been recorded for the British Islands by Spence Bate under the name of *Cyrtophium tuberculatum.*

Dulichia porrecta (Bate).

This easily distinguished species was obtained in a gathering collected by the "Garland" at Muckle Ferry, Dornoch Firth, May 19th, and also in a gathering collected by the steam trawler "St. Andrew" fifty miles south-south-east of Fair Island, October 16th, 1900.

Dulichia monacantha, Metzger.

This *Dulichia* was recorded for the first time as a British species in the "Annals of Scottish Natural History" for January, 1898, p. 55, and afterwards in the Sixteenth Annual Report of the Fishery Board for Scotland (1898), Part III., p. 277. The specimens, which were from the Clyde, comprised a male possessing the well-developed horn-like processes of the first coxal plates and one or two doubtful females. During the recent investigations carried out in the North Sea by the steam trawler "St. Andrew" numbers of small fish were captured, some of which were handed over to me for examination in order to ascertain what they had been feeding on ; among these were a number of Haddocks about four to five inches in length, captured about sixty-five miles south-east by east of Sumburgh Head, Shetland, on September 4th, and it was interesting to find, in nearly all the stomachs examined, specimens of *Dulichia monacantha* ; in several instances the stomachs of the young Haddocks were crowded with nothing else ; both males and females of the Amphipod occurred, but the males appeared to be the most numerous, and unfortunately almost all of them had been more or less damaged by the teeth of the fish. Fragments of *Dulichia porrecta* were also observed, and also of *Aceros phyllonyx,* and a few other forms, but these appeared to be rare. *Dulichia monacantha* occurred also in a bottom tow-net gathering collected about fifty miles south-east of Fair Island on October 19th (1900).

CAPRELLIDÆ.

Phtisica marina, Slabber.

This species was not uncommon in gatherings collected to the east of Fair Island, October 19th, 1900.

Protella phasma (Mont.).

This was occasionally obtained in gatherings collected off Aberdeen in October and November last by the Fishery steamer "Garland."

Caprella (?) *septentrionalis,* Kröyer.

The *Caprellas* which I have ascribed to Kröyer's *C. septentrionalis* were on several occasions obtained in the stomachs of Pollack, *Gadus pollachius,* captured in the salmon-fishers' nets in the Bay of Nigg, but rarely in the stomach of the Cod ; they were observed in the stomachs of these fishes during April and May last year, and I have no notes of their occurrence during any of the other months. An immense number of *Caprellas* were obtained in the stomach of a half-grown Pollack captured on April 20th ; males and females appeared to be equally numerous, the stomach was packed with them. Among other things observed in this stomach were one or two *Idotea pelagica* and fragments of two *Amathilla homari.*

None of these Bay of Nigg *Caprellas*—which I think belong all to the one species—reach the dimensions given by G. O. Sars for *Caprella septentrionalis*; the largest males obtained in the stomachs of the Pollacks measured from 20 to about 22 millimeters in length, while the adult females were only about half that size ; the measurements given by Sars, on the other hand, are "length of adult female reaching 19 mm., of male 27 mm." Some of the Bay of Nigg males and the greater proportion of the females are rather more prominently tuberculated than they are shown to be in Professor Sars's description and figures, but with these exceptions they appear to be identical with *Caprella septentrionalis*, Kröyer. The second gnathopods of the male are large and powerful, they are articulated slightly behind the middle of the second segment of the mesosome, and the propodos is armed with a stout claw; the distal portion of the claw is strongly curved, and the inner margin near the base is furnished with a shallow but distinct tooth ; the palm of the propodos has a dense fringe of short, slender bristles. The surface of the body is thickly besprinkled with minute points, and in certain positions, when viewed with an inch objective, the bases of these points look like hyaline circular depressions, and it is only when the light strikes across the surface of the body that the projecting points can be seen.

One thing which appears to militate against the Bay of Nigg *Caprellas* being really the *C. septentrionalis* of Kröyer is that the distribution of that species seems to be arctic or sub-arctic rather than north temperate; but the same may be said of *Ischyrocerus anguipes*, Kröyer, an undoubted example of which has been found in the Bay of Nigg ; of *Anonyx nugax* (Phipps), which has been obtained near the May Island, Firth of Forth, and in the Cromarty Firth ; of *Byblis gaimardi* (Kröyer), also obtained off the May Island, as well as of other northern forms which have been found on various parts of the Scottish coasts. If the above objection shows anything, it is rather that our knowledge of distribution is not yet complete, and that, as the investigation of our seas is proceeded with, it is highly probable that other forms, whose distribution was wont to be considered more northern or more southern, will yet turn up to swell the list of the Scottish marine fauna.*

An interesting fact in connection with the Bay of Nigg *Caprellas* is that such a large number of them should be found in the stomach of a single fish; it seems as if the Pollack had been able to discriminate between the various organisms which happened to be within its reach and that it had selected the *Caprellas* in preference to the others. But, as is well known, *Caprella* consists of an elongated slender body, to which are articulated still more slender appendages; it can therefore, one would think, afford little nourishment to a hungry fish ; and why the fish should select this creature in preference to more succulent morsels, such as some of the common Gammaridæ, the smaller Eupaguridæ, and others, is somewhat curious.

Sub-order ISOPODA

TANAIDÆ.

Typhlotanais brevicornis (Lilljeborg).

This species was obtained in a gathering collected by the steam trawler "St. Andrew," on November 3rd, about thirteen or fourteen miles northeast from Buckie, in 50 to 55 fathoms. This is the first time that a

---

* It is also to be noted that *Caprella septentrionalis*, Kröyer, has also been recorded from the Clyde on the authority of the late Dr. Robertson of Millport.

species of *Typhlotanais* has been observed in the British seas, and that this should be so is the more interesting from the fact that no fewer than nine species are known to occur off the coast of Norway, but the majority of them are small, and limited in their distribution to moderately deep water, and this may account to some extent for their having escaped notice hitherto. The length of *Typhlotanais brevicornis* is only about 1½ mm.

Paratanais batei, G. O. Sars.

This species was collected by the "Garland" in Loch Etive in 9 fathoms, off Abbot's Island, on March 30th. This is a new station for *Paratanais batei* on the West Coast.

Leptognathia breviremis (Lilljeborg).

This was collected by the "Garland" off Aberdeen in 45 fathoms on November 11th last year.

Leptognathia brevimana (Lilljeborg).

L. brevimana is a more slender species than the last, and is readily distinguished by the structure of the uropods, the outer branches of which appear to be but spiniform prolongations of the outer angles of the basal joints. I have one or two records of this species for the Moray Firth, where it has been taken by the "Garland," as well as off Aberdeen, where it was collected in 45 fathoms along with the next form on November 11th.

Leptognathia (?) *longiremis* (Lillj.), var. (Pl. XVIII., figs. 30–38.)

In the female of this *Leptognathia* the body is elongated and narrow, being about seven times longer than broad (fig. 30). The terminal segment of the metasome has no apparent denticle on its lateral margins as in *Leptognathia longiremis* (Lilleborg). The antennules (superior antennæ) are moderately elongate, being equal to about one-seventh of the length of the body; in the majority of the specimens examined the antennules were four-jointed (fig. 32), but, on the other hand, a number were observed which had the antennules distinctly five-jointed (fig. 33), though in other respects they appeared to be identical with those possessing the four-jointed antennules. In the four-jointed antennules the first joint is scarcely so long as the combined lengths of the next three joints, the second and the last joints are nearly equal, while the third is only equal to about half the length of the preceding joint. In the five-jointed antennules the first joint is about twice as long as the second; the second, in its turn, is about the length of the third joint; the fourth joint is very small, while the end-joint is nearly as long as the entire length of the third and fourth.

The antennæ (inferior antennæ) are very similar to those of *Leptognathia longiremis* (Lillejborg), (fig. 35).

The chelipeds (fig. 36) and the walking legs are very similar to the same appendages in the female *Leptognathia longiremis*.

The inner branches of the uropods, which are elongated, consist of two sub-equal joints, but the outer branches, which are also two-jointed, are scarcely as long as the first joint of the inner branches (fig. 38). Length of the female scarcely 2·5 mm.

In the male the body is considerably shorter and broader than that of the female (fig. 31). The antennules are proportionally longer than those of the female; the first two joints of the peduncle are dilated, but the second is only about half as long as the first one ; the penultimate joint of the flagellum is distinctly longer than either the preceding or the following joints (fig. 34).

The chelipeds of the male (fig. 37) are nearly as robust as those of the female ; they are each provided with a transverse row of slender spines similar to those on the hands of the male chelipeds in *Leptognathia longiremis.*

Habitat.—In the deep water off Aberdeen and in the Moray Firth ; not very rare.

Remarks—This *Leptognathia* is, in its general appearance, as well as in some of its structural details, not unlike *Leptognathia longiremis* (Lilljeborg), but it appears to be a smaller form, and it differs in the terminal segment of the metasome, having no apparent denticles on its lateral margins.

In the male there are one or two marked differences, and especially in the structure of the antennules. In the antennules of the male of *Leptognathia longiremis* the last two joints of the flagellum are about equal to each other in length, and each of them is much longer than either of the two preceding joints, whereas in the form under consideration the end-joint is not much more than half the length of the penultimate one, and its length does not greatly exceed that of the antepenultimate joint. Moreover, the inner branches of the uropods appear to be only two-jointed, whereas in the male of *Leptognathia longiremis* they appear from the drawings of Prof. G. O. Sars to be three-jointed.* Male specimens appear to be very scarce.

Tanaopsis laticaudata, G. O. Sars.

This Isopod, which I have recorded from the Clyde and one or two other places, has had its known distribution still further extended during the past year, having been obtained by the "Garland" in Loch Etive, and also in Loch Eil (off the head of Loch Linnhe). The species was collected in Loch Etive near Abbot's Island in 9 fathoms on March 30th, and in Loch Eil in 30 fathoms on April 3rd, 1900.

Pseudotanais forcipatus (Lilljeborg).

This species occurred along with the *Typhlotanais brevicornis* previously recorded in the gathering collected in 50 to 55 fathoms, north-east of Buckie, on November 3rd. *Pseudotanais forcipatus* is not very rare in the Moray Firth, especially in comparatively shallow water, as at Guillam Bank, in 8 to 10 fathoms.

CIROLANIDÆ.

Cirolana borealis, Lilljeborg.

A specimen of *Cirolana borealis* occurred in a gathering collected by the "St. Andrew" to the eastward of Fair Island on October 12th. The same species has also been obtained by the "Garland" off the coast of Caithness and in the Clyde. Rev. A. M. Norman records it as not uncommon on the "Haddock Ground" near Whalsey Skerries, and in St. Magnus Bay, Shetland.†

* "Crustacea of Norway," vol. ii. (Isopoda), Plate XII.
† *Brit. Assoc. Report,* 1869, p. 288.

Eurydice pulchra, Leach.

Eurydice was taken with the surface tow-net by the "St. Andrew" on October 15th while at anchor in Lerwick Bay, Shetland; only one or two specimens were observed. Rev. A. M. Norman has recorded the same species from St. Magnus Bay.

IDOTHEIDÆ.

Idothea pelagica, Leach.

Specimens of this small Isopod were frequently observed in the stomachs of Pollacks and Flounders, *Pleuronectes flesus,* captured in the salmon-nets in the Bay of Nigg during the past spring and summer (1900); sometimes they occurred in considerable numbers and in a fairly perfect condition. The same Isopod was also occasionally observed in tow-net gatherings collected in the Bay; the specimens observed comprised both males and females—some of the latter with ova.

Idothea emarginata (Fabricius).

This *Idothea* was also of frequent occurrence in the stomachs of fishes captured in the salmon-nets in the Bay of Nigg; it was also obtained in some of the gatherings collected by the "Garland" off Aberdeen in November last.

Idothea linearis (L.).

This species was obtained in the stomachs of Pollacks captured in the salmon-nets in the Bay of Nigg—especially in the stomachs of those captured in April. In one or two of the stomachs of Pollacks captured on the 2nd of the month *Idothea baltica, pelagica, emarginata,* and *linearis* were observed, together with *Paratylus swammerdami, Amathilla homari, Parajassa pelagica, Caprella septentrionalis,* and a post-larval Eel, "*Leptocephalus.*"

ARCTURIDÆ.

Astacilla intermedia (Goodsir). (=*Astacilla affinis,* G. O. Sars.)*

Males and females of this Isopod were obtained in three separate gatherings of Crustacea collected by the "St. Andrew" about fifty miles south-east of Fair Island, in about 65 fathoms, on October 19th, but only a few specimens altogether were observed. Another species—*Astacilla longicornis* (Sowerby)—is described by Rev. A. M. Norman as common in the Shetland Islands.† No specimens, however, were observed in the gatherings collected by the "St. Andrew," but a few were obtained in one of the deep-water gatherings collected by the "Garland" off Aberdeen on October 25th; this gathering was from 60 fathoms. It also occurred in a gathering from 45 fathoms collected on November 11th.

Arcturella dilatata, G. O. Sars.

This somewhat rare species was obtained in one of the gatherings collected to the east of the Fair Island on October 12th. The only other

---

* Sars, "An Account of the Crustacea of Norway," vol. ii., Parts v. and vi. (1897), p. 90, Pl. XXXVII., fig. 2. The Rev. A. M. Norman considers that *A. affinis,* G. O. Sars, is identical with Goodsir's *A. intermedia.*
† *Brit. Assoc. Rept.,* 1869, p. 289.

British record for *Arcturella* that I at present know of is that of the late Dr. Robertson of Millport for the Clyde; he dredged it off Blackwaterfoot, Arran, Firth of Clyde, in 20 fathoms.[*]

MUNNIDÆ.

Paramunna bilobata, G. O. Sars.

This small species was obtained in a gathering collected south-east of Fair Island on October 19th, in about 65 fathoms. This species, though small, is quite distinct, and appears to be extensively distributed around the Scottish coasts. It is one of the species obtained in the stomachs of small Haddocks captured sixty-five miles south-east by east of Sumburgh Head on September 4th, 1900.

Pleurogonium inerme, G. O. Sars.

P. inerme was dredged off Aberdeen by the " Garland " in 45 fathoms on November 7th, 1900.

Pleurogonium spinosissimum, G. O. Sars.

One or two specimens of this distinct species were dredged by the "Garland " in the Sound of Mull in 72 fathoms on March 31st, 1900.

DESMOSOMIDÆ.

Macrostylis spinifera, G. O. Sars.   (Pl. XVIII., fig. 39.)

A single damaged specimen, which is represented by figure 39 on Plate XVIII., was obtained amongst a small quantity of sand collected by the "St. Andrew " about fifty miles south-east of Fair Island on October 19th, 1900 ; nearly all the appendages of the specimen were broken off—even the styliform uropods were damaged. The species was obtained in about 65 fathoms, and is an addition to the British fauna. Prof. Sars mentions that it occurs in a few Norwegian localities, from the Lofoten Islands to Christiania Fjord, and that the *Vana longiremis* of Meinert from the Kattegat belongs also to this species.

Echinopleura aculeata, G. O. Sars.   (Pl. XVIII., fig. 40.)

This Isopod was obtained in the same gathering with the last, and is also an addition to the British Crustacean fauna. Two specimens were observed, and both were damaged ; it is the best specimen of the two which is represented by figure 40 on Plate XVIII. The Norwegian distribution of *Echinopleura* appears to be similar to that of *Macrostylis*, but Prof. Sars remarks that "it seems everywhere to be very rare." The form of the body and the strongly-serrated margins of the segments impart to *Echinopleura* quite a distinctive character.

BOPYRIDÆ.

Phryxus abdominalis (Kröyer).

This Epicarid was moderately frequent on the specimens of *Spirontocaris securifrons* captured by the "Garland" off Aberdeen in October and November last. What appears to be the same species of *Phryxus* was

[*] " Amphipoda and Isopoda of the Firth of Clyde," Part ii., p. 28 (1892), separate copy.

also obtained on *Spirontocaris pusiola*, collected by the "Garland" in 45 fathoms off Aberdeen, November 7th.

Professor G. O. Sars states that *Phryxus abdominalis*, which is the commonest of the Norwegian Bopyrids, is found infesting no fewer than nine different Carides : — *Spirontocaris gaimardi, polaris, pusiola, turgida, spinus*, and *securifrons*, and *Pandalus montagui, borealis*, and *propinquus*, and probably one or two others.* On the other hand, M. Bonnier, who, along with Professor Giard, has devoted a long time to the study of these crustaceans, considers that the identity of the Bopyrids from all these different species of the Caridea with the *Phryxus abdominalis* of Kröyer, which is found on *Spirontocaris gaimardi*, M.-Edw., is at least open to doubt.† I may say, however, that, so far as I have been able to make them out, the species I have recorded above agree very well with the description and figures of *Phryxus abdominalis*, Kröyer, given by Professor Sars in the work referred to.

Sub-Order SYMPODA, Stebbing, 1900 (CUMACEA, Auctorum).

Rev. T. R. R. Stebbing has shown that *Cuma*, so familiar as the name of a genus of Crustaceans, had long before been used as the name of a Molluscan genus, and further, that it was necessary that the sub-ordinal name "Cumacea," which had been derived from *Cuma* and therefore could not stand, must be replaced by some other, and the name he proposes as a substitute is "Sympoda." ‡

Fam. Bodotriidæ, Stebbing

The generic name *Cuma* being set aside as preoccupied, leaves room for the restoration of Goodsir's *Bodotria*, and as *Bodotria* is the typical and the oldest genus in the family with which it is associated, it necessarily forms the basis of the family name.

Genus *Bodotria*, Goodsir (1843).

Bodotria arenosa, Goodsir (*Cuma scorpioides*, G. O. Sars), was obtained in a gathering collected by the steam trawler "St. Andrew," about fifty miles south-south-east of Fair Island on October 16th ; in this species the inner ramus of the uropods is composed of a single piece. *Bodotria scorpioides* (Mont.) (= *Cuma edwardsii*, Goodsir) has been obtained in gatherings collected during the past year in the Dornoch Firth, and off Aberdeen ; in this species, which appears to be more common than the previous one, the inner ramus of the uropods is distinctly biarticulated. *Bodotria pulchella* (G. O. Sars) occurred in gatherings collected by the "Garland" off Aberdeen in 36 to 49 fathoms in October, and in November in 45 fathoms.

Genus *Iphinoë*, Spence Bate (1856).

Iphinoë serrata, Norman, was moderately frequent in one or two of the gatherings collected to the south-eastward of Fair Island in October last.

---

* G. O. Sars, "An Account of the Crustacea of Norway," vol. ii. (Isopoda), p. 217 (1896–99).

† Jules Bonnier, "Contribution a l'étude des Epicarides Les Bopyridæ," *Publications de la Station Zool. de Wimereux*, viii., p. 381 (1900).

‡ Stebbing : On Crustacea brought by Dr. Willey from the South Seas. *A. Willey's Zoological Results* part v., p. 609 (Dec. 1900).

Fam. LAMPROPIDÆ.

Lamprops fasciata, G. O. Sars.

This species was frequent in some tow-net gatherings collected by the "Garland" while at anchor off Ardjachie Point in the Dornoch Firth, between 8.45 p.m. on the 16th and 5 a.m. on the 17th of May, 1900.

Hemilamprops rosea (Norman).

This was moderately frequent in tow-net gatherings collected off Fair Island in October, and it was also observed occasionally in the stomachs of small Haddocks captured about 65 miles south-east by east of Sumburgh Head in September. It was the only species belonging to the Lampropidæ observed in the gatherings.

Fam. LEUCONIDÆ.

Leucon nasicus (Kröyer).

L. nasicus was also one of the more common forms in the gatherings collected off Fair Island by the steam trawler "St. Andrew" in October last.

Eudorellopsis deformis (Kröyer).

This curious small species was occasionally observed in company with *Eudorella marginata* (Kröyer) in the gatherings collected off Fair Island, but neither the one nor the other was very frequent.

Fam. DIASTYLIDÆ.

Diastylis lucifera (Kröyer) and *Diastylus rostrata* (Goodsir) have both been obtained in tow-net gatherings collected in the Bay of Nigg, near Aberdeen, by Mr. H. C. Williamson in July, 1900, while engaged in some special researches in the Bay.

Diastylis tumida (Lilljeborg).

A single specimen of this somewhat rare species was obtained in some material dredged by the "Garland" in 130 fathoms a few miles to the north of Rosehearty, Moray Firth.

Diastylis cornutus (Boeck).

Several specimens of this well-marked species occurred in tow-net gatherings collected in deep water about fifty miles south-eastward of Fair Island on the 12th and 19th October. *Diastylis cornutus* was obtained for the first time as a British species by the Rev. A. M. Norman, who dredged it in Shetland in 1863. Most of the specimens collected off Fair Island were more or less damaged.

Diastylopsis resima (Kröyer).

This species also occurred in the gatherings collected off Fair Island along with *Diastylis cornutus*; male and female specimens of *Diastylopsis resima* were observed, but though the species was moderately frequent, few of the specimens were undamaged, and nearly all were more or less covered with mud. This is the first time that *Diastylopsis resima* has been observed in Scottish waters. *Diastyloides biplicata*, G. O. Sars, was also frequent in the same gatherings.

Leptostylis villosa, G. O. Sars. (Pl. XVIII., fig. 41.)

This species has been added to the British fauna during the past year. The first specimens were obtained in the Firth of Clyde in 1896 in a tow-net gathering from 40 fathoms collected by the "Garland" between Arran and the coast of Ayrshire. It has also been observed in a tow-net gathering collected fifty miles south-east of Fair Island, in about 60 to 65 fathoms, on October 19, by the steam trawler "St. Andrew." *Leptostylis villosa* may be distinguished from its confrères by the peculiar form of the marginal serration of the carapace; figure 41, Plate XVIII., represents a small portion of the margin of the carapace, and shows the peculiar form of the serration.

PSEUDOCUMIDÆ.

Pseudocuma similis, G. O. Sars.

This species has also been lately added to the British fauna from the Firth of Clyde. It was obtained in a gathering collected by the "Garland" in moderately deep water near the mouth of the estuary in December, 1899. *Pseudocuma similis* has also been obtained in gatherings collected by the steam trawler "St. Andrew" off Fair Island in about 80 to 85 fathoms on October 16th, and by the "Garland" off Aberdeen in 42 to 45 fathoms on November 7th and 12th, 1900. Although it is not very difficult to recognise this *Pseudocuma* after the observer has become familiar with its peculiarities, it nevertheless so closely resembles the more common *Pseudocuma cercaria* (v. Beneden), with which it is frequently associated, that it may be readily passed over as belonging to that species. *Pseudocuma similis* has also recently been recorded from the Irish Sea,[*] which seems to indicate that the distribution of this species in British waters may be fairly extensive.

Petalosarsia declivis, (G. O. Sars).

Petalosarsia was obtained in at least three different gatherings collected by the "Garland" off Aberdeen in October and November last year; it occurred in a gathering from 36 to 49 fathoms collected on October 23rd, and in one from 60 fathoms collected on the 25th, and it also occurred in a gathering collected on November 7th from 45 fathoms. The same species was also observed in one of the gatherings collected by the steam trawler "St. Andrew" off Fair Island on October 16th.

NANNASTACIDÆ.

Cumella pygmæa, G. O. Sars.

This small species occurred in the gathering collected by the "Garland" in 45 fathoms off Aberdeen on November 7th. It was also collected in Aberdeen Bay by Mr. H. C. Williamson on August the 9th. The species is very small and easily missed.

[*] A. Scott, in *Lancashire Sea Fisheries Report* for 1900 (March 1901).

Nannastacus unguiculata, Spence Bate.

Male and female specimens of this curious species were captured by the steam trawler "St. Andrew" in Lerwick Bay, Shetland, by means of a surface tow-net put overboard for four hours while the vessel was at anchor. A single male specimen of *Vaunthompsonia cristata* was also obtained in this gathering. *Nannastacus unguiculata* was taken at Shetland in 1863 by the Rev. A. M. Norman; it was also taken in Lerwick Bay in 1867 by David Robertson by means of a surface tow-net,* and in connection with this it is interesting to note, from the point of view of the distribution of the species, that though *Vaunthompsonia cristata* and *Nannastacus unguiculata* are both members of the Shetland marine fauna, and the latter especially not uncommon, yet Professor G. O. Sars in his work on the Norwegian species belonging to the Sympoda, recently published, does not include either of the forms named.

CAMPYLASPIDÆ.

Campylaspis rubicunda (Lilljeborg).

This species was obtained in gatherings collected by the "St. Andrew" off Fair Island on the 12th, 16th, and 19th of October at various depths from 60 to over 80 fathoms.

Campylaspis costata, G. O. Sars.

This species was obtained in the same gatherings with the last, and was, if anything, more frequent. It also occurred in a gathering from Loch Aber (upper end of Loch Linnhe) collected by the "Garland" in 82 fathoms on April 2nd.

Sub-order SCHIZOPODA.

MYSIDÆ.

Mysidopsis didelphys (Norman).

This Schizopod has been obtained during the past year by the "Garland" off Aberdeen, and by the "St. Andrew" off the Fair Island. It was collected off Fair Island in 60 to 70 fathoms on the 19th of October, and off Aberdeen in 60 fathoms on the 25th of the same month. The Rev. A. M. Norman obtained it in 40 to 50 fathoms, five to eight miles east of Balta, Shetland.†

Heteromysis formosa, S. I. Smith.

Hitherto the Firth of Forth is the only Scottish locality whence this Schizopod has been recorded.‡ It is one of the many species which have been added to the Scottish marine fauna as one of the results of the investigations carried on under the direction and by the encouragement of the Fishery Board for Scotland. I have now to record *Heteromysis formosa* for the first time for Loch Fyne and the Moray Firth. It occurred with some other rare things in material dredged by the "Garland" at Tarbert Bank, Lower Loch Fyne, in about 20 fathoms, on

---

* Norman. Last Report on Dredging among the Shetland Islands; *Rept. Brit. Assoc.* for 1868, p. 270. (Pub. 1869).
    † Norman, *op. cit.*, p. 267.
    ‡ *Seventh Annual Report of the Fishery Board for Scotland*, part iii., p. 323.

December 22nd, 1899, and again on the 28th of March, 1900 ; only three specimens altogether were observed, and two of them were slightly damaged. The species was obtained in the Moray Firth district at Station IV. (off Little Ferry), in 7 to 10 fathoms, on April 10th, 1900. This gathering, which was collected by the "Garland," yielded only one or two specimens of the Schizopod.

Erythrops serrata, G. O. Sars.

This species has been obtained during the past year in gatherings collected south-eastward of Fair Island, on the 16th and 19th October, by the "St. Andrew," at a depth of 60 to 80 fathoms. It was also collected by the "Garland" in 60 fathoms, off Aberdeen, on October 25th. *Erythrops serrata*, which is also one of the Schizopods obtained by the Rev. A. M. Norman while dredging among the Shetland Islands,* does not appear to be a rare species in the North Sea.

Macropsis slabberi (Van Beneden).

Though *Macropsis* is frequent enough on some parts of the East Coast of Scotland, and was recorded for the Firth of Forth by Dr. Henderson as long ago as 1884, I have only recently been able to find traces of it on the West Coast. I obtained an undoubted specimen of this species in the stomach of a young Herring—one of several which Dr. Fulton had received from Annan, Solway Firth, and which he had kindly handed over to me for examination. The fact that *Macropsis* was found in the stomach of this Herring from the Solway is fairly satisfactory evidence that the species exists somewhere near the place where the fish was captured. The Herring were captured on July 30th, 1900. *Macropsis slabberi* has also been obtained in a tow-net gathering collected by the "Garland" in the Cromarty Firth in 7 to 8 fathoms on the 30th of November last.

Leptomysis gracilis, G. O. Sars.

This species occurred in one of the bottom gatherings collected south-east of Fair Island by the "St. Andrew" on October 19th ; and it is one of the Schizopods recorded for Shetland by Rev. A. M. Norman in 1869.

Gastrosaccus spinifer (Goes).

Young specimens of *Gastrosaccus spinifer* were moderately frequent in one or two of the gatherings collected by the "Garland" off Aberdeen in October and November, 1900. One of the gatherings in which this species occurred was collected in 60 fathoms, and another in 45 fathoms ; adult specimens were scarce in both gatherings.

Siriella armata (M.-Edw.).

This species was obtained in a bottom tow-net gathering collected by the "Garland" in 7 to 10 fathoms at Station IV., Moray Firth (off Little Ferry), on December 6th, 1900.

---

* Norman, *op. cit.*, p. 270.

Sub-Order MACRURA.

CALLIANASSIDÆ.

Callianassa subterranea (Mont.).

A specimen of this Crustacean, obtained in the stomach of a common Gurnard, *Trigla gurnardus*, captured in the Moray Firth, was forwarded to the Laboratory at Bay of Nigg by Mr. F. G. Pearcey, Naturalist on board the "Garland." The specimen, which, though recognisable, was considerably decomposed, was obtained on June the 14th at Station IX. (about eleven or twelve miles north-east of Lossiemouth). *Callianassa subterranea* has been known for many years as one of the invertebrates of the Moray Firth district. The species was obtained by Thomas Edward of Banff, and it has also been obtained on several occasions by Mr. Sim of Aberdeen, but always in the stomachs of fishes captured in moderately deep water, and where the bottom appears to consist of soft mud.

Eupogebia deltura, Leach.

A specimen of *Eupogebia deltura* was found in the stomach of a Haddock 13½ inches long, captured in the Moray Firth at Station VII. (about five miles east of Tarbat Ness), in February, 1900, and forwarded to the Laboratory at Bay of Nigg by Mr. F. G. Pearcey. *Eupogebia deltura*, which is also one of the Crustaceans obtained in the Moray Firth by the late Thomas Edward of Banff, appears to have a habitat similar to that of *Callianassa*.

CRANGONIDÆ.

Pontophilus spinosus, Leach.

A few fine specimens of this species were obtained by the "Garland" in 40 to 50 fathoms, off Aberdeen, in November last year.

Cheraphilus neglectus, G. O. Sars.

A few specimens of this Crangonid were collected by Mr. H. C. Williamson in the Bay of Nigg in August last year. A fine specimen was also obtained in a gathering collected by the "Garland" in 7 to 8 fathoms at Station I., Moray Firth (between Burghead and the mouth of the River Findhorn), on December 12th. This species resembles *Egion fasciatus* (Risso), but in the first the front of the rostrum is broadly rounded, while in the other the rostrum is truncate and slightly concave at the apex.

HIPPOLYTIDÆ.

Caridion gordoni (Spence Bate).

This species was moderately frequent in a gathering collected by the "Garland" in 60 fathoms, off Aberdeen, on the 25th October last; the majority of the specimens I observed were, however, more or less immature. *Caridion* differs from the closely-allied *Hippolyte* and *Spirontocaris* by the possession of a three-jointed mandible-palp.

Spirontocaris securifrons, Norman.

This Crustacean was moderately common in some gatherings collected by the "Garland," off Aberdeen, in October and November last year.

None of the specimens examined possessed the strong tooth-like projection on the dorsal aspect of the abdomen which is so characteristic of the typical *Spirontocaris spinus* of Sowerby. Though the specimens varied greatly in size and age, it was only in one or two small specimens that there was observed any tendency towards the development of a tooth-like projection on the middle of the posterior margin of the third abdominal segment. It may be that *Spirontocaris securifrons* should only be regarded as a synonym of *Spirontocaris spinus* (Sowerby); and, with the exception of the tooth-like projection on the dorsal aspect of the abdomen alluded to, they are certainly very like each other. My use here of the name is merely to indicate that the form which was common in the gatherings collected by the "Garland" off Aberdeen is the form described by Norman as *Hippolyte securifrons.** The depth from which specimens were obtained ranged from $32\frac{1}{2}$ to 60 fathoms.

Spirontocaris pusiola (Kröyer).

One or two specimens of this small species, as well as of *Spirontocaris gaimardii* (M.-Edw.), were observed in a gathering collected by the "Garland" in about 32 fathoms, off Aberdeen, on October 12th.

PANDALIDÆ.

Pandalina brevirostris (Rathke).

This species was not very rare in the gatherings collected off Aberdeen by the "Garland" in October and November last. Dr. Calman has, in his paper on *Pandalus,*† described certain differences between this species and the others with which it used to be generically associated ; one very obvious difference between *Pandalina,* Calman, and the described species of *Pandalus* is, that *Pandalina* has a very short, nearly straight, rostrum, not unlike that of some of the species of *Spirontocaris,* while *Pandalus* has the rostrum elongated, more or less curved, and slender towards the distal end.

Sub-order BRACHYURA.

CORYSTIDÆ.

Atelecyclus septemdentatus (Mont.).

This species was obtained in a gathering collected by the "St. Andrew" fifty miles south-east of Fair Island at a depth of about 65 fathoms on October 19th. Rev. A. M. Norman describes *Atelecyclus* as a common species among the Shetland Islands. This species was also taken by the "Garland" in 60 fathoms, off Aberdeen, on October 25th.

PORTUNIDÆ.

Portunus puber (Lin.).

Several specimens of the Velvet Crab, *Portunus puber,* were captured by Mr. H. C. Williamson in the Bay of Nigg during the month of June, 1900.

---

* *Brit. Assoc. Rept.,* 1861 ; *Trans. Tyneside Field Club* (1862), p. 267, Pl. XII., figs. 1–7.

† Calman, On the British Pandalidæ ; *Ann. and Mag. Nat. Hist.,* (7), vol. iii., p. 37, Pls. I.–IV., fig. 4 (Jan., 1899).

T

Portunus arcuatus, Leach.

A specimen of this Crab was obtained in the stomach of a Cod captured in the salmon-nets at the Bay of Nigg on June 13th, 1900.

EXPLANATION OF THE PLATES.

PLATE XVII.

(?) *Rhincalanus gigas,* G. S. Brady.

		Diam.
Fig. 1. Foot of fifth pair, female	.	× 51.
Fig. 2. Foot of fifth pair, male, slightly immature		× 51.
Fig. 3. Abdomen and last thoracic segment, female		× 25.
Fig. 4. Abdomen and last thoracic segment, male		× 38.

Cyclopina longicaudata, T. Scott (sp. n.).

Fig. 5. Female, dorsal view	.	× 51.
Fig. 6. One of the female antennules	.	× 63.
Fig. 7. One of the antennæ	.	× 63.
Fig. 8. One of the mandibles	.	× 95.
Fig. 9. One of the maxillæ	.	× 63.
Fig. 10. One of the first maxillipedes	.	× 63.
Fig. 11. One of the second maxillipedes	.	× 63.
Fig. 12. Foot of first pair	.	× 63.
Fig. 13. Foot of fourth pair	.	× 63.
Fig. 14. Foot of fifth pair	.	× 95.

Botryllophilus (?) *ruber,* Hesse.

Fig. 15. Female, dorsal view	.	× 46.
Fig. 16. Male, dorsal view	.	× 72.
Fig. 17. One of the female antennules	.	× 96.
Fig. 18. One of the female antennæ	.	× 145.
Fig. 19. One of the first maxillipedes, female	.	× 145.
Fig. 20. One of the second maxillipedes, female		× 145.
Fig. 21. Foot of the first pair, female	.	× 145.
Fig. 22. Foot of the third pair, female	.	× 145.
Fig. 23. Foot of the fourth pair, female	.	× 145.
Fig. 24. One of the male antennules	.	× 145.
Fig. 25. Foot of the first pair, male	.	× 160.
Fig. 26. Foot of the third pair, male	.	× 72.
Fig. 27. Foot of the fourth pair, male	.	× 160.

(?) *Enteropsis vararensis,* T. Scott (sp. n.).

Fig. 28. Female, side view	.	× 61.
Fig. 29. One of the female antennules	.	× 580.
Fig. 30. One of the female antennæ	.	× 290.
Fig. 31. One of the mandibles	.	× 580.
Fig. 32. One of the maxillæ	.	× 580.
Fig. 33. One of the maxillipedes	.	× 580.
Fig. 34. Foot of the third pair	.	× 195.

Nicothoë astaci, Aud. and M.-Edw.

Fig. 35. One of the female antennules	.	× 190.
Fig. 36. One of the mandibles	.	× 380.
Fig. 37. One of the maxillæ	.	× 380.
Fig. 38. One of the first maxillipedes	.	× 253.
Fig. 39. Abdomen of the female	.	× 80.

PLATE XVIII.

Eucanuella spinifera, T. Scott (gen. et sp. n.).

Fig. 1. Female, dorsal view . . . . × 46.
Fig. 2. One of the female antennules . . . × 76.
Fig. 3. One of the antennæ . . . × 116.
Fig. 4. One of the mandibles . . . × 116.
Fig. 5. One of the maxillæ . . . × 116.
Fig. 6. One of the first maxillipedes . . . × 145.
Fig. 7. One of the second maxillipedes . . × 145.
Fig. 8. Foot of the first pair . . . × 76.
Fig. 9. Foot of the fourth pair . . . × 63.
Fig. 10. Foot of the fifth pair . . . × 63.

Corycæus anglicus, Lubbock (male).

Fig. 11. Male, side view . . . . × 61.

Cancerina confusa, T. Scott (gen. et sp. n.).

Fig. 12. Female, dorsal view . . . . × 38.
Fig. 13. Portion of the same seen from the under-side to show the
 position of the maxillipedes and thoracic feet . . × 38.
Fig. 14. One of the female antennules . . . × 145.
Fig. 15. One of the antennæ . . . × 290.
Fig. 16. One of the mandibles . . . × 290.
Fig. 17. One of the maxillæ . . . × 290.
Fig. 18. One of the first maxillipedes . . . × 193.
Fig. 19. One of the second maxillipedes . . . × 116.
Fig. 20. Foot of the first pair . . . × 116.

Nicothoë astaci, Aud. and M.-Edw.

Fig. 21. Female, dorsal view . . . × 11.
Fig. 22. One of the antennæ . . . × 380.
Fig. 23. One of the second maxillipedes . . × 190.
Fig. 24. Foot of the first pair . . . × 152.
Fig. 25. Foot of the fourth pair . . . × 127.
Fig. 26. Foot of the fifth pair . . . × 190.

Mæra lovéni (Bruzelius).

Fig. 27. One of the first gnathopods . . . × 7.
Fig. 28. One of the second gnathopods . . × 5.

Ischyrocerus anguipes, Kröyer.

Fig. 29. One of the second gnathopods, male . . × 34.

Leptognathia (?) *longiremis* (Lillj.), var.

Fig. 30. Female, dorsal view . . . × 26.
Fig. 31. Male, dorsal view . . . × 38.
Fig. 32. One of the female antennules (four-jointed) . × 154.
Fig. 33. One of the female antennules (five-jointed) . × 154.
Fig. 34. One of the male antennules . . . × 154.
Fig. 35. One of the female antennæ . . . × 154.
Fig. 36. One of the female chelipedes . . . × 123.
Fig. 37. One of male chelipedes . . . × 123.
Fig. 38. One of the uropods, female . . . × 73.

Macrostylis spinifera, G. O. Sars.

Fig. 39. Female, dorsal view (imperfect) . . × 32.

Echinopleura aculeata, G. O. Sars.

Fig. 40. Female, dorsal view (slightly imperfect) . × 48.

Leptostylis villosa, G. O. Sars.

Fig. 41. Portion of serrated margin of cephalon . . × 514.

PLATE XVII.

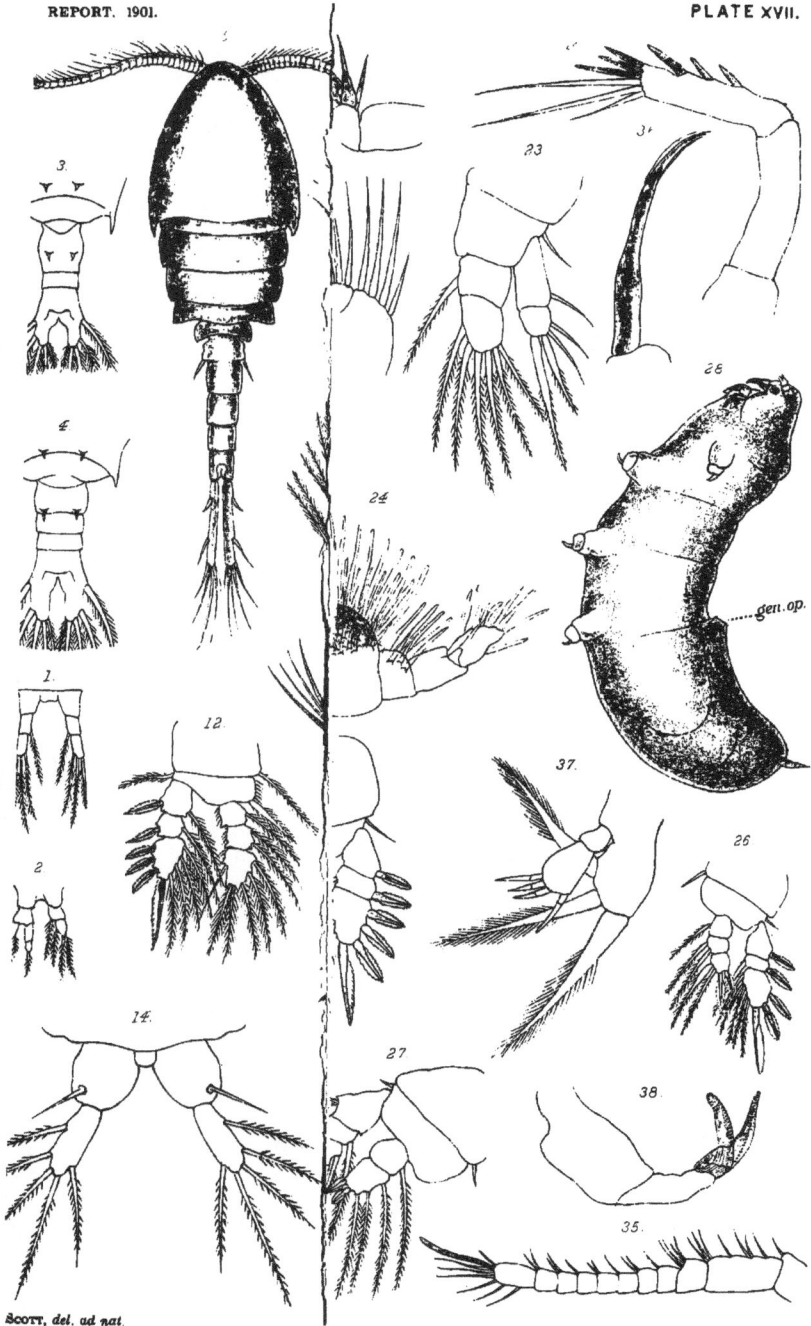

gen.op.

PLATE XVII.

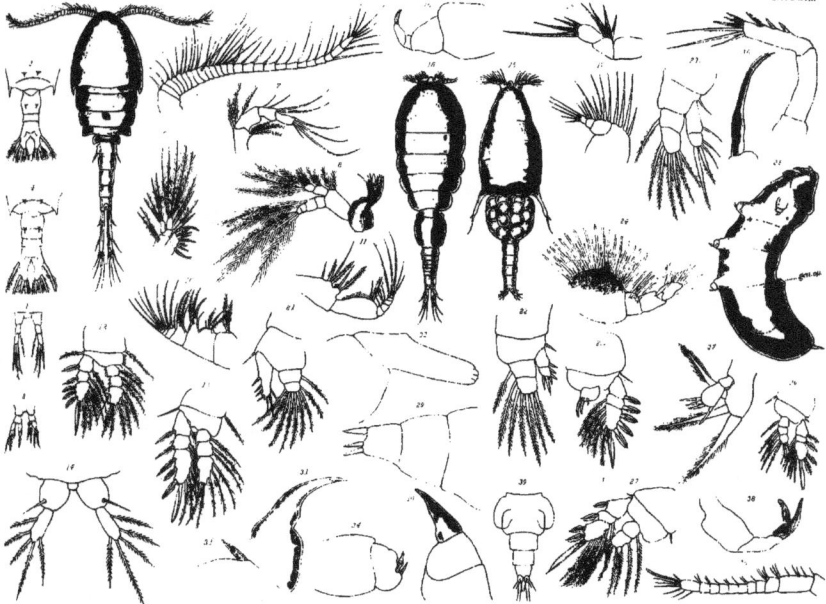

PLATE XVIII.

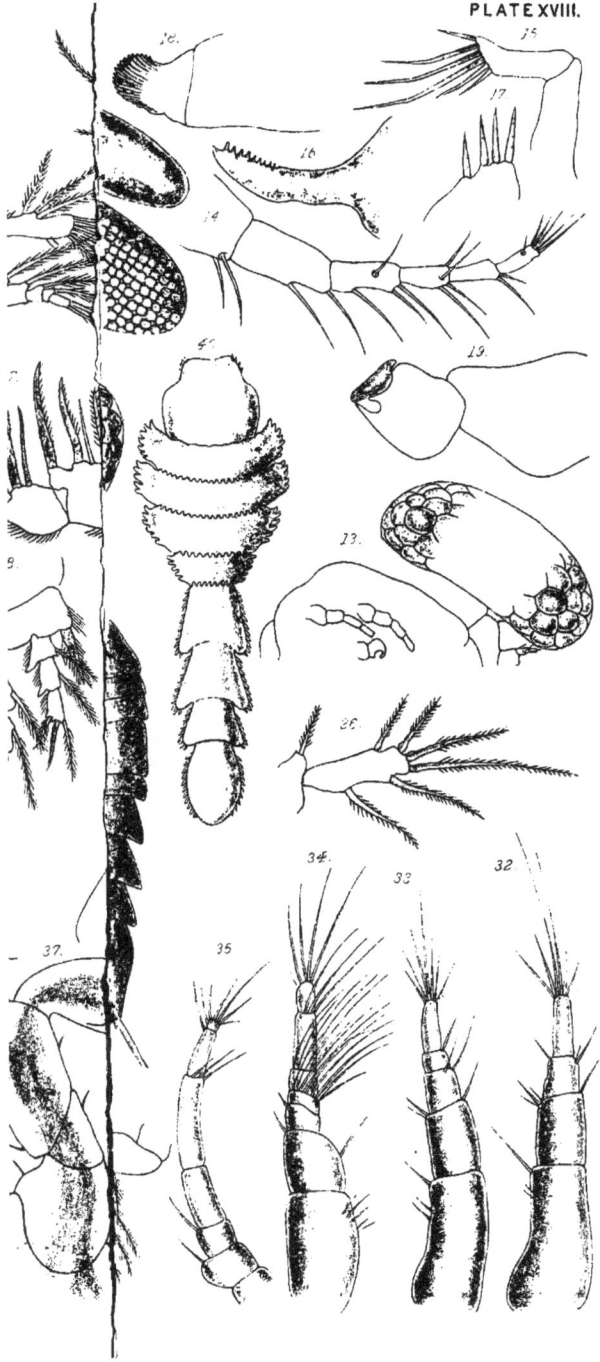

PLATE XVIII.

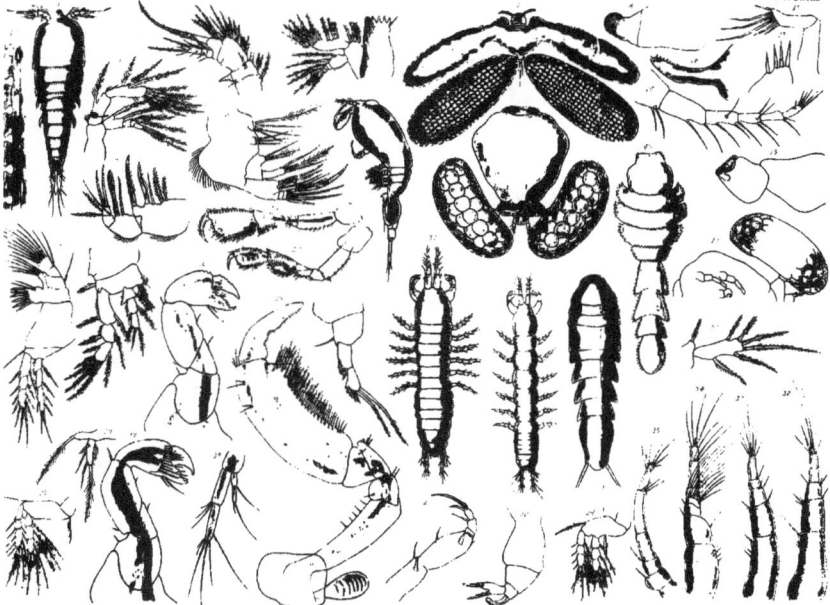

CRUSTACEA FROM TOW-NET AND OTHER GATHERINGS.

·

VIII.—NOTES ON GATHERINGS OF CRUSTACEA COLLECTED BY THE FISHERY STEAMER "GARLAND," AND THE STEAM TRAWLERS "STAR OF PEACE" AND "STAR OF HOPE," OF ABERDEEN, DURING THE YEAR 1901.

By Thomas Scott, F.L.S., Mem. Soc. Zool. de France.

(Plates XXII.—XXV.)

In continuation of my notes on the new and rare Crustaceans which have been obtained in tow-net gatherings, and in gatherings of dredged and other materials collected at different times and in various places during the past year, I have to acknowledge my indebtedness to the Naturalist on board the Fishery steamer "Garland," and to Mr. Dannevig and others who have carried on from time to time a considerable amount of interesting fishery research work during 1901. The collections of Crustacea forwarded to the Laboratory at Bay of Nigg, in connection with these investgations, have in a number of instances proved to be extremely interesting. Several apparently undescribed forms have been obtained, while others, though they may have been recorded elsewhere, have not previously been obtained in Scottish waters. Moreover, the distribution of a number of rare species described in former papers has by these researches been still further extended.

A number of rare Crustaceans—Copepods and others—have occurred in gatherings collected by the "Garland" in the Firth of Forth on the East of Scotland, and in Loch Etive on the West Coast. Other rare forms have been obtained in gatherings collected by the steam trawler "Star of Peace" while working to the east and north of the Shetland Islands; and one or two species, rarely met with, were also captured in deep water (58 to 65 fathoms) about nine to ten miles to the eastward of Aberdeen by the steam trawler "Star of Hope."

There is one point in connection with these investigations which seems to be of peculiar interest, and which it may not be out of place to refer to here; it is this, that localities which have already been subjected to prolonged and careful examination should still continue to yield not only rare but even new forms of life. I do not refer to those minute microscopical species which are difficult to distinguish, and which may easily be overlooked even by those who have acquired a fairly extensive and special knowledge, but species which from their size and shape or colour are sufficiently conspicuous to attract the attention of even the casual observer. Whether these forms, which are turning up in places where they have not been seen before, are recent or new arrivals, or whether they have escaped notice hitherto owing to their distribution being limited to some particular area more or less out of the reach of the dredge or the tow-net, and that, having for some reason left their old haunts, and made their way to a place more accessible, have ultimately been captured, is a question that may not be easily answered. The following two examples will indicate more clearly the aspect of the question concerning the distribution of species to which I refer:—

For many years previous to 1886-87, when arrangements were made by the Fishery Board for the scientific investigation of the Firth of Forth, special efforts had from time to time been put forth by not a few eminent naturalists to study the invertebrate fauna of the Forth estuary, and a considerable number of species of Crustacea, including several Schizopods, were made known to science, yet none of these investigators appears to have noticed *Erythrops goesii*, G.O.S. (= *E. erythrophthalmus*, Goes). Now this Schizopod, though probably one of the largest of the species belonging to the genus *Erythrops*, is under half an inch in length, and might on account of its small size be overlooked by an ordinary observer, but, like its confrères, it possesses eyes of such a brilliant red colour as to make the little creature quite conspicuous even in a crowded tow-net gathering, especially if the gathering be examined soon after it is collected ; it is therefore scarcely likely that this *Erythrops* would have escaped being noticed by naturalists so experienced as Goodsir, Henderson, Leslie, and others, had it been present in any of their collections. Other and less conspicuous Schizopods were recorded by these early investigators : then why not this one? When, on the other hand, we turn to the work of the Fishery steamer "Garland," as described in the various Annual Reports of the Board, we find that mention is made of *Erythrops* as early as October 3rd, 1888, when it was obtained in a bottom tow-net gathering collected at Station V.,* while in a paper on the fauna of the Firth of Forth published in 1889 *Erythrops goesii* is described as "frequent all over the Forth from Inchkeith to May Island,"† and as being new to Britain ; but since that time it has been found to be of moderately common occurrence, especially in the part of the Firth described above. The question which naturally suggests itself here is—Was the recognition of *Erythrops* in the Firth of Forth in 1888, and every year since, the result of a recent migration of the species, or had it simply been overlooked by former observers?

The second example is even more interesting than the one just referred to—viz., the occurrence of *Calocaris macandreæ* in the Firth of Forth. *Calocaris* was for a long time considered to be a rare species and to have a very limited distribution ; subsequent investigations have shown, however, that its distribution is not so restricted as it was formerly believed to be ; but till as recently as the past summer it had never been known to occur in the Firth of Forth, notwithstanding the fact that the Crustacean fauna of that estuary has been very carefully examined by the various methods of tow-netting, dredging, and trawling, as well as by the examination of the organisms contained in the stomachs of fishes captured within its limits. About the end of May of the present year (1901), when the Fishery steamer "Garland" was engaged in carrying on some special work, a number of specimens of the Crustacean referred to were obtained, along with several other organisms, at Station III.—to the east of Inchkeith —in a small-mesh net which was being employed for the capture of small fishes ; and it was also about the same time obtained in the stomach of a Long Rough Dab, *Drepanopsetta platessoides*, from about the same place. Other specimens of *Calocaris* were obtained by Mr. Pearcey, the Naturalist on board the "Garland," in the stomachs of large Witch Soles, *Pleuronectes cynoglossus*, captured on the 28th of June and 15th July at Station V.—to the west of May Island—in 25 to 27 fathoms. These various captures of *Calocaris* within such a short time would almost warrant the belief that this Crustacean was not uncommon in the Firth of Forth ; and should that be found to be really the case, the fact that no trace of the species had been noticed by any previous observer is of considerable interest. Of course, if the species be of true fossorial habits— and any evidence we have concerning it seems to support such a con-

---

* Seventh Annual Report of the Fishery Board for Scotland (1889), Part III., p. 57.
† Op. cit., p. 322.

clusion—it might easily escape capture by the dredge or the trawl by burrowing deep into the soft mud which forms the floor of the more central portion of the estuary of the Forth; but its occurrence in the stomachs of the Long Rough Dab and the Witch Sole indicates that it does not always remain in hiding, but occasionally comes to the surface of the mud in which it burrows, and though it may have escaped being captured by the dredge or the trawl, it apparently does not always escape the fishes that happens to be on the look-out for food, yet we can find no reference to its ever having been observed even in the stomachs of fishes taken within the estuary. *Calocaris* is considered to be somewhat sluggish in its habits, because specimens are occasionally found overgrown with a small zoophyte; but such habits should make its capture by trawl or dredge of more easy accomplishment. Moreover, it is not such a small species as to be easily passed over, and it is sufficiently distinct that any one with a fair knowledge of the Crustacea would be likely to recognise it as different from the more common forms, yet the fact remains that not one of the many students who have investigated the Crustacean fauna of the Firth of Forth appears to have obtained any evidence to lead them to regard it as even of doubtful occurrence within the limits of the estuary.

Whether *Calocaris* be a recent introduction or not, there is apparently no doubt as to its having now a right to be reckoned amongst the Crustacean fauna of the Forth.

The number of Crustacea recorded in the present paper is scarcely so large as in that published last year, but there are included several species apparently undescribed, and others which are new or rare in the Scottish seas. The following are the more interesting of the species recorded :—

> *Xanthocalanus* (?) *borealis*, G. O. Sars.
> (?) *Phœnna zetlandica* T. Scott (sp. n.)
> *Scolecithrix* (?) *brevicornis*, G. O. Sars.
> *Platypsyllus minor*, T. Scott (gen. et sp. n.).
> *Nereicola concinna*, T. Scott (sp. n.).
> *Stenhelia confusa*, T. Scott (sp. n.).
> *Ameira tenuicornis*, T. Scott (sp. n.).
> *Ameira propinqua*, T. Scott (sp. n.).
> *Pseudomesochra longifurcata*, T. Scott (gen. et sp. n.).
> *Leptopontia curvicauda*, T. Scott (gen. et sp. n.).
> *Normanella attenuata*, A. Scott.
> *Fultonia hirsuta*, T. Scott (gen. et sp. n.).
> *Monstrilla longiremis*, Giesbrecht.
> *Thaumaleus thompsoni*, Giesbrecht.
> *Pseudopsyllus elongatus*, T. Scott (gen. et sp. n.).
> *Acontiophorus ornatus* (Brady and Robertson).
> *Cancerilla tubulata*, Dalyell.
> *Salenskya tuberosa*, Giard and Bonnier.

It will be observed from the above list that five new genera and ten new species are described in the present paper. There were a few other interesting organisms observed, such as *Arca pectunculoides* and *Cadulus subfusiformis* (two species of Mollusca), but valves only of the first, and a recently dead specimen of the other, were obtained.

The following is a detailed description of the more interesting of the species of Crustacea observed in the numerous gatherings examined during the year :—

CRUSTACEA.

Sub-Class ENTOMOSTRACA.

Order I.—COPEPODA.

CALANIDÆ.

Calanus hyperboreus, Kröyer.

1838.   *Calanus hyperboreus*, Kr., Danske Selsk. Afh., vol. 7
        p. 310, t. 4.

A few specimens were obtained in a gathering collected about 22 miles
to the north of Shetland on May 17th, 1901. One or two specimens
were also found in another gathering collected to the east of the
Shetland Islands on the 22nd of the same month, and as both these
localities are within the British area, this northern *Calanus* is entitled
to be regarded as a member of the copepod fauna of the British Islands.
This appears to be the only *Calanus* in the North Atlantic or Arctic seas
in which the last segment of the thorax has the postero-lateral margins
distinctly angular. The size of specimens appears to vary a good deal,
but one or two of the largest of those now recorded measured one-fifth
of an inch (5 mm.) in length.

Rhincalanus (?) gigas, Brady.

I have again to record this copepod from the Moray Firth. It does
not appear to differ much except in size from the form described by
Dr. Brady under the name given above.

Pseudocalanus elongatus, Boeck.

I include this common species in these notes in order to mention the
occurrence of a form somewhat smaller than the ordinary one. These two
forms have been observed in the Firth of Forth and in the Moray
Firth as well as off the Aberdeenshire coast. They were first noticed a
good many years ago, and they are still occasionally noticed. Though
the two forms have been carefully examined no important difference has
been observed between them.

Stephus minor, T. Scott.

1892.   *Stephus minor*, T. Scott, 10th Ann. Rept. Fishery Board
        for Scot., pt. iii., p. 245, pl. vii., figs. 1, 2, 10–13.

This species occurred in a bottom gathering from Smith Bank, Moray
Firth, collected on February 15th, 1901, at a depth of about 24 fathoms.
Stephus minor, though apparently widely distributed, seems to be a rare
species, as seldom more than one or two specimens are obtained in any
single gathering.

Stephus scotti, G. O. Sars.

1896.   *Stephus gyrans*, T. Scott (not Giesbrecht), 15th Ann.
        Rept. of the Fishery Board for Scot., pt. iii., p. 146, pl. ii.,
        fig. 9, pl. iii., figs. 17, 18.

Recently, when re-examining some copepoda collected in the Firth of
Forth in 1892, I obtained a single female specimen of this species.

This is the first time the species has beeen observed on the east coast, but on the west coast it has been taken in Loch Fyne and in the Sound of Mull; it appears to be a very scarce species. Prof. G. O. Sars informs me *in lit.* that this species has long ago been found by him off the Norwegian coast, and that, though the female resembles somewhat the female of *Stephus gyrans*, Giesbrecht, the male (which I have not seen) differs very considerably from the male of the form described by that author. He also informs me that the name he proposes for this species is *Stephus scotti*, a full description of which will be found in vol iv of his work on the Crustacea of Norway, now in course of publication.

Pseudocyclopia caudata, T. Scott.

> 1894. *Pseudocyclopia caudata*, T. Scott, 12th Ann. Rept. Fishery Board for Scot, pt. iii., p. 236, pl. v., figs. 1–8.

This is a smaller and rarer species than *Pseudocyclopia crassicornis*; it has been taken in the Firth of Clyde as well as in the Firth of Forth, and I now record its occurrence in the deep water (60 fathoms) about 10 miles off Aberdeen, where it was collected on July 30th, 1901. In this species the caudal furca are distinctly more elongate than in the other described species.

Ætidius armatus, Brady.

The female specimen of a copepod which I regard as identical with the species described by Dr. Brady under the above name in his Report on the Challenger Copepoda, occurred in a mid-water tow-net gathering collected off the east side of the Shetland Islands on May 22nd, 1901. Professor Sars, in his new work on the Copepoda of Norway, remarks that the *Pseudocalanus armatus* of Boeck is probably identical with the *Ætidius armatus* described by Professor G. S. Brady, and if so, the name of the species would stand as *Ætidus armatus* (Boeck).

Bradyidius armatus (Vanhöffen).

> 1878. *Pseudocalanus armatus*, Brady (not Boeck), Brit. Copep., vol. i., p. 46, pl. iv., figs. 1–11.
>
> 1899. *Bradyidius armatus* (Vanhöffen). *Vide* Das Tierreich, p. 32.
>
> (?) 1884. *Undinopsis Bradyi*, G. O. Sars, in Sp. Schneider's Rept. of Invertebrata from the Kvænangen Fjord.

A set of drawings showing the structural differences between the male and female of this species, with descriptive notes, was published in the 14th Annual Report of the Fishery Board for Scotland. It is a species that is of frequent occurrence in the Clyde, where it was first obtained by Dr. Brady many years ago, and it is also found on other parts of the Scottish coasts—its distribution extending to the north of the Shetland Islands, where it was collected on the 17th of May of the present year (1901). *Undinopsis Bradyi*, G. O. Sars, may be identical with this species, but it differs somewhat in the structure of the fifth thoracic feet of the male.

Scolecithrix hibernica, A. Scott.

> 1896. *Scolecithrix hibernica*, A. Scott, Ann. and Mag. Nat. Hist., (6), vol. 18, p. 362, pl. xvii., figs. 1–9; pl. xviii., figs. 1–9.

This species was obtained in Loch Etive in about 60 fathoms on September 17th, 1901; Loch Etive is a new station for this *Scolecithrix*.

(?) *Scolecithrix brevicornis*, G. O. Sars. Pl. XXV., figs. 1 and 2.

> 1900. *Scolecithrix brevicornis*, G. O. Sars, The Norw. N. Polar Exped. (1893–96), p. 49, pl. x.

Description of the Male.—Length 1·5 mm. (about $\frac{1}{17}$ of an inch). Body viewed from above, elongate-oval; equal to rather more than two-thirds of the entire length, widest behind the middle. The anterior somite is equal to nearly twice the entire length of the next three: abdomen, slender, scarcely half as long as the thorax. Antennules moderately short, scarcely reaching to the first abdominal segment, the first seven joints very short, the eighth to the (?) twelfth coalescent, the remaining joints somewhat similar to those of the female.

The fifth thoracic feet were slightly damaged, and the following description of them is to some extent imperfect. The left leg is composed of four joints, the first is swollen but scarcely as long as the next, the third and fourth, which are sub-equal, are also rather shorter as well as being more slender than the second. The right leg consists of three (? or four) joints, the first and second are moderately elongate, the third is somewhat shorter and narrower, alongside of third joint and articulated with it to the end of the second is a slender branch-like appendage equal in length to the third joint (fig. 2).

Habitat.—Collected about sixty miles to the east of the Shetland Islands on May 22nd, 1901. In the same gathering were obtained *Metridia longa, Xanthocalanus* (?) *borealis*, and one or two other rare copepods. The copepod which I have provisionally ascribed to *Scolecithrix brevicornis* agrees with the female described by Professor G. O. Sars in its general form, in the proportional lengths of the thoracic segments, and in the comparatively short antennules. Sars did not obtain the male of his species, and there is therefore some uncertainty as to whether our specimen belongs to that species or not. This gathering was from moderately deep water.

Xanthocalanus (?) *borealis*, G. O. Sars. Pl. XXII., figs. 8 and 9.

> 1900. *Xanthocalanus borealis*, G. O. Sars, Crust. of Norw. N. Polar Exped., p. 49, pl. xi.

A female specimen of a *Xanthocalanus* was obtained in a tow-net gathering collected to the eastward of the Shetland Islands on May 22nd, 1901. It was thought at first that this specimen might belong to the *Xanthocalanus minor*, Giesbrecht, but on a careful comparison of it with the description and figures of that species and with the description and figures of G. O. Sars' *Xanthocalanus borealis* it was found to agree much better with the latter than with the former species. Prof. G. O. Sars, in the portion of his new work on the Crustacea of Norway, just published,* gives a figure of the fifth pair of thoracic feet of a slightly immature female, which agrees fairly well with the drawing of the fifth pair of the Shetland specimen (fig. 9).

This specimen (fig. 8) measures 2·89 mm. (rather less than an eighth of an inch) in length. The cephalothorax is moderately robust, and when viewed from above the width is seen to be equal to more than a third of its entire length, the sides are evenly rounded, and the posterior thoracic segment is produced on each side into acute angular processes which reach beyond the middle of the first abdominal segment. A minute seta springs from each side of the second-last thoracic somite.

---

*Crustacea of Norway, by G. O. Sars, vol. iv. (Copepoda), parts iii. and iv. (1902), p. 46, pl. xxxi., xxxii.

The abdomen is narrow and short, being scarcely more than a fifth of the entire length of thorax and abdomen.

The caudal furca are very short. No furcal setæ are shown in the figure, as they had all been broken off.

The fifth pair of thoracic feet are of moderate size, the first joint is equal to rather more than half the length of the second one, and is slightly gibbous on the interior aspect, the inner margin is also densely fringed with minute hairs; the second joint is slightly distorted, being bent inward somewhat abruptly near the middle; this joint is armed with three moderately stout setiferous terminal spines, and is also furnished with a few minute setæ on the lateral aspect and near where the joint is bent as shown in the drawing (fig. 9).

The occurrence of *Xanthocalanus borealis* in the neighbourhood of the Shetland Islands adds another species to the copepod fauna of Scotland. The distribution of this species appears to be somewhat restricted, as, with the exception of a single young female obtained in a gathering collected north of the New Siberian Islands, Prof. Sars has only found it in Stavanger Fjord, and in a few other places off the west coast of Norway, but its occurrence in this Shetland gathering seems to indicate that it may after all have an extensive distribution.

Phænna zetlandica, sp. n.   Pl. XXII., figs. 5–7.

A male specimen of a (?) *Phænna*, which I have provisionally named *P. zetlandica*, was obtained in the same gathering with the *Xanthocalanus* just described, agrees in some respects very closely with *Phænna spinifera,* Claus, and may probably be only a form of that species. The specimen (fig. 5) measures nearly two and a half millimetres ($\frac{1}{10}$ of an inch) in length. The thorax is moderately robust, and when seen from above is broadest behind the middle. The cephalo-thoracic segment, which is about equal to half the entire length of the animal, tapers gradually to the broadly rounded forehead, the next three thoracic segments are short. The abdomen is narrow, and its length is equal to little more than a third of that of the cephalo-thorax; it consists of five segments, the genital segment is slightly longer than any of the others, the last is very short; the caudal furca are very short (the furcal hairs are not shown, as they had accidentally been broken off).

The antennules, which appear to be the same on both sides, do not reach to the end of the abdomen, the first six joints are short, the seventh, eighth, and ninth are partly coalescent, the tenth to the fifteenth joints, which are of moderate length, are sub-equal, the seventeenth joint is also nearly equal to these in length, but the remaining joints are rather shorter; the antennules are only sparingly setiferous, as shown by the drawing (fig. 5).

The fifth pair of feet, in which both branches are developed, are very similar to those of *Phænna spinifera*, Claus. The right branch is elongated and slender and composed of four joints, the first three are of nearly equal length, but the last is about one and a half times the length of the preceding joint and considerably attenuated so as to resemble a spine rather than a joint (fig. 6). The left branch is rather longer than the right one and apparently five-jointed; the terminal portion of this branch consists of two appendages, the inner one being short and moderately broad, rounded at the end and fringed with setæ, the other is narrow and longer than the inner one and forms with it a kind of finger and thumb-like arrangement, as shown in figure 7.

The fifth thoracic feet of this Shetland specimen are seen to differ somewhat from the fifth pair of the male of *Phaenna spinifera*, Claus, especially in the terminal part of the left branch, when compared with the figure of the male fifth pair in Dr. Giesbrecht's work *; and I have therefore provisionally retained it under a separate specific name.

CENTROPAGIDÆ.

Metridia longa (Lubbock).   Pl. XXII., figs. 1–4.

One or two specimens of Lubbock's *Metridia (Calanus) longa* were found in a gathering collected about 22 miles to the north and in another collected about 60 miles to the east of the Shetland Islands; the first was collected on May 17th, and the other on May 22nd, 1901.   One of the largest specimens taken in the first gathering measured 4 mm., or about one-sixth of an inch in length.   The male and female specimens represented by the drawings (figs. 1. to 4) were obtained in the second of the two gatherings mentioned above; this female, which is scarcely so large as the largest specimen found in the first gathering, measured about 3·7 mm. and the male 2·8 mm. in length.   The fifth pair of thoracic feet of the male and female, as represented by the figures 3 and 4, are practically identical with the figures of the same appendages given by Dr. Giesbrecht.†

Metridia lucens, Boeck.

This species was moderately frequent in the same gatherings with the last, as well as in gatherings from the Moray Firth and from other parts of the Scottish coasts.

PSEUDOCYCLOPIDÆ.

Pseudocyclops obtusatus, G. S. Brady.

One or two specimens of this curious species occurred in the washings of some dredged material from the north-west end of Inchkeith, Firth of Forth, collected on May 23rd, 1901. *Pseudocyclops obtusatus* has already been recorded from the Forth estuary, but has not been very often met with.

PONTELLIDÆ.

Acartia bifilosa, Giesbrecht.

Acartia bifilosa was obtained along with the two species of *Metridia* mentioned above in the gatherings from the north and east of Shetland, and also in gatherings from the Firth of Forth recently collected.

CYCLOPIDÆ.

Oithona (?) *setiger*, Dana.

Several specimens of an *Oithona* which I ascribe to *Oithona setiger*, Dana, occurred in a tow-net gathering from the Firth of Forth collected on April 22nd, and also in a similar gathering collected to the east of the Shetland Islands on the 22nd of May, 1901.   In these specimens the rostrum, which projects forward instead of downward as in *O. helgolandica*, Claus (*O. similis*, Giesbrecht), tapers gradually to an acute

---

* Fauna u. Flora d. Golfes v. Neapel, vol. 19, p. 293, t. 12, fig. 5.
† Pelagischen Copepoden des Golfes von Neapel, pl. 33, figs. 20 and 23.

point, and not abruptly as in *O. plumifera*, Baird. This same *Oithona* was recorded from the Firth of Forth in 1891 under Dana's name—*Oithona setiger.**

ASCIDICOLIDÆ.

Doropygus normani, G. S. Brady.

1898. *Doropygus normani*, G. S. Brady, Mon. Brit. Copep., vol. i., p. 136, pl. xxxii., figs. 1–14.

This large and distinct species was obtained in some material dredged in 8 fathoms off the North Craig, Firth of Forth, on July the 4th, 1901. Though the branchial chamber of the larger ascidians is the usual habitat of this species, it sometimes happens that the test of the ascidian is ruptured by the dredge, and the copepods that may be contained within the branchial chamber are then set free. Probably this may explain the reason why the specimens of a somewhat peculiar type of copepod, which I now describe, were obtained "free" amongst the same dredged material from the North Craig in which the *Doropygus* occurred.

~~Platypsyllus~~, T. Scott (gen. nov.).

Body flat and sub-ovate. Antennules rudimentary. Antennæ (?) obsolete. Mouth consisting of a small suctorial tube. Mandibles, maxillæ, and maxillipeds (?) obsolete. No thoracic feet observed. Abdomen scarcely distinguishable from the thorax. Ovisacs two, elongated.

Platypsyllus minor, T. Scott (sp. nov.). Pl. XXV., figs. 15–16.

Description of Female.—Length 1·7 mm. (nearly $\frac{1}{15}$ of an inch) in length. Body, seen from above, flat, oblong-ovate, greatest width near the posterior end, but the form varies somewhat in different specimens. Colour (after a short immersion in alcohol), opaque-white. Antennule obsolete or nearly so, reduced to a minute lobe on each side of the forehead, and bearing one or two extremely minute setæ. Antennæ obsolete. Mouth suctorial, and consisting of a small trumpet-shaped tube (fig. 16). Mandibles and other mouth appendages wanting. Thoracic feet also wanting. Abdomen indistinct from thorax. Ovisacs two, elongated, and containing numerous small ova; each ovisac originates from a small lateral angular process at the posterior end of the body (fig. 15).

Habitat.—Vicinity of North Craig, Firth of Forth, dredged in 8 fathoms on July 4th, 1901. No males have been observed.

Remarks.—The first specimens of this curious copepod observed were without ovisacs, and from their shape, their colour, and the apparent entire absence of appendages, there was at first considerable doubt regarding them, but ere long a specimen turned up with two long ovisacs attached to it, and then their true character was revealed.

From the simple and unarmed structure of these copepods it is fairly evident, I think, that if they are not commensals of some ascidian they must receive from some other host the shelter and protection necessary to organisms apparently so helpless as these animals seem to be.

Nereicola concinna, T. Scott (sp. nov.). Pl. XXV., figs. 8–14.

Description of the Female.—Length, 1·6 mm. (about $\frac{1}{16}$ of an inch) Body considerably dilated, rather more than one and a half times longer

* Ninth Annual Report of the Fishery Board for Scotland, Part III., p. 301.

than broad, widest in the middle. The cephalo-thoracic segment scarcely distinct, being indicated by a simple constriction ; thoracic segments all coalescent. Abdomen very small, and apparently consisting of two somites ; the first somite is short but moderately broad ; the second is also short, and tapers abruptly to the slightly bilobed extremity. Caudal furca extremely small. Ovisacs (two) large (fig. 8).

The antennules are short, moderately slender, and five-jointed ; the first and the last three are sub-equal in length, but the second is about one and a half times the length of the third ; all the joints are sparingly setiferous (fig. 9).

The antennæ are short and three jointed, the end joint is armed with one small marginal and three or four stout terminal spines, which are slightly hooked (fig. 10) ; they are not provided with secondary branches.

The mandibles are large and elongated ; proximally they are somewhat dilated, but they taper gradually to the distal extremity, where they are armed with two rows of short but stout tooth-spines as shown in the drawing (fig. 11). No maxillæ could be observed.

The first and second maxillipeds are very stout but of a somewhat rudimentary structure ; the terminal claws are very short, but stout and tooth-like (figs. 12, 13). No thoracic feet were observed.

Habitat.—Parasitic on specimens of a marine annelid, *Eulalia viridis* Œrsted, dredged by the "Garland" in 55 to 65 fathoms in Loch Etive, west coast of Scotland, on September 17th, 1901. Several specimens, including old and young, were observed ; one or two of the specimens occurred still adhering to fragments of the annelid, on which they appeared to be able to take a very firm hold.

At first I was not sure but that these Loch Etive specimens might belong to the same species as those found by Professor M'Intosh on *Nereis cultrifera*, Grube,[*] on the shores of the Channel Islands. I therefore sent a specimen to him for his opinion as to whether it was the same as those he had discovered ; in replying to me he pointed out certain differences observed by him, and also very kindly sent me an example of the form from the Channel Islands so that I might more easily observe the differences he referred to. He stated further that the form found by him was the *Nereicola ovata* described by Keferstein in 1860.[†] I am also indebted to Professor M'Intosh for the name of the annelid from Loch Etive on which the copepods described here were obtained.

The difference in the form of immature specimens of *Nereicola concinna* from that of similar specimens of *N. ovat* is even more marked than in the adults (fig. 14).

HARPACTICIDÆ.

Eucanuella spinifera, T. Scott.

1901.   *Eucanuella spinifera*, T. Scott, 19th Ann. Rept. Fishery
         Board for Scotland., pt. iii., p. 245, pl. xviii., figs. 1–10.

This species, described in Part iii. of the 19th Annual Report, has again been observed in a gathering of bottom material collected to the east of the Shetland Islands in 60 to 70 fathoms on May 22nd, 1901. *Eucanuella* is apparently a deep-water species.

---

[*] On a Crustacean parasite of *Nereis cultrifera*, Grube, by W. C. MacIntosh, M.D., *Micr. Journ.*, vol. x., N.S., p. 39, pl. v.

[†] Zeit. f. w. Zool., bd. xii. (1860), taf. xliii., f. 1–4, p. 461.

Ectinosoma melaniceps, Boeck, (?) var.   Pl. XXII., figs. 10-16.

　1864.   *Ectinosoma melaniceps*, Boeck, Overs. Norg. Copep., p. 20.

A few specimens of an *Ectinosoma* obtained in some dredged material from Station VI., Firth of Forth, have such a general resemblance to *Ectinosoma melaniceps*, Boeck, that though they differ in some details of structure they may after all be only a form of that species. The following description will indicate a few of the more important points of difference :—

The antennules are six-jointed, the basal joint appears to be the largest, being nearly twice the length of the next one, but the others are comparatively small (fig. 11). The antennæ appear to be similar to those of *Ectinosoma melaniceps*. The mandibles and mandible-palps are slender (fig. 12). The other mouth organs and swimming feet are somewhat similar to the same appendages in *E. melaniceps* (figs. 13, 14).

The fifth pair of the present form have the basal joints not very broad, the inner produced part scarcely reaches to the middle of the secondary joints, and is abruptly truncate at the apex ; a short and a moderately long seta spring from the apex, the inner seta being the longest. The secondary joints are sub-cylindrical, and about one and a half times longer than broad ; they are furnished with three terminal setæ, the middle one, which springs from a slightly produced lobe, is considerably longer than the other two ; a small lateral hair is observed between the elongated middle seta and the outer one as shown in the figure (fig. 15). No males were observed.

Stenhelia ima, G. S. Brady.

　1872.   *Canthocamptus imus*, Brady, Nat. Hist. Northumb. and Durham, vol. iv., p. 432, pl. xix., figs. 1-5.

This species occurred very sparingly in washings from dredged material collected near Inchkeith, Firth of Forth, on July 4th, 1901. Though *Stenhelia ima* is apparently widely distributed, I have not found it to be very common.

Stenhelia intermedia, T. Scott.

　1897.   *Stenhelia intermedia*, T. Scott, 15th Ann. Rept. Fishery Board for Scotland, pt. iii., p. 169, pl. ii., figs. 10-21.

This somewhat rare species was dredged in Loch Etive in about 60 fathoms on September 17th, 1901.

Stenhelia hirsuta, I. C. Thompson.

　1893.   *Stenhelia hirsuta*, I. C. Thompson, Trans. L'pool Biol. Soc., vol. vii., p. 20, pl. xxi., fig. 2, d.e.f. (separate reprint).

Specimens of *Stenhelia hirsuta* were occasionally observed in gatherings of dredged material from the Firth of Forth collected in July, 1901.

(?) *Stenhelia hispida*, G. S. Brady.   Pl. XXIV., figs. 19-26.

　1880.   *Stenhelia hispida*, Brady, Mon. Brit. Copep., vol. ii., p. 32, pl. xlii., figs. 1-14.

Description of the Female.—Length about 1 mm. ($\frac{1}{25}$ of an inch). The body is in general appearance somewhat similar to *Stenhelia ima* (fig. 19).

Antennules scarcely reaching to the end of the cephalo-thoracic segment, eight-jointed; the first four joints are large, but the last four are very small (fig. 20).

Antennæ short and moderately stout; secondary branches small and three-jointed (fig. 21).

Mandibles robust, with a broadly truncate biting edge which is armed with several small and somewhat irregular teeth; palp well developed, two-branched, but the posterior branch is very small (fig. 22).

Second maxillipeds stout, and armed with a moderately strong terminal claw (fig. 23).

First pair of thoracic feet stout; the proximal joint of the inner branches scarcely reach beyond the ends of the outer ones, second joint small, the third is about twice the length of the second, while the second and third together are scarcely equal to half the length of the first joint; the outer branches are composed of three sub-equal joints (fig. 24). The second, third, and fourth pairs are slender and moderately elongated, and the branches are all three-jointed; figure 25, which represents the fourth pair, shows that the inner branches are only slightly longer than the outer ones.

The fifth pair are foliaceous, the basal joints being broadly subtriangular, and furnished with five moderately stout but not very long setæ on the somewhat rounded apex. The secondary joints are subrotund, and scarcely reach beyond the apex of the basal joints; they are each provided with five setæ of unequal lengths, the second and third, counting from the inside, being considerably longer than the other three (fig. 26).

The caudal furca are very short.

Habitat.—This species was obtained in some dredged material from Station VII., Firth of Forth (between Fidra and the Bass Rock), on July 9th, 1901; it appears to be somewhat rare. No males were observed.

Remarks.—The *Stenhelia* just described, and which I have referred to *Stenhelia hispida*, G. S. Brady, while differing in a few particulars from the species named, agrees very well with it in several important particulars. The structure of the antennules, for example, is almost identical with that of the antennules in *Stenhelia hispida* as described and figured by Professor G. S. Brady, and the first and fifth pairs of feet are also nearly alike in both.

Stenhelia confusa, T. Scott (sp. n.). Pl. XXII., figs. 17–25.

Description of the Female.—Length about 9 mm. (about $\frac{1}{28}$ of an inch). Body moderately stout, tapering slightly towards the posterior end; rostrum prominent (fig. 17).

Antennules shorter than the cephalo-thoracic segment, moderately stout, and composed of eight joints; the first, second, and end joints are the longest, while the fifth and sixth are very small (fig. 18). The approximate proportional lengths of the various joints are shown in the formula—

Proportional lengths of the joints,	.	22	.	14	.	8	.	6	.	3	.	4	.	7	.	14
Numbers of the joints,	.	.	.	1	.	2	.	3	.	4	.	5	.	6	.	7	.	8

All the joints with the exception of the first are moderately setiferous.

The antennæ are somewhat similar to those of *Stenhelia ima*, Brady (fig. 19).

The mandibles are also somewhat similar to those of the same species, but the branches of the mandible-palp appear to be shorter (fig. 20).

The second maxillipeds resemble the same organs in *Stenhelia hispida*, G. S. Brady (fig. 21).

The first pair of thoracic feet are moderately stout, the inner branches are about one and a half times the length of the outer branches, first joint is equal to the entire length of the second and third, while the second is equal to about two-thirds the length of the end joint ; the joints of the outer branches are sub-equal (fig. 22).

The remaining three pairs of feet do not differ very materially from the same appendages of *Stenhelia ima*, except that they are scarcely so elongated (fig. 23).

The fifth pair is foliaceous, but comparatively short ; the secondary joints do not extend much beyond the produced inner portion of the basal joints, which in outline is broadly triangular, and furnished with three setæ of unequal length on the bluntly rounded apex, while two dagger-like spines spring from the inner margins ; the secondary joints are also somewhat triangular, but they are rather narrower than the produced inner portion of the basal joints ; the apex of the secondary joints, which is somewhat truncate, bears three small setæ, and three small hairs spring from the outer margin (fig. 24).

Caudal furca slender, and about as long as the last abdominal segment (fig. 25).

Habitat.—The species occurred very sparingly in some washings of dredged material from Station III., Firth of Forth (to the east of Inchkeith), collected on June 7th, 1901. No males were observed.

Remarks.—This species resembles *Stenhelia hirsuta*, I. C. Thompson, in some respects, and especially in the structure of the antennules, and in the length of the caudal furca ; but in that species the inner branches of the first pair of thoracic feet are long and slender, much more so than the present form. The two species differ also in the form of the fifth pair. And though the structure of the antennules of *Stenhelia confusa* bears a certain resemblance to those of *Stenhelia hirsuta*, it differs very distinctly in this as well as in some other respects from almost every other species of the genus.

Ameira tenuicornis, T. Scott (sp. n.). Pl. XXIV., figs. 1–9.

Description of the Female.—Length ·67 mm. (about $\frac{1}{37}$ of an inch). Body moderately slender and sub-cylindrical ; rostrum small (fig. 1).

Antennules slender and elongated, exceeding in length the cephalo-thoracic segment, and composed of eight joints ; the first and second joints are sub-equal ; the third and fourth are also sub-equal, but smaller than the preceding joints (fig. 2). The approximate proportional lengths of the various joints are shown by the formula. All the joints with the exception of the first one are sparingly setiferous.

Proportional lengths of the joints,	.	24	.	25	.	18	.	15	.	10	.	9	.	6	.	9
Number of the joints,	- - - - -	1	.	2	.	3	.	4	.	5	.	6	.	7	.	8

The antennæ are very slender, and the secondary branches are small and two-jointed, the end joint being the smallest (fig. 3).

The mandibles are of moderate size, narrow-cylindrical, and obliquely truncate at the apex, which is armed with a few minute spinules ; the basal portion of the mandible-palp is small but slightly dilated, and carries two branches ; the marginal branch is small and one-jointed and furnished with a few setæ ; the end joint is long and very slender, and is minutely serrated at the extremity (fig. 4).

The second maxillipeds (posterior foot-jaws) are moderately stout and armed with strong terminal claws (fig. 5).

The first pair of thoracic feet are elongated and slender, and especially the inner branches, the first joint being about as long as the entire length of the outer branches; the other joints are small, but the end one is about twice the length of the penultimate joint (fig. 6). The following three pairs have the outer branches long and slender; the inner branches are also slender, but they are shorter than the outer, as shown by the figure (fig. 7) which represents one of the fourth pair.

The fifth pair are small and somewhat foliaceous, the basal joints are sub-triangular and provided with about four setæ on the rounded apex. The secondary joints are subovate, and the inner margins are fringed with minute hairs, while one or two moderately long setæ spring from the apex, and one or two others from the outer margin (fig. 8).

Caudal furca shorter than the last abdominal segment (fig. 9).

Habitat.—Dredged at Station VI., Firth of Forth (off St. Monans), in July 1901; only one or two specimens were obtained, but no males were observed.

Remarks.—This species is readily distinguished by the elongated antennules and the long and slender first pair of feet; it differs in both of these appendages from *Ameira longipes*, Boeck, as well as from the other described species of *Ameira* known to me.

Ameira propinqua, T. Scott (n. sp.). Pl. XXIV., figs. 10–18.

Description of the Female.—Length about ·6mm. (nearly $\frac{1}{40}$ of an inch). Body slender, sub-cylindrical, the cephalo-thoracic segment about equal to the entire length of the next three segments, rostrum very small (fig. 10).

Antennules slender and rather longer than the cephalo-thoracic segment, eight-jointed; the second joint is the longest, the first and third are sub-equal and about two-thirds the length of the second; the remaining joints are small (fig. 11). The formula shows approximately the proportional lengths of the various joints :—

Proportional lengths of the joints,	12 · 17 · 12 · 4 · 4 · 5 · 3 · 5
Numbers of the joints, · ·	1 · 2 · 3 · 4 · 5 · 6 · 7 · 8

Antennæ elongate and moderately stout, secondary branches small, slender, and one-jointed (fig. 12).

Mandibles cylindrical and not very broad, the truncate apex is armed with a stout spine on the outer angle and a few small spiniform setæ, as shown by the drawing (fig. 13). The mandible-palp is of moderate size, the basal joint is provided with a single one-jointed and terminal branch.

The second maxillipeds (posterior foot-jaws) are small but with well-developed terminal claws, which are rather longer than the joints to which they are articulated (fig. 14).

The first four pairs of thoracic feet are slender and elongated. In the first pair the inner branches are narrow and considerably longer than the outer branches, the length of the first joint is equal to that of the second and third combined, but the second and third joints are sub-equal in length; a single small seta springs from the inner margins of the first and second joints, while the end joints are provided with three terminal hairs, the middle one being the longest. The outer branches, which are composed of three sub-equal joints, reach to a little beyond the end of the first joint of the inner branches (fig. 15). The inner branches of the next three pairs are considerably shorter than the outer branches, which are slender and elongated (fig. 16). In all the four pairs the outer and inner branches are three-jointed.

Iu the fifth pair the basal joints are broadly foliaceous and sub-triangular in outline, with the apex truncate and provided with four spiniform setæ, the outermost of which is very small ; the secondary joiuts are long, narrow, and cylindrical, being about four times longer than broad, the extremity, which is obliquely truncate, carries several setæ, the two inner ones being longer than the others (fig. 17).

The caudal furca are shorter than the last abdominal segment; the furcal setæ are elongated (fig. 19).

Habitat.—Station VI. (off St. Monans), Firth of Forth, dredged on July 8th, 1901. The species is apparently very rare. (For drawings of what may be the male of this species, *see* Pl. XXII., figs. 36–42, and Pl. XXIII., fig. 1.)

Remarks.—In some respects *Ameira propinqua* comes rather near to *Ameira longiremis*, T. Scott, the fifth feet especially being very similar to those of that species, as well as to those of *Ameira longipes*, Boeck, but the structure of the antennules and of the first pair of feet separate it distinctly from both the species named.

Pseudotachidius coronatus, T. Scott

> 1898. *Pseudotachidius coronatus*, T. Scott, 16th Ann. Rept. Fishery Board for Scot., pt. iii., p. 267, pl. xiii., figs. 12–26 ; pl. xv., figs. 1–4.

This distinct species has been obtained in a gathering of small crustacea dredged in Loch Etive in 55 to 65 fathoms on September 17th, 1901. *Pseudotachidius coronatus* has not previously been recorded out of the Clyde district.

Pterinopsyllus insignis, G. S. Brady.

> 1868. *Lophophorus insignis*, G. S. Brady, Mon. Brit. Copep., vol. i., p. 122, pl. xiii., figs. 1–10. (See also *op. cit.*, vol. iii., p. 23, where the generic name is changed to *Pterinopsyllus*, —"*Lophophorus*" being preoccupied.)

This fine species was obtained in the same gathering as the *Pseudotachidius* just recorded, and appears to be the first record of it from the West of Scotland. It has in previous years been obtained in the Firth of Forth and the Moray Firth. Although *Pterinopsyllus insignis* and *Pseudotachidius coronatus* have a general resemblance to one another they may be readily distinguished by the difference in the lengths of the antennules—those of the first-named species being distinctly longer and more slender than in the other.

Mesochra macintoshi, T. and A. Scott.

> 1895. *Mesochra macintoshi*, T. and A. Scott, Ann. and Mag. Nat. Hist., (6), vol. xv., p. 53, pl. vi., figs. 1–7.

Mesochra macintoshi, which was first observed amongst a number of peculiarly slender copepods collected on the south shore of the Firth of Forth near Musselburgh, has lately been obtained at Station VI. (off St. Monans). It is apparently a rare species, but being one of those forms which live upon the bottom it may from its small size be easily overlooked.

Pseudomesochra, T. Scott (gen. nov).

This genus is somewhat intermediate between *Mesochra*, Boeck, and *Cletodes*, G. S. Brady. The antennules (anterior antennæ) are composed

of six joints.   The secondary branches of the antennæ (posterior antennæ)
are small and two-jointed.   Mandibles stout, mandible-palp well
developed and provided with two branches,   Other mouth organs similar
to those in *Mesochra* and *Cletodes*.   The first four pairs of thoracic feet
have the outer branches three- and the inner branches all two-jointed.
Fifth somewhat rudimentary and composed of a single lamelliform joint.
Ovisacs apparently double.

Pseudomesochra differs from *Mesochra, Cletodes*, and allied genera
chiefly in the structure of the mandible-palp and fifth pair of feet.

Pseudomesochra longifurcata, T. Scott (sp. nov.) Pl. XXIV., figs. 27–35.

Description of the Female.—Length about ·5mm. ($\frac{1}{50}$ of an inch).
Body moderately stout, tapering slightly towards the posterior end ;
abdomen not distinct from thorax ; rostrum small ; caudal furca elongated,
being nearly equal to the entire length of the last three abdominal
segments (fig. 27).

Antennules short, moderately stout, and composed of six joints, the
first two and the last being each considerably longer than any of the
other three (fig. 28).   The approximate proportional lengths of the
various joints are shown by the formula :—

Proportional lengths of the joints, - 23 · 20 · 8 · 5 · 4 · 15

Numbers of the joints,   -   -   -   1 · 2 · 3 · 4 · 5 · 6

The antennæ are of moderate size ; secondary branches two-jointed and
provided with several marginal and terminal setæ (fig. 24).

Mandibles robust and having the biting end armed with several stout
teeth ; mandible-palp well developed and the basal part furnished with
two branches (fig. 30).

Second maxillipeds short and moderately stout, but the terminal
claw is rather feeble, and fringed with a few minute hairs (fig. 31).

The first two pairs of thoracic feet slender.   In the first pair both
branches are about the same length ; the joints of the inner branches are
sub-equal, and a single seta springs from the inner margin and three from
the end of the second joint, the proximal joint appears to be unprovided
with setæ or spines (fig. 32).   In the second, third, and fourth pairs the
inner branches in each are rather shorter than the outer branches, and
the end joints somewhat longer than the proximal ones, and in these
three pairs of feet the inner branches are more setiferous than the inner
branches of the first pair (fig. 33).   The fifth pair, which are small and
somewhat rudimentary, appear to be composed of a single lamelliform
joint, bearing three long apical setæ (fig. 34).

Habitat.—Upper Loch Etive, where it was dredged by the "Garland"
in over 60 fathoms on September 17th, 1901.   It appears to be a rare
species, as only a single specimen (a female) was observed, but as the
species is a very small one it may be easily overlooked.

Leptopontia, T. Scott (gen. nov.).

Body slender and cylindrical.   Antennules (anterior antennæ) seven-
jointed, slender.   The secondary branches of the antennæ are entirely
wanting or reduced to a single seta.   Mandibles slender and moderately
elongate ; mandible-palp also slender and one-branched.   Other mouth
organs somewhat as in *Mesochra*, Boeck.   The first four pairs of thoracic
feet slender, outer branches three- and inner branches all two-jointed.
The inner branches of the first pair elongated—the end joint being the
shortest, the outer branches short, the inner branches of the other three
pairs very short, and the outer elongated.   Fifth pair small, foliaceous,
two-branched ; secondary branches (or joints) minute.

Leptopontia curvicauda, T. Scott (sp. n.). Pl. XXII., figs. 26-35.

Description of the Female.—Length ·6mm. (about $\frac{1}{42}$ of an inch). Body slender, cylindrical; cephalo-thoracic segment moderately short, scarcely three times the length of the next one; rostrum short (fig. 26).

The antennules are slender and longer than the cephalo-thoracic segment, seven-jointed, the first joint is longer than any of the others, the second and third are sub-equal, the fourth, fifth, and sixth joints are each about half the length of the third, while the end joint is about twice the length of the one next to it (fig. 27). The formula shows approximately the proportional lengths of the different joints :—

Proportional lengths of the joints, -	28	18	17	8	7	7	14
Numbers of the joints, -	1	2	3	4	5	6	7

Antennæ slender, consisting of two elongated joints; the secondary branches are represented by a single minute hair (fig. 28).

Mandibles slender and elongated, the biting end is obliquely truncated and armed with a number of small teeth; the mandible-palp is also elongated, and one branched, the basal joint is moderately stout, and about twice the length of the single-jointed branch, and is furnished with a few small setæ (fig. 29).

The posterior foot-jaws (fig. 30) are not unlike those of *Tetragoniceps incertus,* T. Scott. The inner branches of the first pair of thoracic feet are slender, elongated, and two-jointed, the end joint being only about one-third the length of the other; the outer branches, which are three-jointed, are scarcely half the length of the inner branches (fig. 31). In the second, third, and fourth pairs the outer branches are long and slender, and three-jointed; the joints of the outer branches of the second pair are sub-equal in length, but in the outer branches of the third and fourth pairs the first and second joints are sub-equal and rather longer than the end joints. The inner branches of the same three pairs are short and two-jointed; in the second pair the inner branches are rather longer than the first joint of the outer branches; but in the third and fourth pairs the inner branches are shorter than the first joint of the outer branches, as shown by the drawing (figs. 32, 33).

In the fifth pair the basal joints are moderately large and foliaceous, the inner produced part is broadly rounded, and bears three apical setæ, and the secondary joints, which are very small, also carry three setæ (fig. 34).

The last abdominal segment is armed on the median dorsal aspect with a backward-pointing tooth as shown in figures 26 and 35.

The caudal furca, which are nearly as long as the last abdominal segment, become gradually attenuated towards the distal extremity; and in all the specimens that have been obtained the furca are distinctly recurved as shown in the habitus drawing (fig. 1).

Habitat.—Dredged in the Firth of Forth at Station VI. (off St. Monans) on July 8th, 1901; not very common.

Remarks.—Odd specimens of this species have been observed from time to time for a considerable while past, but the first specimens were put aside as being probably immature forms of some species already described. I am now, however, convinced that they are distinct *Leptopontia* in some respects resembles both *Mesochra,* Boeck, and *Tetragoniceps,* G. S. Brady, in its structural details, but in the absence of a secondary branch to the antennæ, in the elongate and slender form of the mandibles, and the one-branched mandible-palp, it does not agree very well with either of the genera named, which are its nearest allies.

Laophonte similis (Claus).

This species was obtained in the same material with the last, collected off St. Monans ; and it was also taken from the swimmerets of a Spider Crab (*Hyas arenarius*) from the Bay of Nigg on May 23rd, 1901.

Laophonte curticauda, Boeck.

Specimens of this species were obtained on several occasions on the swimmerets of the common Shore Crab (*Carcinus mœnas*), but whether its occurrence on the swimmerets of the Crab was accidental or whether the copepod is associated with the Crab as a commensal, is a question that will require further research to determine. It may be mentioned, however, the *Laophonte* was obtained as described on almost ever Crab examined.

Normanella attenuata, A. Scott.   Pl. XXIII., figs. 2-4.

1896.   *Normanella attenuata*, A. Scott, Rept. for 1895, on Lancashire Sea-Fisheries Laboratory, p. 47, pl. iv., figs. 8-20.

This species, which was first discovered in a gathering collected one mile off Spanish Head, Isle of Man, from a depth of 16 fathoms, is now added to the fauna of the Firth of Forth ; it was obtained in dredged material from Station VI. (off St. Monans), at a depth of about 10 fathoms. *Normanella attenuata* is a slender species, measuring about a millimetre in length. The rostrum is very small ; the antennules are slender, somewhat elongated, and nine-jointed, The outer branches of the first four pairs of thoracic feet are all three-jointed, but the inner branches of the first three pairs are two-jointed, while in the fourth pair the inner branches, like the outer ones, are three-jointed. The inner branches of the first pair, which are considerably longer than the outer ones, have the end joints short (fig. 3), and armed at the apex with an elongate spine, and two setæ of unequal length. The inner branches of the second, third, and fourth pairs are shorter than the elongate outer branches.

The fifth pair are broadly foliaceous and of moderate size, as shown by the drawing (fig. 4).

In the female the first segment of the abdomen is larger than those which follow, being composed of two coalescent joints, as shown by the habitus figure (fig. 2).

In this species the thoracic portion of the body is rather shorter than the abdomen, and gives to the animal a more than usually slender appearance.

Though, from its occurrence in the Forth, *Normanella attenuta* would appear to have a moderately extensive distribution, it does not seem to be very common. This species differs from *Normanella dubia*, Brady and Robertson, in its general conformation, by its elongated antennules, and by the form of the fifth pair of thoracic feet ; but it differs more particularly in that the inner branches of the fourth pair are three-jointed. In this respect it disagrees with the generic definition of *Normanella*, and may for this, and perhaps one or two other reasons, require to be removed to another genus.

Cletodes longicaudata, Brady and Robertson. Pl. XXIII., figs. 26–33.

1875.   *Cletodes longicaudata*, Brady and Robertson, Brit. Assoc. Rept., p. 196.
1880.   *Cletodes longicaudata*, Brady, Mon. Brit. Copep., vol. iii., p. 92, pl. lxxix., figs. 13–19.

A perfect specimen—a female with ova—was obtained in a gathering from the west end of Station III., Firth of Forth, collected on June 7th, 1901.   As the species is somewhat rare I give a short description of it, along with figures of some of the principal appendages.   The female specimen referred to measured about ·84 mm. ($\frac{1}{30}$ of an inch) in length from the forehead to the end of the caudal furca, but the caudal furca, which are very long, are about equal to a fourth of the entire length of the animal; the rostrum is of moderate length (fig. 26).

The antennules are short, moderately stout, and composed of five joints.   Professor G. S. Brady describes the antennules of his specimen as six-jointed, but this difference may be due to a slight local variation. The fourth joint is very small and carries an asthetask or sensory filament (fig. 27).   The formula shows approximately the proportional lengths of all the joints :—

Proportional lengths of the joints,	6	·	13	·	20	·	4	·	23
Numbers of the joints,	1	·	2	·	3	·	4	·	5

The antennæ (fig. 28) are of moderate size, but the secondary branches are small and one-jointed.

In the first four pairs of thoracic feet, the outer branches, which are all three-jointed, are all of them somewhat similar in structure ; the first joint is slightly longer than the second, while the end joint is distinctly longer than either of the other two; the marginal armature consists also of setæ rather than spines.   The inner branches of all the four pairs are slender and two-jointed, the first joints being very short.   In the first pair the end joints of the inner branches extend somewhat beyond the outer branches and are furnished with two long terminal setæ (fig. 29).   In the second pair the end joints of the inner branches are, like those of the first pair, long and slender, but they do not reach to the extremity of the outer branches; they bear two long terminal setæ (fig. 30).   In the third and fourth pairs, the inner branches are considerably shorter than the outer and bear one sub-terminal and two apical setæ (fig. 31).

The fifth pair are moderately large ; the secondary branches are elongated, narrow-cylindrical, and about four times longer than broad ; they are armed with two terminal and three strong marginal setæ, as shown in the figure ; the inner produced portion of the basal joint is nearly as long as the secondary joint, and provided with three moderately elongated setæ near the distal end ; a long slender seta also springs from the outer aspect of the basal as shown (fig. 32).

Cletodes longicaudata does not appear to be a very common species anywhere, although it seems to be widely distributed.

Fultonia, T. Scott (gen. nov.).

This genus is somewhat like *Cletodes*, G. S. Brady, in general appearance.   The abdomen in not distinctiy separated from the thorax. The antennules are moderately short and composed of about eight joints. The antennæ are each furnished with a small secondary branch.   The mouth organs are similar to the same appendages in *Cletodes*.   The outer branches of the first four pairs of thoracic feet are all three-jointed, while the inner branches of the first pair are composed of two, and of the second, third, and fourth pairs of three joints.   The fifth pair are nearly

2 F

as in *Cletodes*. The copepods for which I have instituted this genus do not correspond to any described form known to me. The genus is named in compliment to Dr. T. Wemyss Fulton, Superintendent of the Scientific Work of the Fishery Board for Scotland. •

Fultonia hirsuta, T. Scott (sp. nov.). Pl. XXIII., figs. 5–12.

Description of the Female.—Length about ·63 mm. ($\frac{1}{40}$ of an inch). Body somewhat slender and sub-cylindrical, the posterior margins of abdominal segments fringed with short projecting hairs, which give to this part of the animal a peculiarly hirsute appearance, caudal furca short, rostrum small (fig. 5).

Antennules rather slender, and shorter than the cephalo-thoracic segment, eight-jointed; the first three joints are moderately large, but the others are short; all the joints, with the exception of the first, are more or less setiferous; an elongated and slender asthetask springs from the upper distal angle of the fourth joint (fig. 6). The approximate proportional lengths of all the joints are shown by the formula—

Proportional lengths of the joints, . 10 · 14 · 11 · 6 · 5 · 7 · 6 · 6
Numbers of the joints . . . . 1 · 2 · 3 · 4 · 5 · 6 · 7 · 8

The antennæ are similar to those of *Cletodes*; the secondary branches are very small, and one-jointed, and furnished with a single terminal hair (fig. 7).

Mouth organs somewhat similar to those of *Cletodes*; the second maxillipeds, which are moderately elongated, are provided with long slender terminal claws (fig. 8).

The thoracic feet are all moderately slender; the outer branches of the first four pairs and· the inner branches of the second, third, and fourth pairs are all three-jointed, while the inner branches of the first pair are only two-jointed. All the inner branches are short; those of the first pair are just about half the length of the outer branches and carry three small terminal setæ or spines (fig. 9). The inner branches of the other three pairs are scarcely more than a third of the length of the outer branches (fig. 10). The outer branches of the same three pairs are rather more elongated than those of the first pair.

In the fifth pair the basal consists of a narrow plate articulated to the last thoracic segment, and bearing one or two small marginal setæ; the exterior extremity of the basal joint is produced into a very narrow lobe which forms the base of a small seta. The secondary joints are narrow and sub-cylindrical, and about four times longer than broad, and furnished with about half-a-dozen small marginal and terminal setæ (fig. 11).

Habitat.—Station VI., Firth of Forth (off St. Monans), dredged in 13 to 15 fathoms, on May 22nd, 1901.

Remarks.—This species somewhat resembles *Cletodes irrasa*, T. Scott, in its hirsute appearance, but in that species the antennules are six-jointed, the inner branches of all the first four pairs of thoracic feet are only two-jointed, while the caudal furca are moderately long and slender. The species does not resemble any described form known to me.

Nannopus palustris, G. S. Brady. Pl. XXIII., figs. 13–25.

1880. *Nannopus palustris*, Brady, Mon. Brit. Copep., vol. ii., p. 101, pl. lxxvii., figs. 18–20.

This curious and rather interesting copepod, which Professor G. S. Brady discovered living in brackish water pools in a salt march at Seaton Sluice, Northumberland, and which has subsequently been found in similar situations in various parts of the British Islands as well as on the Continent, was during the preceding summer (1901) obtained near Newburgh on the Ythan, Aberdeenshire. Both the males and females of *Nannopus palustris* were obtained in the brackish water pools at this Aberdeenshire Station, and as no description or drawing of the male has yet, so far as I know, been published, the following notes on both of the sexes may be of interest.

The body of the female seen from the dorsal aspect is moderately stout anteriorly, but tapers gradually and evenly towards the posterior end (fig. 13); the female specimen represented by the drawing measures about ·9 mm. ($\frac{1}{28}$ of an inch) in length.

The antennules, which are short and stout, are composed of five joints, the penultimate joint being very small; the last four joints are densely setiferous, but the sensory filament (asthetask) which springs from the end of the third joint is comparatively short and slender (fig. 15).

The antennæ are very small, though comparatively moderately stout ; the secondary branch is small and one-jointed and provided with a few apical setæ (fig. 16).

The mandibles are moderately stout, their biting end is somewhat truncate and provided with a few stout teeth ; the mandible-palp consists of a one-jointed and somewhat dilated appendage articulated near the base of the mandible and provided with a few short setæ (fig. 17).

The second maxillipeds are small, the terminal claws are moderately stout and carry a few minute hairs on their inner margins (fig. 18).

In the first four pairs of thoracic feet the outer branches, which are moderately stout, are all three-jointed and armed with elongate though somewhat slender spines on their outer margins; the inner branches of the first, second, and third pairs are considerably shorter than the outer, and each composed of two joints, but the inner branches of the first pair are rather smaller than those of the second and third (fig. 19). The inner branches of the fourth pair are rudimentary and consist of a single minute joint which carries a very small and a moderately elongated seta (fig. 20).

The fifth pair resemble somewhat closely the fifth pair in *Enhydrosoma curvatum*, Brady and Robertson, the basal joint is very broad and short, they form short lamelliform plates along the postero-ventral margin of the last thoracic segment, four moderately stout plumose setæ spring from the edge of the inner half of each basal joint; the secondary joints are small and sub-rotund, and each furnished with five moderately long setæ round the distal margin (fig. 21).

The caudal furca are very short.

Description of the Male.—The male differs little from the female except in the following points ; it is, when seen from above, distinctly narrower than the female, especially towards the anterior end (fig. 14). The antennules are shorter and less prominent, the basal joints are also more dilated, but the antennules taper quickly towards the distal end, the penultimate joint is considerably swollen and assumes a utricule-like form, while the end joint is very small (fig. 23).

The mouth organs in the male appear to be very similar to those of the female. The thoracic feet are also similar in both sexes, except that the inner branches of the third pair are provided with a short but stout terminal spine slightly hooked at the end, and a single plumose seta instead of a moderately long slender spine and two plumose setæ as in the third pair in the female (fig. 24). The fifth pair in the male resemble those of the female, except that the secondary joint appears to be almost obsolete (fig. 25); the appendages to the first abdominal segment in the male are small and carry three setæ—two moderately long and one short.

Dr. Eugene Canu has given a very full series of figures of the female, but he refers to the male as being unknown.*

Dactylopus coronatus, T. Scott.

> 1894. *Dactylopus coronatus*, T. Scott, 12th Ann. Rept. of the Fishery Board for Scot., pt. iii., p. 255, pl. ix., figs. 12–20.

This species was dredged off the North Craig, Inchkeith, Firth of Forth, in 8 fathoms, on July 4th, 1901. The first specimens from which the species were described were also obtained in the Firth of Forth, but in the vicinity of the Bass Rock, near the mouth of the estuary. The Firth of Forth appears to be the only Scottish locality where this *Dactylopus* has been hitherto obtained.

Cylindropsyllus lævis, G. S. Brady.

> 1880. *Cylindropsyllus lævis*, G. S. Brady, Mon. Brit. Copep., vol. iii., p. 30, pl. lxxxiv., figs. 1–8.

A male specimen of this large and somewhat remarkable species was obtained in some bottom material from Smith Bank, Moray Firth, on February 15th, 1901.

MONSTRILLIDÆ.

The Family Monstrillidæ is represented in the copepod fauna of our seas from the English Channel to the Shetland Islands; a few of the species appear to be somewhat local in their distribution, while others are co-extensive with the seas that surround our shores. The family comprises the two genera *Monstrilla*, Dana, and *Thaumaleus*, Kröyer, and both are represented in the marine fauna of Scotland. Specimens belonging to this curious group of copepods have been captured at odd times in the Firth of Forth almost every year since 1888.† The first specimens obtained were ascribed to *Cymbasoma rigida*, I. C. Thompson, but they were shortly afterwards submitted to Mr. Gilbert C. Bourne, who was preparing a revision of the various forms which had recently been observed in the British seas, and his opinion of these specimens from the Forth estuary was that they were identical with *Monstrilla helgolandica*, Claus.‡ Other specimens have been obtained at odd times which appeared to belong to the same species, and also one or two belonging to a different species, and which were subsequently identified as the true *Monstrilla rigida* of I. C. Thompson. Two apparently adult specimens of a *Monstrilla* were obtained in some tow-net gatherings collected by the "Garland " on the 24th of July, 1901, and as they seem to differ from those previously mentioned as occurring in the Forth estuary, I give here a short description of them—they appear to be identical with the form described by Dr. Giesbrecht under the name of *Monstrilla longiremis*.

Specimens of *Monstrilla* have also in recent years been obtained in the Firth of Clyde, and though usually they have occurred very sparingly, yet on one or two occasions large numbers have been obtained in a single tow-net gathering. In a small gathering of material, in which there was

---

* Les Copép. du Boulonnais, p. 166, pl. iv., fig. 6-21 (1892).
† Annual Report of the Fishery Board for Scotland, Part III., p. 316 (1889).
‡ Notes on the genus *Monstrilla*, Dana ; Quart. Journ. Micr. Sci., (2), vol. 30, p. 515 pl. xxxvii., figs. 14, 15 (1890).

a good deal of fibrous matter, collected in Loch Fyne on September 29th, 1899, over eighty specimens of *Monstrilla* were obtained, while in another, collected on the 28th of November, twenty-seven specimens were found.* All these specimens, which appear to belong to the one species, I have recorded as *Monstrilla* (?) *danæ*, Claparède.† It may be remarked, however, that Dr. Giesbrecht seems to think that the species described by Claparéde is not a *Monstrilla*, but should be placed in the genus *Thaumaleus*.‡

In a tow-net gathering collected in Lerwick Harbour in October last year (1900), a specimen belonging to the group of copepods under consideration was obtained, but owing to some doubt concerning its identification it was left over for further examination. I now find that this specimen belongs to the genus *Thaumaleus* of Kröyer, and a short description of it follows that of *Monstrilla*.

But besides the specimens alluded to above, others have been obtained in the Moray Firth, which have still to be examined; but the study of the group is a somewhat difficult one, and the more so as some of the descriptions and figures of the earlier writers are sometimes wanting in that fulness necessary for certain identification.

Monstrilla longiremis, Giesb. Pl. XXV., figs. 3, 4.

  1892. *Monstrilla longiremis*, Giesbrecht, Pelagischen Copepoden des Golfes von Neapel, p. 589, pl. 46, figs. 10, 14, 22, 37, 41.

The *Monstrilla* which I now record from the Firth of Forth was obtained in a bottom tow-net gathering collected at Station V. (to the west of May Island) on July 24th, 1901. The antennules (first antennæ) are moderately elongated, but with the exception of an articulation near the base, the joints are very indistinct; the setæ were imperfect, and did not show the branched structure exhibited in Dr. Giesbrecht's figure of the antennule of *Monstrilla longiremis*. In the general form of the body and of the thoracic appendages the Forth specimen agrees very well with the species named. The abdomen consists of four segments, but the articulation between the first and second is not very marked, the third and fourth are short and distinct. The furca are of moderate size, and are each provided with five setæ as shown in the figure; no trace of a sixth seta could be observed.

The fifth feet consist each of a short, one-jointed sub-cylindrical branch, the proximal half of which is somewhat delated interiorly; each branch is furnished with three apical setæ, the inner one being much shorter than the other two, and a fourth seta springs from the inner margin as shown in the figure.

From these descriptive notes on this specimen from the Firth of Forth there seems to be little doubt that it is identical with the species described by Dr. Giesbrecht under the name of *Monstrilla longiremis*. The Forth specimen certainly does not show the branching setæ exhibited in Dr. Giesbrecht's figure; but this is not very surprising when it is remembered how delicate these long branching setæ are, and the friction they may be subjected to while in the tow-net. The specimen I have described is a female; no male was observed.

* Eighteenth Annual Report of the Fishery Board for Scotland, Part III., pp. 398-99, pl. xiii., figs. 15-20 (1890).
 † This species has also recently been observed in a gathering from the Firth of Forth collected at Station V. in 1901.
 ‡ Pelagischen Copepoden des Golfes von Neapel, p. 578 *et seq.*

Dr. Giesbrecht, in his great work on the Pelagic Copepoda of the Gulf of Naples, seems to be in doubt as to whether the Forth specimens considered by Mr. Bourne as belonging to the *Monstrilla helgolandica*, Claus, can really be identical with that species, and is rather inclined to ascribe them to his *M. longiremis*; the chief difficulty in the way of accepting this conclusion, however, is that Bourne, in describing the specimens, states that they possessed six furcal setæ, whereas in *M. longiremis* there are only five; unfortunately the mounted preparations from which the original description and figures were prepared were accidentally destroyed, and I am therefore unable to throw any light on the subject; but the occurrence of the specimen of *Monstrilla longiremis* just recorded seems to lend some support to the doubt expressed by Dr. Giesbrecht. Moreover it is interesting to note that none of the species mentioned by Mr. Bourne appear to have been provided with five furcal setæ, the number observed being either three or six.

Thaumaleus thompsoni, Giesb.   Pl. XXV., figs. 5, 6.

> 1892.   *Thaumaleus thompsoni*, Pelagischen Copepoden des Golfes
> von Neapel, p. 584, pl. 46, figs. 7, 27, 31, 36, 40.

A male specimen of a Copepod, which is apparently identical with *Thaumaleus thompsoni*, Giesbrecht, was obtained in a tow-net gathering collected in Lerwick Harbour, Shetland, on October 15th, 1901. In the male of this species the first of the two abdominal somites is shorter than the second, while in the female the first abdominal segment is, according to Dr. Giesbrecht, distinctly more dilated than the other. The caudal furca in the male are each provided with three setæ (fig. 5), but the female, according to Giesbrecht, has four.

One of the more obvious differences between the genus *Thaumaleus* and *Monstrilla* is that in the first the abdomen is composed of not more than two segments, exclusive of the caudal furca, whereas in *Monstrilla* the abdomen is composed of three, and sometimes of four, segments. Moreover, in *Thaumaleus* the number of hairs on the caudal furca is usually three or four, while *Monstrilla*, on the other hand, is provided with five or six furcal setæ.

LICHOMOLGIDÆ.

Lichomolgus furcillatus, Thorell.

This species was obtained in the washings of some dredged material collected at the north end of Inchkeith on May 23rd, 1901.

Lichomolgus hirsutipes, T. Scott.

> 1893.   *Lichomolgus hirsutipes*, T. Scott, 11th Ann. Rept. of the
> Fishery Board for Scotland, pt. iii., p. 206, pl. iv., figs.
> 1-12.

This was dredged off the North Craig, Firth of Forth, on July 4th, 1901; the species was described from specimens collected in the vicinity of the Bass Rock at the mouth of the estuary in 1893; it appears to be a rare species.

Pseudanthessius liber (Brady and Robertson).

> 1875.   *Lichomolgus liber*, B. and R., Brit. Assoc. Report, p. 197.

This species was obtained in the same gathering with the last. It appears to be a more common and more widely distributed species than *Lichomolgus hirsutipes*.

Pseudopsyllus, T. Scott (gen. nov.).

Somewhat like *Clausia,* Claparède, in general appearance ; the abdomen scarcely distinct from the thorax ; antennules composed of six short but stout joints. The antennæ are somewhat similar to those of *Hersiliodes,* Canu. Second maxillipeds apparently two-jointed, and armed with extremely long and powerful terminal claws ; other mouth organs unknown. The first four pairs of thoracic feet have both branches three-jointed as in *Hersiliodes.* Fifth pair short, and composed of a single lamelliform joint. Male unknown.

Pseudopsyllus elongatus, T. Scott (sp. nov.). Pl. XXIV., figs. 36–42.

Description of the Female.—Length 1·4 mm. ($\frac{1}{18}$ of an inch). Body elongate-narrow ; when viewed from above the anterior thoracic portion is slightly broader than the abdomen, but the distinction between thorax and abdomen is not very marked (fig. 36). The cephalo-thoracic segment is about equal to the entire length of the next four ; rostrum short.

Antennules very short and stout and composed of six sub-equal joints ; moderately setiferous and provided with three sensory filaments—one on the fourth joint, one on the fifth, and one on the end of the last joint (fig. 37). The formula shows approximately the lengths of the different joints :—

Proportional lengths of the joints,	15 · 11 · 9 · 12 · 10 · 13
Numbers of the joints, -	1 · 2 · 3 · 4 · 5 · 6

The antennæ are small but moderately stout, the first joint is the largest, the second and third are small, while the fourth is nearly as long as the second and third combined ; the exterior angle of the joint extends forward to near the middle of the end joint and terminates in a small spine ; the end joint seems, for the reason just stated, to arise from slightly beneath the penultimate one, somewhat similar to the structure of the same appendages in species of *Hersiliodes* (fig. 38).

The second maxillipeds—the only mouth organs obtained—are robust and armed with long and powerful terminal claws (fig. 39).

The first four pairs of thoracic feet are stout, moderately elongated, and with both branches three-jointed and of nearly equal length. In the outer branches of the first pair (fig. 40) a stout and moderately long spine springs from the outer distal angle of the first and second joints, while the end joint carries three similar spines on the outer margin and apex ; a plumose seta springs from the inner margin of the second joint, and five from the inner margin and apex of the last joint ; the first and second joints of the inner branches are each provided with one plumose seta on the inner margin, while the end joint carries two marginal and two apical setæ, in addition an elongate spine which springs from its outer distal angle. In the fourth pair the armature of the first and second joints of the outer branches is similar to that of the first and second joints of the outer branches of the first pair except that the marginal spines are not so elongated ; but the armature of the end joint differs from that of the same end joint of the first pair in that it carries only one small spine on the distal half of the outer margin and a moderately long but slender sub-apical spine, while round the inner margin and apex there are six instead of five plumose setæ ; the armature of the inner branches differs from that of the inner branches of the first pair, the second joint is provided with two setæ on its inner margin instead of one, while the end joint bears only three apical setæ, instead of four setæ and an elongate spine as in the first pair (fig. 41).

The fifth pair consist each of a single one-jointed lamelliform branch, sub-cylindrical in outline and fully twice as long as broad; they are each furnished with a single seta on the outer margin, while two setæ and a small spine spring from the truncated apex—the spine being articulated at the inner angle; a seta also springs from the exterior angle of the last thoracic segment, to which the fifth foot is articulated (fig. 42).

The caudal furca are moderately broad and nearly as long as the last abdominal segment (fig. 36).

Habitat.—Dredged at Station VII., Firth of Forth (between Fidra and the Bass Rock), on July 9th, 1901. Only a single female specimen was observed.

Remarks.—The specimen described above has such a close general resemblance to *Clausia cluthae*, T. and A. Scott,[*] that it was at first considered to belong to the same genus, but when closely examined it is found to differ in several important points, *i.e.* the structure of the antennæ and the form and armature of the second maxillipeds. The inner branches of the first four pairs of thoracic feet are all three-jointed, and thus differ very distinctly from those of *Clausia*.

I do not know of any genus or species to which this copepod could be ascribed.

ASTEROCHERIDÆ.

Asterocheres violaceus (Claus).

This somewhat rare species was obtained in a bottom gathering collected about 60 miles to the east of Shetland (or 180 miles north-east of Buchan-ness), on May 22nd, 1901. This species has been taken by my son, Mr. Andrew Scott, in the Irish Sea,[†] and it has also been obtained in the Firth of Clyde.[‡]

Rhynchomyzon purpurocinctum (T. Scott).

This distinct and widely distributed species was observed in some material dredged at the north end of Inchkeith on May 23rd, 1901. The specimens from which the species was described were also obtained in the Firth of Forth, but nearer the mouth of the estuary. *Rhynchomyzon purpurocinctum*, though widely distributed, is not very common, and occurs only very sparingly.

Acontiophorus ornatus (Brady and Robertson).

> 1875. *Ascomyzon ornatum*, Brady and Robertson, Brit. Assoc. Rept., 1875, p. 197.
> 1880. *Acontiophorus armatus*, G. S. Brady, Mon. Brit. Copep., vol. iii., p. 71, pl. lxxxvii., figs. 8–15.

A few specimens of this fine species were obtained in the washings of some dredged material collected near North Craig, Firth of Forth, on July 4th, 1901. This is the first time *Acontiophorus ornatus* has been observed in the Firth of Forth. It is a moderately large species, being not only robust in form but reaching a length of about one and a half millimetres.

---

[*] Ann. and Mag. Nat. Hist., (6), vol. xviii., p. 1, pl. 1, figs. 1–12.

[†] Report for 1895 on the Lancashire Sea-Fisheries Laboratory, p. 54, pl. v., figs. 16-26 described under the name of *Ascomyzon thompsoni*), pub. 1896.

[‡] Sixteenth Annual Report of the Fishery Board for Scotland. Part III., p. 270 (1898).

Cribropontius Normani (Brady and Robertson).

>1875. *Dyspontius Normani*, Brady and Robertson, Brit. Assoc. Report (1875), p. 197.
>
>1880. *Artotrogus Normani*, G. S. Brady, Mon. Brit. Copep., vol. iii., p. 63, pl. xci., figs. 12-15; pl. xcii., fig. 14; pl. xciii., fig. 10.
>
>1897. *Bradypontius Normani*, T. Scott, 15th Ann. Rept. Fishery Board for Scotland, pt. iii., p. 154, pl. ii., figs. 1, 2; pl. iii., figs. 1-11.
>
>1899. *Cribropontius Normani*, Giesb., Die Asterocheriden des Golfes von Neapel, p. 86, pl. 7, figs. 40-47.

This fine species was obtained in the same gathering with the one last recorded, and is the first time it has been noticed in the Firth of Forth. As will be observed from the synonymy, the genus-name has undergone several changes. Moreover, the colour of the living specimens, as shown by the drawings given in the 15th Annual Report of the Fishery Board for Scotland, is somewhat singular, and more ornate than is usually met with amongst the copepod-fauna of our seas; unfortunately, however, the bright colouration is very evanescent in specimens preserved in alcohol.

Bradypontius magniceps (G. S. Brady).

>1880. *Artotrogus magniceps*, Brady, Mon. Brit. Copep., vol. iii., p. 61, pl. xciii., figs. 1-9.
>
>1895. *Bradypontius magniceps*, Giesb., Sub-fam., gen., and sp. of the Copepod Fam. Ascomyzontidæ, Thorell; Ann. Mag. Nat. Hist., (6), vol. 16, p.

This, which is also a moderately large species, was obtained in a gathering of dredged material collected at the north-west end of Inchkeith on May 23rd, and in another gathering collected near North Craig on July 4th, 1901; one or two of the female specimens obtained in the last gathering were carrying ovisacs.

Bradypontius magniceps, though apparently widely distributed, is not very common, very few specimens being found in any single gathering.

Cancerilla tubulata, Dalyell. Pl. XXV., fig. 7.

>1857. *Cancerilla tubulata*, Dalyell, The Powers of the Creator, vol. i., p. 233, pl. xxii., figs. 1-5.
>
>1892. *Cancerilla tubulata*, Canu, Les Copépodes du Boulonnais, p. 255, pl. xxix., figs. 5-13.

A female specimen (fig. 7) of *Cancerilla tubulata* was obtained in a gathering of crustacea collected in the deep water (60 fathoms) off Aberdeen on August 2nd, 1901. The species, which appears to be rare in the British seas, has been recorded by Mr I. C. Thompson from the Irish Sea*, and Dr. Canu has described it in his work on the Copepods found near Boulogne.† This is the first time I have met with it notwithstanding that a careful look-out for it has been kept up for several years.

In 1893, T. and A. Scott recorded *Caligidium vagabundum*, Claus, from the Moray Firth‡. This copepod Canu in 1892 § and more

---

* Additions to the Copepoda of Liverpool Bay, *Trans.* L'pool Biol. Soc., vol. ix. (1898) p. 101.
† Les Copépodes du Boulonnais, p. 255, pl. xxix., figs. 5-13 (1892).
‡ Ann. and Mag. Nat. Hist., (6), vol. xii., p. 343 (October, 1893).
§ Les Copépodes du Boulonnais, p. 256.

recently Giesbrecht* have described as the male of *Cancerilla*. There is in some respects a similarity in the structure of *Caligidium* with that of *Cancerilla*, but I am not sure that the relationship between the two has been fully established. Dr. Edward Graeffe in his fauna of the Gulf of Trieste retains *Caligidium vagabundum*, Claus, under its distinctive name, without any reference to its sexual relationship with *Cancerilla*. †

HERPYLLOBIIDÆ.

Salenskya tuberosa, Giard and Bonnier. Pl. XXV., figs. 17–22.

> 1895.   *Salenskya tuberosa*, Giard and Bonnier, Contrib. á l'étude des Epicarides ; Bull. Scient., vol. xxv., p. 472, pl. xiii.

A few specimens of this remarkable form were found fixed between the thoracic legs of *Ampelisca spinipes*, Boeck, dredged near North Craig, Firth of Forth, on July 7th, 1901. They agree very closely with the figure of Giard and Bonnier, who obtained a single female and three " pygmy males " on a specimen of the same amphipod from Le Croisie.

The female (fig. 17) measures about ·84 mm. ($\frac{1}{3T}$ of an inch) in length, and it is about as broad as long ; one or two of the females carried two globular ovisacs, each one being nearly as large as the copepod itself. No appendages are present.

What seem at first sight to be the males, but, as Hansen has shown (Choniostomatidæ, p. 19), are really the larvæ—the adult males being degenerate like the females—(fig. 18) measure about ·15 mm. ($\frac{1}{168}$ of an inch). The anterior segment of the body is comparatively greatly dilated, the remaining segments are small. The antennules are very short, and three (or four) jointed, and furnished with two terminal setæ and a club-shaped appendage represented in the figure (fig. 19). Two pairs of limbs which represent the first and second maxillipeds are shown in figures 12 and 13.

My son, when dissecting the larvæ, was able to make out two pairs of thoracic feet ; each foot appeared to be composed of a two-jointed basal part and a single one-jointed branch, which was armed with two small spines on the exterior margin, and four plumose on the inner margin and apex.

This description will be found to differ (possibly by reason of age) from the character shown by Giard and Bonnier, and my figures show a further difference in a greater segmentation of the hinder part of the abdomen.

It is very probable that the parasite is congeneric, and quite possible that it is identical with *Rhizorbina ampliscæ*, Hansen, described from *Ampelisca lævigata*, Lilljeborg, by Hansen, in 1892 (Entomol. Meddelelser, ii., pp. 207–234, pl. iii.), a memoir which I have not yet been able to consult. ‡

Order OSTRACODA.

A considerable number of Ostracoda have been observed in the dredged material examined during the year, but as they apparently all belong to described species which for the most part are more or less generally distributed, I will only refer to the two following which appear to be somewhat rare.

---

* Fauna u. Flora d. Golfes u. Neapel, vol. xxv., Astrocheriden, pp. 95 and 112, pl. x. figs. 1–11.

† Fauna d. Golfes v. Triest, *Arbeit. el Zoolog. Institute* zu Wien, t. xiii., haft 1, p. 43 (1900).

‡ I desire to express my indebtedness to Prof. D'Arcy W. Thompson for the identification of this interesting crustacean, as well as for other information concerning it.

Sarsiella (?) *capsula*, Norman. Pl. xxv., figs. 27–32.

Three specimens of an Ostracod—a *Sarsiella*—were obtained in some sand collected about twenty-two miles to the north of the Shetland Islands on May 17th, 1901. The sand had been for a time immersed in formaline, and this may probably be the reason why the shells of these specimens are comparatively soft and, when seen from the side (fig. 27), show a somewhat even surface instead of being solid, and with the surface more or less corrugated as seen in typical specimens (see Part II. of Monograph published by Professor G. S. Brady and the Rev. A. M. Norman*).

The specimen represented by the drawing (fig. 27) is that of a female, and the shell when seen from the side is somewhat obliquely rotundate, the length being rather greater than the height, and measuring about 1·2 mm. by ·9 mm. respectively; the produced posterior projection is finely ciliated, the ventral margin is also fringed with delicate hairs. The posterior end is slightly compressed and bounded by an indistinct ridge which extends obliquely across from the dorsal to the ventral margins. The surface of the shell is not rugose as in the typical *Sarsiella capsula*, but is ornamented with numerous minute pits.

In the dorsal view of the shell given in figure 28, the valves are open to some extent on the ventral aspect; this was due to the soft structure of the shell (the shell could not be mounted dry in the usual way, but had to be kept in water under a cover-glass while being figured).

The various appendages of the contained animal resemble the drawings given in the Monograph by Brady and Norman already referred to.

The secondary appendages of the antennæ are rudimentary. Each consists of a rounded tubercle bearing *two* small spiniferous setæ (fig. 30A). The caudal lamina is provided with six spines on the one side, but only five on the other; on this side there is no trace of a sixth spine, nor any indication that a sixth had been present but had been broken off.

In this specimen two ova were observed.

The shell of the male differed little, if at all, from that of the female, except that it was slightly smaller. The appendages also appeared to be similar, except that the secondary branches of the antennæ were, as in closely allied species, more developed and fitted for grasping, and that the caudal lamina was only provided with five spines on both sides.

The first joint of the antennules in the male is rather longer than the next, and the second, third, and fourth joints gradually decrease in length, the fifth joint is almost obsolete, so much so that it is difficult to make out whether there is really a fifth joint or not. There are no setæ on the first joint; the second joint is furnished with a small seta near the middle of the upper margin, and another on its lower and its upper distal angles; the third joint bears one seta on the upper and two on the lower distal angles, while the fourth and (?) fifth joints carry several terminal setæ as shown in the drawing (fig. 29); some of these hairs have an annulated structure, but this is not shown in the drawing.

The antennæ of the male are provided with a number of long plumose setæ, and the secondary branches, though somewhat rudimentary (fig. 30), are moderately elongated and are apparently four-jointed—the two middle joints being very small; there is also a minute terminal appendage.

The first maxillæ (fig. 31) are similar to those of the female in structure and armature.

* A Monograph of the Marine and Fresh-water Ostracoda of the North Atlantic and of North-western Europe, Part II., *Sci. Trans.* Roy. Dublin Soc., vol. v. ser. ii. (January, 1896), p. 677, pl. lx., figs. 1–4, 18.

The caudal lamina of the male appears to have only five spines on both sides (fig. 32), no trace of a sixth could be observed.

The male described and figured here does not agree with *Nematohamma obliqua*, Brady and Norman, the structure of the antennæ and of the secondary branches of the antennules differs considerably from that of the same appendages in *N. obliqua*.

No male of *Sarsiella capsula* appears to have been described hitherto.

Conchœcea elegans, G. O. Sars. Pl. xxv., fig. 33.

A specimen of this species was dredged 180 miles north-east from Buchan-ness (about sixty miles to the east of the Shetland Islands) on May 22nd, 1901. Another specimen was obtained in the stomach of a whiting captured in 65 fathoms about 10 miles off Aberdeen on the 19th of the same month.

Order BRANCHIOPODA.

POLYPHEMEDÆ.

Genus *Podon*.

Three species of *Podon* have been described from the North Sea viz. *Podon polyphemoides*, Leuckart, *Podon leuckartii*, G. O. Sars, and *Podon intermedius*, Lilljeborg, and two of these—the first and the third—have sometimes been included in lists of British Crustacea ; there is a probability, however, that *Podon leuckartii* has sometimes been mistaken for *P. polyphemoides*, and as I have, with the assistance of Professor Lilljeborg's great work on Swedish Cladocera recently published, been enabled to recognise *Podon leuckartii* in some tow-net gatherings from the Firth of Forth and also from the Moray Firth, I will here indicate what seem to be the more obvious differences between this species and *Podon intermedius,* which is occasionally observed in the Firth of Clyde, and between both of these and *Podon polyphemoides*.

Podon leuckartii (G. O. Sars). Pl. XXV., figs. 23, 24.

The specimen represented by the drawing measures about a millimetre in length. The lower branches of the antennæ are composed of three, the upper of four joints as in the other two species referred to above ; the joints of the lower branches are sub-equal in length, and the first two bear each one and the last four terminal setæ, the first joint of the upper branches is very small, but other three are larger and sub-equal in length, and are provided with the same number of setæ as the lower branches (fig. 24).

The caudal spines are strong and slightly curved, and are rather longer than the caudal spines of *Podon intermedius*.

Habitat.—Firth of Forth and the Moray Firth.

The species does not appear to be rare on the east of Scotland : it has probably been mistaken for *Podon polyphemoides*.

Podon intermedius, Lilljeborg. Pl. XXV., figs. 25, 26.

> 1853. *Podon intermedius*, Lilljeborg, De Crust. ex ordinibus tribus`: Cladocera, Ostrocoda, Copepoda, in Scania occurr., sec. ii. ; da Crust. Marina, Ord. Clad., p. 161.

The specimen represented by the drawing measures about 1·5 mm. The antennae (second antennæ) are somewhat similar in structure to those of *Podon leuckartii,* and the lower branches are provided with the same number of setæ; but the upper branches differ in being furnished with an additional seta on the penultimate joint—one of the setæ springs from the middle of the joint and the other from its distal end—this branch, therefore, carries seven instead of six setæ (fig. 26).

The caudal spines are moderately stout, and straight, but smaller than those of *Podon leuckartii* (fig. 25).

Habitat.—Firth of Clyde, not very rare. It may be readily distinguished from *Podon leuckartii* by having an additional seta on the upper branches of the antennæ and by the caudal spines being smaller and straight.

Podon polyphemoides, Leuckart—a species considerably smaller than the other two—is, like *Podon intermedius,* provided with seven setæ on the upper branches of the antennæ, but the end joints of both branches are distinctly shorter than the preceding joints; the supplementary seta on the penultimate joint of the upper branches springs from near the distal end instead of near the middle of the joint. The caudal spines are also smaller. Probably *Podon leuckartii* has sometimes been mistaken for this species.*

Order EDRIOPHTHALMA.

Sub-Order AMPHIPODA.

Many species belonging to the Amphipoda have been observed in tow-net gatherings, in dredged material, and in the stomachs of fishes examined during the past year, but only some of the rare forms are recorded here.

PONTOPOREIIDÆ.

Bathyporeia norvegica, G. O. Sars, occurred in a tow-net gathering collected in Aberdeen Bay on September 4th, 1901. *Argissa hamatipes* (Norman) was also observed in some of the gatherings collected off Aberdeen during the past year.

PHOXOCEPHALIDÆ.

The only species belonging to this family which may be noted is the *Phoxocephalus oculatus,* G. O. Sars; it was obtained in a tow-net gathering collected about 22 miles to the north of the Shetland Islands on May 17th, 1901.

AMPELISCIDÆ.

Several species of the Ampeliscidæ have been observed, not only in gatherings collected with the tow-nets and dredge, but also in the stomachs of fishes, with whom they appear to be a favourite kind of food. The following species were observed:—*Ampelisca macrocephala* has occurred in gatherings from the Firth of Forth, from off Aberdeen, and from the Shetland district. *Ampelisca assimilis* has been obtained in gatherings and in fishes' stomachs from the Firth of Forth and Collieston, Aberdeenshire. *Ampelisca spinipes* was obtained in dredgings from the Firth of Forth (with parasites attached) and in the

---

* All the three species of *Podon* mentioned above have recently been obtained in some tow-net gatherings from the Firth of Clyde collected for the most part in April and May and in July and August 1901; *P. leuckartii* was obtained in the spring gatherings, but not in those collected later; while the other two species were observed in these later gatherings only.

stomachs of fishes. *Ampelisca brevicornis* (A. Costa), occurred in fishes' stomachs from the Shetland district and the Firth of Forth ; and *Haploops tubicola* was also obtained in the Forth estuary and off Aberdeen.

STEGOCEPHALIDÆ.

Stegocephaloides (?) *christianiensis* (Boeck). Pl. XXV., figs. 34–40.

Stegocephaloides christianiensis has again been observed in Loch Fyne, where it appears to be generally, though sparingly, distributed. Three specimens of what seems to be a variety of this species was dredged by the " Garland " in the Sound of Mull in 72 fathoms on March 31st, 1900. The fourth pair of coxal plates in these specimens are scarcely so broad in proportion to their length as in the typical forms (fig. 34). The basal joints of the last pereiopods do not terminate so acutely, but have the ends slightly rounded (fig. 38). The last epimeral plates of the metasome do not appear to be minutely notched on the low distal angles (fig. 39) ; and the telson seems to be slightly broader in proportion to its length (fig. 40). The other parts are similar to those of *Stegocephaloides christianiensis*.

AMPHILOCHIDÆ.

The somewhat rare *Amphilochus tenuimanus*, Boeck, was obtained in the stomach of a small Whiting from Station III., Firth of Forth, in April. *Amphilochoides intermedius* was dredged at the same station on May 25th. *Gitana sarsii*, Boeck, was also obtained in this gathering, along with a few other species noticed in the sequel.

METOPIDÆ.

Metopa pusilla, G. O. Sars, was obtained in some dredged material from the west end of Station III. collected in May, and in similar material from Stations V. and VII. collected in April—all from the Firth of Forth. *Metopa pollexiana* was obtained in gatherings collected off Aberdeen and to the east of the Shetland Islands ; it was also dredged at the west end of Station III., Firth of Forth, on May 23rd, 1901, and *Sthenometopa* (*Metopa*) *robusta* (G. O. Sars) was also obtained in the same gathering, as well as off the North Craig—also in the Firth of Forth—in 8 fathoms, in July.

DEXAMINIDÆ.

The somewhat rare *Guernia coalita* (Norman) has been dredged in the Firth of Forth, where it has been previously observed ; it was also dredged at Smith Bank in the Moray Firth, in 24 fathoms, on February 15th, and in Loch Linnhe in 48 fathoms on September 12th, 1901.

GAMMARIDÆ.

Mœra loveni (Bruzel.) was obtained in the stomach of a Witch-sole, *Pleuronectes cynoglossus*, captured at Station V., Firth of Forth, by the " Garland " on June 28th, 1901 ; this is one of the rarer species in the Firth estuary. *Megaluropus agilis*, Norman, was obtained in gatherings of dredged material from Stations IV. and V., Firth of Forth, on the 22nd and 24th April ; while *Cheiroccrates sundevalli* (Rathke) and *Cheiroccrates assimilis* (Lilljeborg) were dredged at the north-west end of Inchkeith, Firth of Forth, on May 23rd, 1901.

LILLJEBORGIIDÆ.

Lilljeborgia kinahani (Bate) occurred in some dredged material collected at Stations III. and VI., Firth of Forth, in May and June ; the species appears to be somewhat rare in the Firth of Forth.

PHOTIDÆ.

The following species belonging to this family were observed during the year. *Leptocheirus pilosus*, Zaddach, was dredged at the north-west end of Inchkeith on May 23rd. (This is the form described by G. O. Sars, having a six-jointed accessory appendage to the antennules.) *Protomedeia fasciata*, Kröyer, was obtained on several occasions in the stomachs of fishes.

Megamphopus cornutus, Norman, was dredged at Smith Bank, Moray Firth, on February 15th, and also at Stations III. and V., Firth of Forth, in April and May, 1901 ; it was also observed in the stomachs of small Whiting captured off Aberdeen on September 3rd.

COROPHIIDÆ.

I record two species belonging to this family. *Corophium affine*, Bruzelius, a male specimen, was obtained in the stomach of a small Whiting collected about ten miles off Aberdeen, in 55 to 60 fathoms, on September 3rd, 1901. *Unciola planipes*, Norman, was obtained in 78 fathoms about 110 miles north-east of Buchan-ness on May 15th, 1901.

DULICHIIDÆ.

Three species of *Dulichia* have during the past year been obtained in the Firth of Forth. *Dulichia porrecta*, Bate, was dredged at the west end of Station III. on May 23rd, and *Dulichia falcata*, Bate, at Station V. on April 24th. *Dulichia monacantha*, Metzger, which is new to the crustacean fauna of the Forth, was obtained in dredged material from Stations V. and VII. in April; all three species have also been observed in stomachs of fishes examined during the year.

Sub-Order ISOPODA.

Typhlotanais brevicornis (Lilljeborg). This minute isopod, which is only about 1·5 mm., was dredged in 50 to 55 fathoms, about 14 miles off Buckie, on November 3rd, 1900. So far as I know, this is the first representative of the genus *Typhlotanais* which has been recorded from the British seas.

Idothea granulosa, Rathke. Pl. XXV., fig. 41.

Two specimens of a small *Idothea* captured in the Bay of Nigg on March 23rd, 1901, agree with the *Idothea granulosa* of Rathke, as figured in Sars' Isopoda of Norway. The specimen represented by the drawing is a female with ova, and measures about 12·7 mm. (½ an inch) in length.

Sub-Order SYMPODA.

The following, amongst other species belonging to this sub-order, have been observed :—*Iphinoë serrata*, Norman, was obtained in a gathering collected to the east of Shetland on the 15th of May. *Diastylis rostrata* and *lucifera* were dredged at Station V., Firth of Forth, in April, the one on the 24th and the other on the 26th.   *Diastylis tumida*, Lilljeborg, was obtained in a bottom tow-net gathering from 60 fathoms, about ten miles off Aberdeen, collected on August 21st, and *Pseudocuma similis*, G. O. Sars, was dredged at Smith Bank, Moray Firth, on February 15th, and at Station V., Firth of Forth, on April 24th; this is an addition to the crustacean fauna of the Forth.   The two species belonging to the Cumellidæ, *Cumella pygmæa*, G. O. Sars, and *Nannastacus unguiculatus*, Bate, were obtained in some bottom mud brought up in a tow-net about 22 miles to the north of the Shetland Islands on May 17th. *Cumella* was also observed in a gathering collected in Uyea Sound between Unst and Uyea Island on the 18th, and also in some dredged material from the west end of Station VI., Firth of Forth, collected on the 22nd of the same month.   In this last gathering the somewhat rare but widely distributed ~~Pseudocuma~~ *pulchella*, G. O. Sars, was also obtained.

Order PODOPHTHALMA.

Sub-Order SCHIZOPODA.

In the gatherings forwarded by the "Garland" from the Firth of Forth during the past year, the schizopod *Erythrops goesii* was, as usual, of frequent occurrence.   Previous to the investigations instituted by the Fishery Board this species had not been recognised as a member of the British fauna.   *Erythrops elegans*, G. O. Sars, has also been obtained in the Forth estuary during the past year, and the curious copepod-parasite *Aspidoecia normani*, Giard and Bonnier, was found adhering to a specimen of both *Erythrops elegans* and *goesii*.   *Siriella armata* (M. Edw.) was obtained in the Firth of Forth in a bottom tow-net gathering from Station IV. collected on April 22nd, and *Siriella crassipes* in a gathering from Station V. collected on April 24th. *Mysidopsis angusta*, G. O. Sars, was obtained in the same gathering with *Siriella crassipes*, and in a gathering from Smith Bank, Moray Firth. *Rhoda raschii* (M. Sars) was, as usual, frequent in many of the gatherings forwarded to the Laboratory.   It will be observed that the Rev. Mr. Stebbing has restored Mr. Sim's generic name *Rhoda* for the group to which this species belongs.[*]

Sub-Order MACRURA.

Pasiphæa sivado (Risso) was obtained in 95 fathoms in Loch Linnhe on April 2nd, 1900, but the tow-net gathering in which it occurred was not examined till the following March.   A specimen of *Pandalus montagui* (with *Phryxus* (?) *abdominalis* attached) was found in a gathering from Station III., Firth of Forth, collected on May 9th.   A specimen of *Spirontocaris securifrons*, Norman, which also had a *Phryxus* (?) *abdominalis* attached to its abdomen, occurred in a bottom tow-net gathering collected in 78 fathoms, 110 miles north-east of Buchan-ness, on May 15th, while *Spirontocaris pusiola*, with what appeared to be the same species of parasite, was obtained in a gathering collected in the Forth to the west of Queensferry on April 26th.   *Crangon* (*Cheraphilus*) *nanus* was obtained at Station V. on April 26th, and *Calocaris macandreœ* was obtained at the same station as well as at Station III.; this Macrurid

---

[*] Arctic Crustacea : Bruce Collection, by the Rev. T. R. R. Stebbing, *Ann. and Mag. Nat. Hist.* (7), vol. v. (Jany. 1900), p. 10.

was found in the stomachs of the Long-rough Dab and the Witch Sole as well as amongst the contents of the small-mesh trawl-net. The *Trachelifer* stage of *Jaxea nocturna*, Nardo, was collected in abundance in the surface tow-net in Tobermory Bay, Sound of Mull, after dark, on September 9th, 1901. The same organism has lately been recorded by my son from the Barrow Channel, near Barrow-in-Furness.† The occurrence of *Trachelifer* at three different places seems to indicate a somewhat extended distribution for this crustacean.

ADDITIONAL NOTE.

Cancerina confusa, T. Scott, 19th Ann. Rept. of the Fishery Board for
    Scotland, pt. iii., p. 252, pl. xviii. figs. 12–20.

The copepod described under this name is identical with *Selioides bolbroei*, Levinsen (Vidensk Meddel. Naturh. Forening Kjöbenhaven. 1878, p. 373, *Crust. Copep.* parasit.) ; see also *Seloides, op. cit.* (1877). This copepod is said to be parasitic on *Harmothoe imbricata* (Lin.).

A marked peculiarity in this copepod is the position occupied by the ovisacs, as shown by the drawing in the Report of the Fishery Board for Scotland mentioned above. I have to thank the Rev. A. M. Norman for drawing my attention to the paper by Levinsen, and also for permitting me to examine the specimens he had received from that author.

Eurynotus insolens, T. and A. Scott, Ann. and Mag. Nat. Hist. (7), vol.
    i., p. 188 (1898) ; and *Eurynotopsyllus insolens*, T. Scott,
    19th Ann. Rept. of the Fishery Board for Scotland, pt. iii.,
    p. 256.

I am obliged to Dr. Steuer of Trieste for pointing out to me that this species is probably identical with *Eunicicola Clausii*, Kurz.—a parasite on a species of *Eunice*—one of the Annelida.

† Fifteenth Annual Report of the L.M.B.C., 1901, p. 13.

EXPLANATION OF THE PLATES.

PLATE XXII.

Metridia longa (Lubbock).   Diam.

Fig. 1. Female, dorsal view   × 18·6.
Fig. 2. Male, dorsal view   × 23.
Fig. 3. Foot of fifth pair, female   × 77.
Fig. 4. Foot of fifth pair, male   × 77.

Phænna, zetlandica, sp. n.

Fig. 5. Male, dorsal view   × 22.
Fig. 6. Foot of fifth pair   × 77.
Fig. 7. Terminal part of fifth pair of feet   × 525.

Xanthocalanus (?) *borealis*, G. O. Sars.

Fig. 8. Female, dorsal view   × 18·6.
Fig. 9. Foot of fifth pair   × 115.
    2G

Ectinosoma melaniceps, Boeck, var

Fig. 10.	Female, side view	.	.	.	×	77.
Fig. 11.	Antennule	.	.	.	×	175.
Fig. 12.	Mandible and palp	.	.	.	×	210.
Fig. 13.	First maxilliped	.	.	.	×	262.
Fig. 14.	Second maxilliped	.	.	.	×	350.
Fig. 15.	Foot of fifth pair	.	.	.	×	210.
Fig. 16.	Part of abdomen and caudal furca	.		×		102.

Stenhelia confusa, sp. n.

Fig. 17.	Female, side view	.	.	.	.	×	61.
Fig. 18.	Antennule	.	.	.	.	×	233.
Fig. 19.	Antenna	.	.	.	.	×	233.
Fig. 20.	Mandible and palp	.	.	.	.	×	162.
Fig. 21.	Second maxilliped	.	.	.	.	×	525.
Fig. 22.	Foot of first pair	.	.	.	.	×	233.
Fig. 23.	Foot of fourth pair	.	.	.	.	×	175.
Fig. 24.	Foot of fifth pair	.	.	.	.	×	350.
Fig. 25.	Part of abdomen and caudal furca	.	.		×	87.

Leptopontia curvicauda, nov. gen. et sp.

g. 26.	Female, side view	.	.	×	154.
ig. 27.	Antennule	.	.	×	350.
Fig. 28.	Antenna	.	.	×	525.
Fig. 29.	Mandible and palp	.	.	×	700.
Fig. 30.	Second maxilliped	.	.	×	630.
Fig. 31.	Foot of first pair	.	.	×	525.
Fig. 32.	Foot of second pair	.	.	×	525.
Fig. 33.	Foot of fourth pair	.	.	×	350.
Fig. 34.	Foot of fifth pair	.	.	×	700.
Fig. 35.	Part of abdomen and caudal furca		×	350

Ameira propinqua, sp. n.

Fig. 36.	Male, side view	.	.	.	×	38.
Fig. 37.	Antennule, male	.	.	.	×	117.
Fig. 38.	Antenna	.	.	.	×	175.
Fig. 39.	Mandible and palp	.	.	.	×	262.
Fig. 40.	Foot of first pair	.	.	.	×	175.
Fig. 41.	Foot of third pair	.	.	.	×	105.
Fig. 42.	Part of abdomen and caudal furca	.		×	103.

PLATE XXIII.

Ameira propinqua, n. sp.

| Fig. | 1. | Foot of third pair, male . | | × | 183. |

Normanella attenuata, A. Scott.

Fig.	2.	Female, side view	.	.	.	×	77.
Fig.	3.	Foot of first pair .	.	.	.	×	262.
Fig.	4.	Foot of fifth pair	.	.	.	×	262.

Fultonia hirsuta, nov. gen. et sp.

Fig.	5.	Female, side view	.	.	×	103.
Fig.	6.	Antennule	.	.	×	262.
Fig.	7.	Antenna .	.	.	×	525.
Fig.	8.	Second maxilliped	.	.	×	525.
Fig.	9.	Foot of first pair .	.	.	×	467.
Fig.	10.	Foot of fourth pair	.	.	×	350.
Fig.	11.	Foot of fifth pair	.	.	×	262.
Fig.	12.	Part of abdomen and caudal furca	.		×	154.

Nannopus palastris, Brady.

Fig. 13. Female, dorsal view	×	92.
Fig. 14. Male, dorsal view	×	92.
Fig. 15. Antennule, female . . .	×	350.
Fig. 16. Antenna ,, . . .	×	350.
Fig. 17. Mandible and palp, female . . .	×	350.
Fig. 18. Second maxilliped, ,, . . .	×	700.
Fig. 19. Foot of first pair, ,, ...	×	262.
Fig. 20. Foot of second pair, ,, ...	×	262.
Fig. 21. Foot of fourth pair, ,, ...	×	210.
Fig. 22. Foot of fifth pair	×	262.
Fig. 23. Antennule, male	×	262.
Fig. 24. Foot of third pair, male	×	262.
Fig. 25. Foot of fifth pair and appendage to first abdominal segment .	×	262.

Cletodes longicaudata, Brady and Robertson.

Fig. 26. Female, side view . .	×	102.
Fig. 27. Antennule . .	×	350.
Fig. 28. Antenna . . .	×	350.
Fig. 29. Foot of first pair . .	×	262.
Fig. 30. Foot of second pair . .	×	262.
Fig. 31. Foot of fourth pair . .	×	262.
Fig. 32. Foot of fifth pair . .	×	131.
Fig. 33. Part of abdomen and caudal furca	×	131.

Ameira reflexa, T. Scott, var.

Fig. 34. Female, side view . .	×	77.
Fig. 35. Antennule . .	×	350.
Fig. 36. Antenna . . .	×	525.
Fig. 37. Mandible and palp . .	×	1050.
Fig. 38. Second maxilliped . .	×	525.
Fig. 39. Foot of first pair . .	×	350.
Fig. 40. Foot of fourth pair . .	×	131.
Fig. 41. Foot of fifth pair . .	×	131.
Fig. 42. Part of abdomen and caudal furca	×	315.

PLATE XXIV.

Ameira tenuicornis, sp. n.

Fig. 1. Female, side view . .	×	103.
Fig. 2. Antennule . .	×	262.
Fig. 3. Antenna . . .	×	350.
Fig. 4. Mandible and palp . .	×	630.
Fig. 5. Second maxilliped, enlarged .	×	.
Fig. 6. Foot of first pair . . .	×	210.
Fig. 7. Foot of fourth pair . .	×	175.
Fig. 8. Foot of fifth pair . .	×	350.
Fig. 9. Part of abdomen and caudal furca	×	210.

Ameira propinqua, sp. n.

Fig. 10. Female, side view . .	×	103.
Fig. 11. Antennule . .	×	350.
Fig. 12. Antenna . . .	×	350.
Fig. 13. Mandible and palp . .	×	525.
Fig. 14. Second maxilliped . .	×	700.
Fig. 15. Foot of first pair . .	×	467.
Fig. 16. Foot of fourth pair . .	×	262.
Fig. 17. Foot of fifth pair . .	×	350.
Fig. 18. Part of abdomen and caudal furca	×	275.

Stenhelia hispida, Brady.

Fig. 19. Female, side view	.	× 51.
Fig. 20. Antennule	. . . . .	× 175.
Fig. 21. Antenna . .	. . . . .	× 262.
Fig. 22. Mandible and palp	. . . .	× 420.
Fig. 23. Second maxilliped	. . . .	× 350.
Fig. 24. Foot of first pair	. . . .	× 175.
Fig. 25. Foot of fourth pair	. . . .	× 175.
Fig. 26. Foot of fifth pair and appendage to first abdominal segments .		× 175.

Pseudomesochra longifurcata, nov. gen. et sp.

Fig. 27. Female, side view	. .	× 103.
Fig. 28. Antennule	. .	× 305.
Fig. 29. Antenna .	. .	⤳ 305.
Fig. 30. Mandible and palp	. .	× 305.
Fig. 31. Second maxilliped	. .	× 700.
Fig. 32. Foot of first pair	. .	× 315.
Fig. 33. Foot of fourth pair	. .	× 315.
Fig. 34. Foot of fifth pair	. .	× 525.
Fig. 35. Part of abdomen and caudal furca		× 175.

Pseudopsyllus elongatus, nov. gen. et sp.

Fig. 36. Female, dorsal view		× 46.
Fig. 37. Antennule	.	× 262.
Fig. 38. Antenna .	.	× 262.
Fig. 39. Second maxilliped		× 174.
Fig. 40. Foot of first pair		× 174.
Fig. 41. Foot of fourth pair		× 117.
Fig. 42. Foot of fifth pair		× 262.

———

PLATE XXV.

Scolecithrix (?) *brevicornis*, G. O. Sars.

Fig. 1. Male, dorsal view	. . . . .	× 38.
Fig. 2. Fifth pair of thoracic feet (slightly damaged) .	.	× 77.

Monstrilla longiremis, Giesbrecht.

Fig. 3. Female, dorsal view .	.	× 19.
Fig. 4. Fifth pair of thoracic feet	.	× 51.

Thaumaleus thompsonii, Giesbrecht.

Fig. 5. Male, dorsal view .		× 19.
Fig. 6. Fifth pair of thoracic feet		× 525.

Cancerilla tubulata, Dalyell.

~~Female~~ Fig. 7. ~~Male~~, dorsal view	. . .	× 51.

Nereicola concinna, T. Scott, sp. n.

Fig. 8. Female, dorsal view .		× 25.
Fig. 9. Antennule	. .	× 262.
Fig. 10. Antenna .	. .	× 350.
Fig. 11. Mandible .	. .	× 420.
Fig. 12. First maxilliped	.	× 262.
Fig. 13. Second maxilliped	.	× 350.
Fig. 14. An immature (?) female .		× 47.

✗ *Platypsyllus minor*, T. Scott, gen. et. sp. n.

Fig. 15. Female, dorsal view . . . . . × 34.
Fig. 16. Siphon . . . . (greatly enlarged).

Salenskya tuberosa, Giard and Bonnier.

Fig. 17. Female, dorsal view . . . . × 35.
Fig. 18. Male, dorsal view . . . . × 235.
Fig. 19. Antennule . . . . . × 1050.
Fig. 20. First maxilliped . . . . × 788.
Fig. 21. Second maxilliped . . . . × 787.
Fig. 22. Foot of second pair . . . . × 1050.

Podon leuckartii, G. O. Sars.

Fig. 23. Female, side view . . . . . × 38.
Fig. 24. Antenna . . . . × 62.

Podon intermedius, Lilljeborg.

Fig. 25. Female, side view . . . . × 38.
Fig. 26. Antenna . . . . . × 62.

Sarsiella (?) *capsula*, Norman.

Fig. 27. Female, side view . . . . . . × 38.
Fig. 28. Female, dorsal view . . . . . . × 38.
Fig. 29. Antennule, male . . . . . . × 77.
Fig. 30. Antenna ,, ... × 77.
Fig. 30A. Secondary appendage of antenna, female . (greatly enlarged).
Fig. 31. First maxilla, male . . . . . . × 154.
Fig. 32. Post-abdomen, male . . . . . . × 77.

Conchœcia elegans, G. O. Sars.

Fig. 33. Side view . . . . × 34.

Stegocephaloides christianiensis, Boeck.

Fig. 34. Female, side view, slightly immature . . . . × 15.
Fig. 35. Antennule . . . . × 77.
Fig. 36. Antenna . . . . × 31.
Fig. 37. First gnathopod . . . . . . × 31.
Fig. 38. Seventh pereiopod . . . . . × 33.
Fig. 39. Last epimeral plate of metasome . . . × 31.
Fig. 40. Telson . . . . × 115.

. *Idothea granulosa*, Rathke.

Fig. 41. Female, dorsal view . . . . × 5

PLATE XXII.

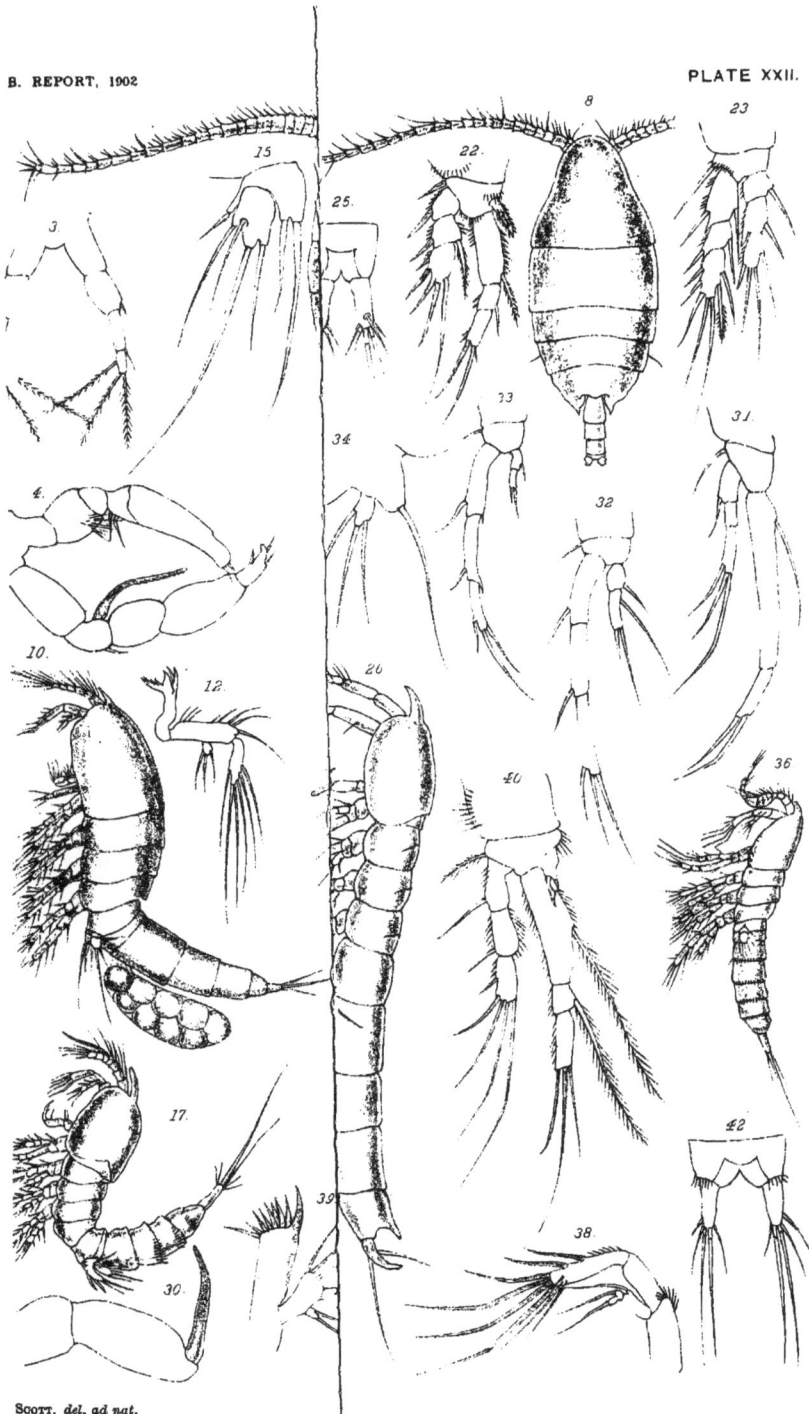

PLATE XXII.

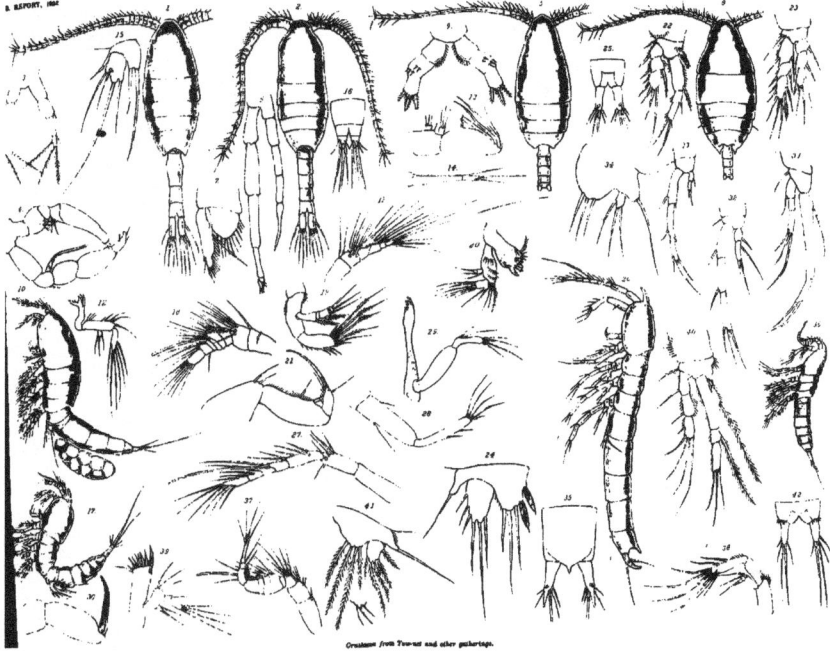

Crustacea from Tow-net and other gatherings.

PLATE XXIII.

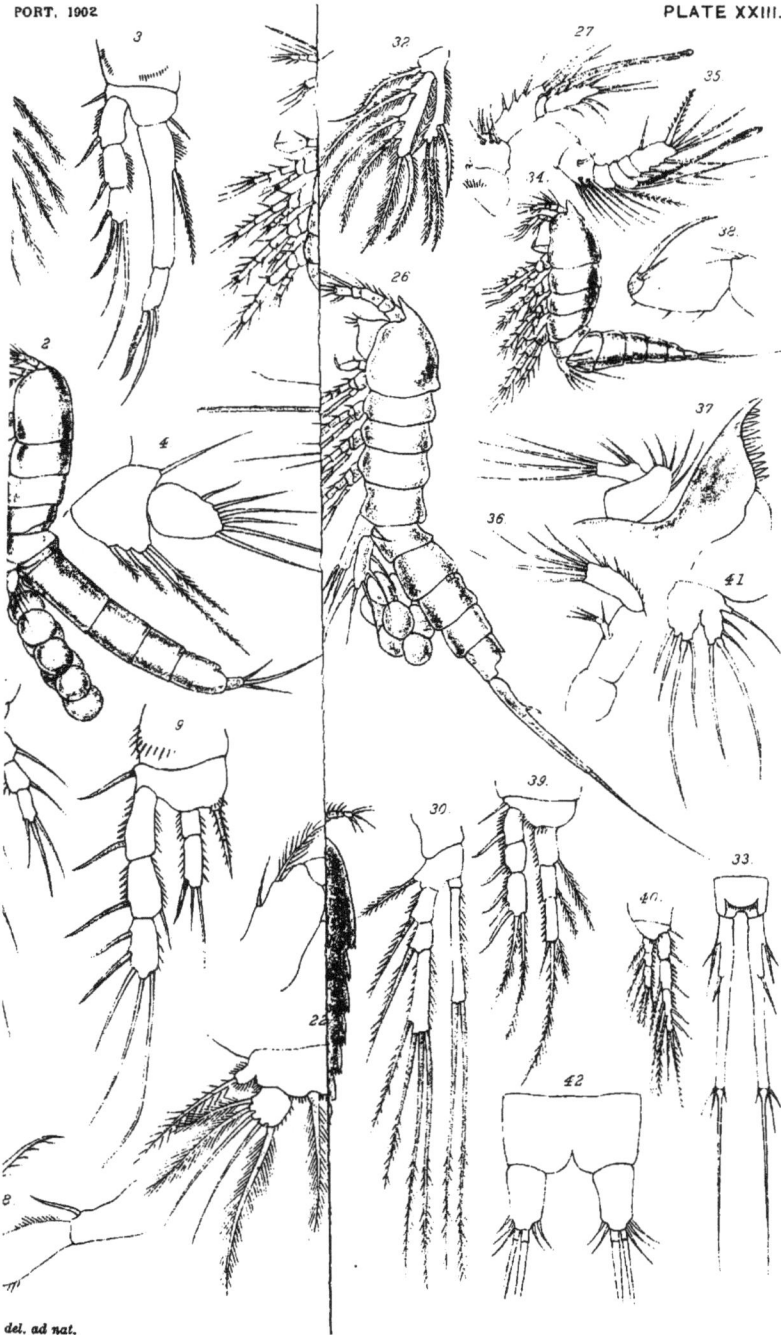

del. ad nat.

PLATE XXIII.

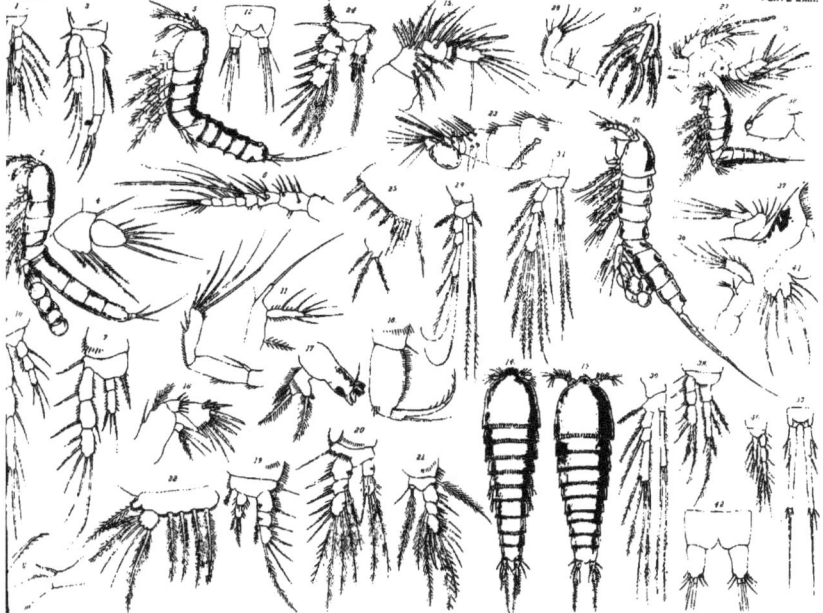

PLATE XXIV.

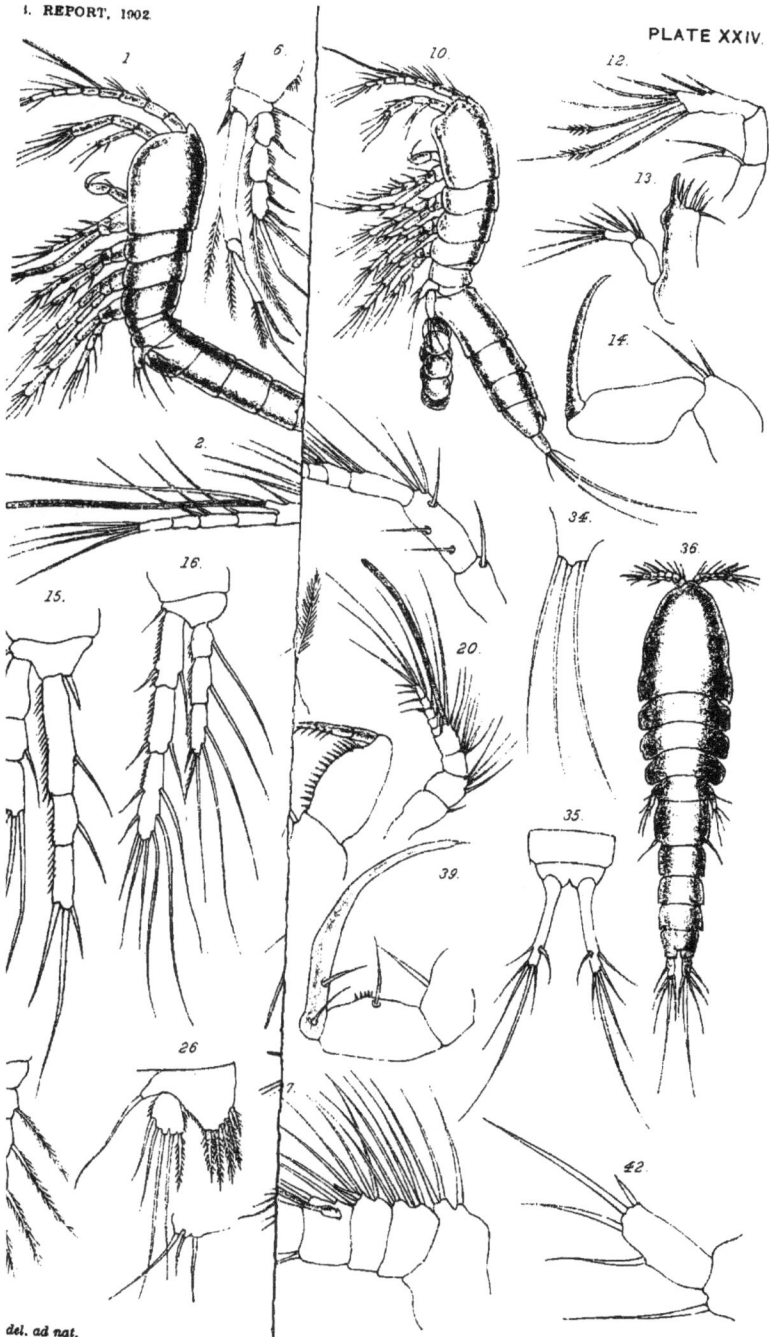

PLATE XXIV.

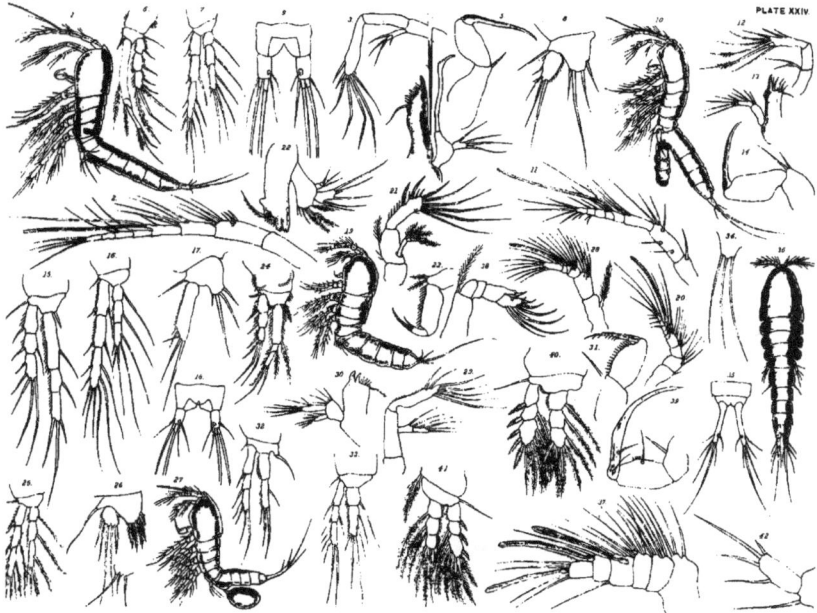

PLATE XXV.

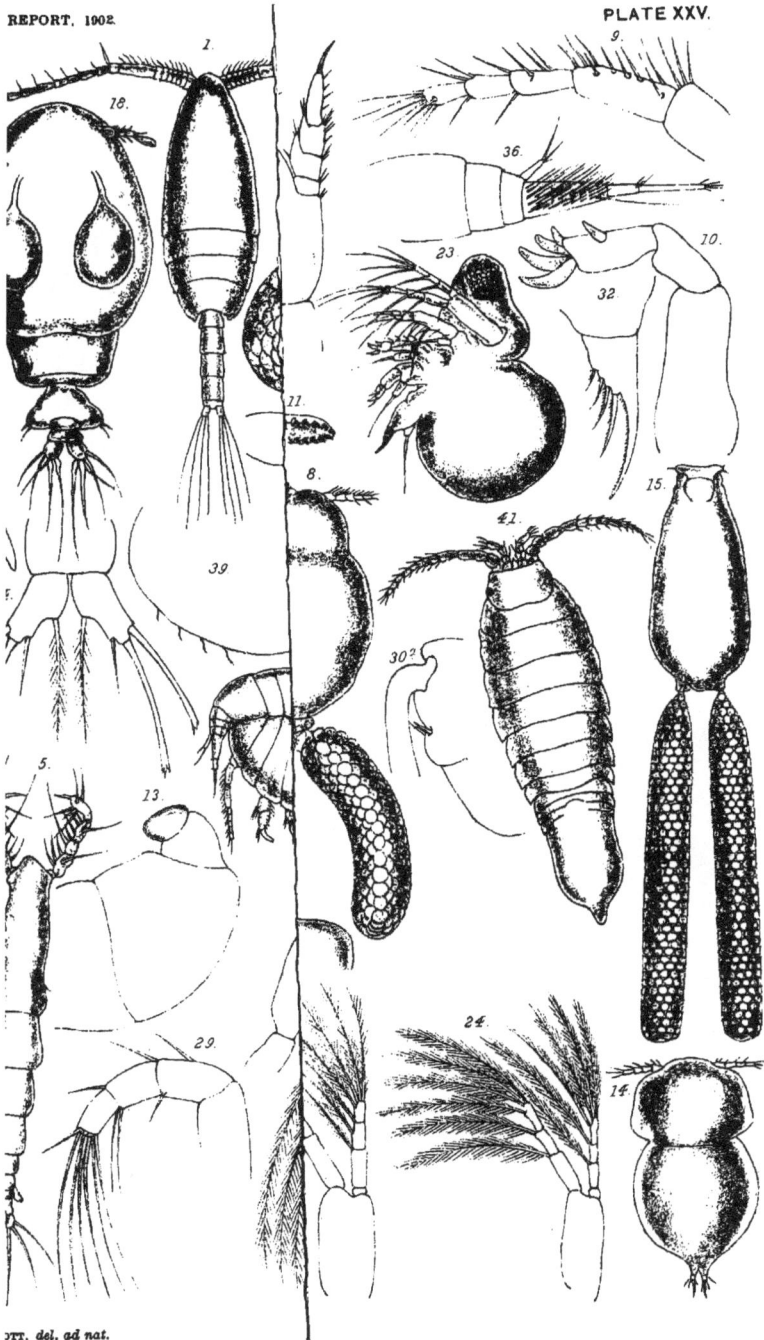

PLATE XXV.

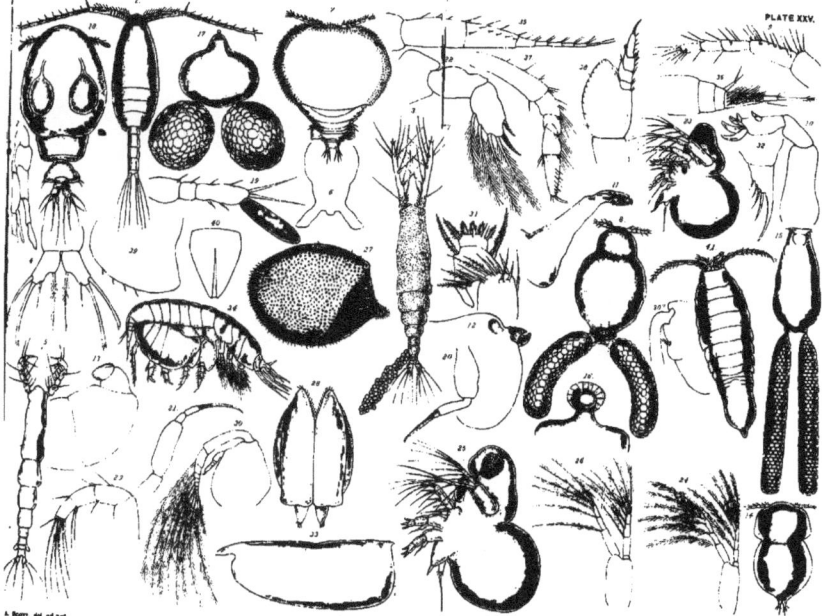

Crustacea from Tow-net and other gatherings.

SOME NEW AND RARE CRUSTACEA
LLECTED AT VARIOUS TIMES IN CON-
CTION WITH THE INVESTIGATIONS OF
E FISHERY BOARD FOR SCOTLAND.

— —

from *Twenty-first Annual Report of the Fishery Board for
Scotland—Part III.—Published July 20, 1903.*]

[*Reprint from Twenty-first Annual Report of the Fishery Board for Scotland—Part III.—Published July 20, 1903.*]

II.—ON SOME NEW AND RARE CRUSTACEA COLLECTED AT VARIOUS TIMES IN CONNECTION WITH THE INVESTIGATIONS OF THE FISHERY BOARD FOR SCOTLAND.

By THOMAS SCOTT, LL.D., F.L.S., Mem. Zool. Soc. de France.

(Plates II.–VI.)

Though the larger, or, as they are sometimes called, the higher Crustacea of the British seas are now fairly well known, our knowledge of the smaller forms is still very defective ; and as these smaller species constitute an important part of the food of small and young fishes, their study becomes important from a fisheries' point of view as well as from the point of view of the naturalist.

A large number of gatherings of small crustaceans has been examined during the past year. These gatherings have been collected at various times, and some of them several years ago ; some of the earlier gatherings were not examined or only partially examined at the time they were collected, as other work requiring more immediate attention had to be done. The re-examination of these collections has yielded several new forms, besides a number of rare and interesting species already described. The new species all belong to the Copepoda—a group which from their great abundance, their wide distribution in all our seas and estuaries, and their nutritive qualities, is of immense value as food for fishes.

In a recent paper on the food of fishes* it is shown that the young of almost all the food-fishes live very largely on small crustaceans, and a considerable proportion of them belong to the Copepoda.

Among the species recorded in the present paper the following appear to be undescribed :—

Ameira pusilla, T. Scott (sp. nov.).
    ,,   *ambigua,* T. Scott (sp. nov.).
Delavalia minutissima, T. Scott, (sp. nov.).
Tetragoniceps pygmæus, T. Scott (sp. nov.).
Laophonte gracilis, T. Scott (sp. nov.).
Cletodes neglecta, T. Scott (sp. nov.).
Enhydrosoma gracile, T. Scott (sp. nov.).
Enhydrosoma minutum, T. Scott (sp. nov.).
Dactylopus littoralis, T. Scott (sp. nov.).
Dactylopus vararensis, T. Scott (sp. nov.).
Dactylopus mixtus, T. Scott (sp. nov.).
Paranthessius dubius, T. Scott (gen. et sp. nov.).

While among the other species recorded in the sequel there are several that are of special interest. For example, *Parastephos pallidus* (G. O. Sars) is a copepod new to the British fauna.

Eucalanus crassus, Giesbrecht, is a copepod which was added to the British fauna a few years ago, and is now recorded from a new station.

Stephos scotti, G. O. Sars, is also recorded from a new station, and the male is described for the first time from Scottish specimens.

---

* *Twentieth Annual Report Fishery Board for Scotland*, Part III., pp. 486-538 (1902).

The number of Crustacea described in the present paper is scarcely so large as in some of those previously published.

I am, as formerly, indebted to my son Mr. Andrew Scott, A.L.S., for the drawings required to illustrate the new and rare species described here; and the arrangement of the species is similar to that adopted in previous papers.

CRUSTACEA.

Sub-Class ENTOMOSTRACA.

Order I.—COPEPODA.

CALANOIDA.

Genus *Eucalanus.*

Eucalanus crassus, Giesbrecht.

> 1881. *Eucalanus crassus*, Giesb., Atti. Acc. Lincei Rend., ser. 4, vol. 4, lem. 2, p. 333. *See also* Pelagisch. Copep. Golfes von Neapel, pp. 131–152, *t.* 11 and 35.

Several specimens of this species were captured about 10 miles off Aberdeen on November 6th, 1901. This is the first time that *Eucalanus crassus* has been taken so far south on the east side of Scotland. The species has been several times captured in the Moray Firth,* and it has also been collected along with *Eucalanus elongatus* about fifty miles south-east of Fair Island.† Dr. R. Norris Wolfenden records it from the Faröe Channel, where he has also taken several other interesting species.‡

Genus *Stephos*, T. Scott (1892).

Stephos scotti, G. O. Sars. Pl. ii., figs. 1–4.

> 1897. *Stephos gyrans*, T. Scott (not Giesbrecht), 15th Ann. Rept. Fishery Board for Scotland, pt. iii., p. 146, pl. iv., fig. 9; pl. iii., figs. 17–18.
>
> 1892. *Stephos scotti*, G. O. Sars. An Account of the Crustacea of Norway, vol. iv., p. 63, pl. xliii.

This species was first observed in some material dredged in 1896 in Loch Gair—a small lagoon opening into Loch Fyne. Only a single female was obtained on this occasion, and, as it had a somewhat close resemblance to *Stephos gyrans*, Giesbrecht, it was ascribed to that species. Additional female specimens were subsequently obtained not only in other parts of the Clyde area but also in the Firth of Forth, but till quite recently no males had been observed among Scottish specimens. Prof. G. O. Sars, however, had already obtained both sexes of the species in Norwegian waters, and had found that the males especially differed considerably from the males of *Stephos gyrans*, and, therefore, in vol. iv. of his great work on the Crustacea of Norway described it under the name given above.

It happened that I had a gathering of small Crustacea which had been collected in 1894 in an old quarry near Granton to which the tide has access during high water. This gathering, which was examined during

* *Eighteenth Annual Report of the Fishery Board for Scotland*, Part III., p. 382 (1900).
† *Nineteenth Annual Report of the Fishery Board for Scotland*, Part III., p. 237 (1901).
Journ. Marine Biol. Assoc., vol. vi., No. 3 (January, 1902), p. 361.

the past summer, was found to contain many copepods, some of them being rare forms; several specimens of *Stephos scotti* were obtained in this gathering, and they included both males and females. The dissections of both species corresponded exactly with Prof. G. O. Sars' description and figures in the work referred to.

The two drawings (pl. ii., figs. 1 and 2) represent an adult female and male from the Granton quarry gathering; the only obvious external difference between them is in the structure of the fifth pair of feet, separate drawings of which are represented by figures 3 and 4. A full description of the species, with drawings, is given by Prof. Sars in the work already mentioned.

Genus *Parastephos*, G. O. Sars (1902).

Parastephos pallidus, G. O. Sars. Pl. ii., figs. 5–10.

1892. *Parastephos pallidus*, G. O. Sars, Crustacea of Norway, vol. iv., p. 65, pl. xliv.

The genus *Parastephos* was recently instituted by Prof. G. O. Sars for a single male specimen of a copepod found many years before at Sjerjehavn, west coast of Norway, in about 100 fathoms, where the bottom was soft and muddy.

During the past summer, while examining a bottom tow-net gathering collected near the head of Loch Fyne in November, 1901, I observed several specimens, both males and females, of this interesting species; but though most of the male specimens were adult, the females, with the exception of one, were more or less immature. The drawings of the male prepared by Prof. G. O. Sars are perfectly characteristic, but the figures given here may be useful to those who have not seen the drawings of the learned author referred to.

The species is a moderately large one, the adult male (fig. 6) is very nearly two millimetres in length, while the adult female (fig. 5) is some- what larger, being 2·19 mm. (about $\frac{1}{11}$ of an inch). The description of the fifth feet of the male may be best given in Prof. G. O. Sars' own words:—"Last pair of legs in the male largely developed and very asymmetrical, right leg slender and terminating in a strong denticulated claw, left leg much coarser, with the antepenultimate joint the largest' (fig. 10). The distal portion of the right leg can apparently be folded completely back upon the proximal portion as shown in the figure.

The female antennules resemble those of the male; they are equally elongated and composed of twenty-four joints, the second and eighth joints are each nearly as long as the combined lengths of the two joints which immediately follow them. The antennules are only sparingly setiferous, and are furnished with several small sensory filaments (fig. 7). The fifth pair in the female appear also to be asymmetrical; in the only adult specimen obtained the left leg of the fifth pair is considerably longer than the right one, but this appears to be the only difference (fig. 9).

In the female, each of the first three abdominal segments expands posteriorly into a ridge which is fringed with fine hairs, as shown in fig. 2.

In the adult male represented by the drawing, the second pair of thoracic feet (fig. 8) are alike on both sides, the outer branches being both three-jointed and the inner two-jointed. Prof. G. O. Sars, in his descrip- tion of the only male specimen he had, states that the outer branch of the right foot of the second pair was only two-jointed; but such a difference is rather unusual among the Calanoida, and probably his specimen may have been defective, as no such difference was observed in Clyde specimens. The Clyde is, so far, the only known British habitat for this interesting species.

Genus *Pseudophœnna*, G. O. Sars (1900).

Pseudophœnna? *typica*, G. O. Sars.   Pl. ii., figs. 11–15.

> 1902. *Pseudophœnna typica*, G. O. Sars.   An Account of the Crustacea of Norway, vol. iv., p. 44, pl. xxix., xxx.

A single male specimen of a Calanoid, which I have referred, though somewhat doubtfully, to *Pseudophœnna typica*, G. O. Sars, was obtained in a bottom tow-net gathering of Crustacea collected last year near the head of Loch Fyne.   The specimen agrees very closely with *Pseudophœnna typica* in its general outline and in the structure of the various appendages so far as these can be made out, except that the fifth feet slightly differ from the drawing given in the work of Prof. G. O. Sars referred to above, but not so much in their general structure as in the apical part of the right leg (fig. 15).

This Loch Fyne specimen measures fully one and a half millimetres ; the thorax is moderately stout, but the abdomen is slender (figs. 11 and 12). The antennules, which reach to about the distal end of the thorax, appear to be composed of twenty-one joints.   The basal joints, from the third to the seventh, are shorter than the others ; the right is elongated and appears to be indistinctly articulated near the distal end.   The antennules are only sparingly setiferous, but they are well supplied with sensory filaments as shown in the drawing (fig. 13).   The species will not be satisfactorily determined till more specimens of both sexes are obtained.

In his note on the distribution of this Calanoid, Sars states that he has obtained it at several places, from Christiania Fjord to Vardö, and that it is a true bottom form, it is therefore probable that the species may not be rare in the deeper water off the Scottish coast.

Genus *Pseudocyclops*, G. S. Brady (1872).

Pseudocyclops obtusatus, Brady and Robertson.   Pl. vi., figs. 13–15.

> 1873. *Pseudocyclops obtusatus*, B. and R., Ann. and Mag. Nat. Hist. (4), vol. xii., p. 128, pl. viii., figs. 4–7.
> 1878. *Pseudocyclops obtusatus*, Brady, Mon. Brit. Copep., vol. i., p. 84, pl. xii., figs. 1–13.

Although the distribution of this species seems to be extensive, it does not appear to be anywhere very common.   The female represented by the drawing (fig. 13) was obtained during the past summer by washing the filters at the Hatchery, Bay of Nigg.   The species, which is fairly well marked, has been described by Brady and Robertson in the Annals and Magazine of Natural History, and by Prof. G. S. Brady in his Monograph of the British Copepoda.

In this species the rostrum is of a somewhat broadly triangular form and the antennules (fig. 14) are short and moderately stout, and are apparently composed of seventeen joints and are furnished with numerous plumose setæ ; the basal joint also carries two moderately long sensory filaments.

The outer branches of all the thoracic feet are armed with stout dagger-like spines on their outer aspect.   The inner branches of the fifth pair are considerably shorter than the outer ones, and the end joints terminate abruptly, as shown in the figure ; moreover, the marginal setæ on the inner edges of both the inner and outer branches have the basal half distinctly thicker than the distal portion, so much so as to be observable without dissection (fig. 15).

This species has also been obtained in the Moray Firth and in the Firth of Forth as well as in the Clyde, but seldom more than one or two specimens have been noticed in any single gathering.

Genus *Labidocera*, Lubbock (1853).

Labidocera wollastoni, Lubbock.
> 1857.   *Pontella wollastoni*, Lubb., Ann. Nat. Hist. (2), vol. 20, p. 406, pl. 10, 11.

This somewhat rare species was captured in Loch Fyne with the surface tow-net at Station XIII. (off Largymore), October 9, 1901. It also occurred in a bottom tow-net gathering collected at Station XIII., near the mouth of the Clyde estuary, on November 11, 1901. In the gathering at Station VIII. there were also obtained *Candacia pectinata, Metridia lucens, Parapontella brevicornis,* and other forms.

Fam. HARPACTICIDÆ.

Genus *Ectinosma*, Boeck (1864).

Ectinosma curticorne, Boeck.   Pl. vi., fig. 1.
> 1885.   *Ectinosoma curticorne*, Boeck, Abhandl. Natur. Vereins zu Bremen, ix. Bd., p. 194, t. vi., figs. 1–12.
> 1895.   *Ectinosoma curticorne*, T. and A. Scott, Trans. Linn. Soc., vol. vi. (Zool.), p. 430, pl. 36, figs. 22, 30, 34, *et. seq.*

This is a marine species, and though recorded from several British localities it does not appear to be anywhere very common ; it is, however, more frequently met with amongst the fronds and roots of algæ in the littoral zone than in off-shore waters. It is usually of a brownish colour, and there is also usually a dark-coloured blotch at the bases of the antennules, such as is observed in *Bradya minor*, but in that species the outline of the blotch is more distinctly defined. Specimens of this species have been obtained in gatherings collected some years ago in shallow water near Musselburgh and Granton, Firth of Forth, but which have only recently been examined. In this species, as in one or two others, the furcal joints are each provided at the apex with a short but stout cone-shaped spine and with two other short setæ which are moderately stout and spiniform, as shown in the drawing (pl. vi., fig. 1). There are also, as in other Harpactids, one or two elongated terminal setæ, the principal of which is moderately stout.

A few other *Ectinosomas* were obtained in the Musselburgh gatherings along with *E. curticorne*, two of which may be referred to here, viz. *E. gracile* and *E. herdmani*. *Ectinosoma gracile*, T. and A. Scott, is a small and slender form which was first discovered near St. Monans, Firth of Forth, and has since been found sparingly in several British localities. This species was moderately frequent in one of the gatherings from Musselburgh. *Ectinosoma herdmani*, T. and A. Scott, though a larger form than the last, is also moderately slender, and was also first observed near St. Monans. Most of the *Ectinosomas* require careful examination, but this is one of a few that are comparatively easily identified. This species was one of the more common Harpactids in the Musselburgh gatherings ; it is readily noticed by its elongated slender form and the opaque white colour it assumes when preserved in spirit.

H

Genus *Ameira*, Boeck.

Ameira pusilla, T. Scott, sp. nov.　Pl. v., figs. 1–10.

Description of the Female.—Body elongated and slender, resembling generally a small *Canthocamptus* or *Attheyella* ; length about three and a half millimetres (about $\frac{1}{70}$ of an inch), rostrum short (fig. 1).

The antennules are only moderately elongated and composed of seven joints, the second joint is considerably longer than any of the others, and the antepenultimate one is apparently the smallest, as shown by the drawing (fig. 2).

The antennæ, which are moderately stout (fig. 3), are furnished with a small uniarticulated secondary branch.

The mandibles are of a narrow cylindrical form and are armed with several small teeth or spinules on the obliquely truncate apex ; the palp is of moderate size and is composed of a somewhat dilated basal joint bearing two small one-jointed branches (fig. 4).

The second maxillipeds are small, two-jointed, and armed with a small but moderately stout terminal claw (fig. 5).

In the first pair of thoracic feet the inner branches, which are three-jointed, are very long, but this is owing to the elongation of the first and third joints, the middle joint being a short one, the first joint reaches to about the extremity of the three-jointed outer branches, while the third joint is fully half the length of the first and twice the length of the second joint ; the inner branches are also slender in proportion to the length (fig. 6).

The inner branches of each of the following three pairs are all shorter than the outer ones, which are somewhat elongated, and both branches in all the three pairs, and especially of the third and fourth, are moderately slender (figs. 7 and 8).

In the fifth pair the inner produced portion of the basal joint is broadly sub-cylindrical, and does not reach to the end of the secondary joint ; it appears to be provided with four apical setæ, the two inner ones being moderately short and spiniform, but the two others are longer.　The secondary joint is also somewhat cylindrical in form, and rounded at the ends, its breadth being nearly equal to half the length ; it appears to be provided with only three apical setæ, arranged as shown in the drawing (fig. 9), the middle one being very long and slender and the inner one also slender and elongated, but the outer is short.

The furcal joints are fully half as long as the last segments of the abdomen, and the principal tail setæ are very long and slender (fig. 10).

Habitat.—Off Musselburgh, Firth of Forth, in shallow water, but not very common.

Remarks.—This species has a close general resemblance to some forms of *Canthocamptus*, not only in its general configuration but also in some of its appendages ; this is especially noticeable in the structure of the first thoracic feet, which do not differ much from *Canthocamptus staphylinus* or *C. northumbricus* ; this pair is also somewhat similar to those of *Stenhelia ima*, but the terminal joints of the inner branches are proportionally considerably longer.　The structure of the antennæ and mandibles shows its relationship with *Ameira*, but it differs in the structure of the first and fifth pairs of thoracic feet from any species previously described, so far as these are known to me.

Ameira ambigua, T. Scott, sp. nov.　Pl. v., figs. 11–19.

Description of the Female.—This form, which somewhat resembles *Ameira longipes* in general appearance, is comparatively small ; the specimen represented by the drawing (fig. 11) measures only ·56 mm. (about $\frac{1}{44}$ of an inch). In this form the rostrum is small and the furcal joints short.

The antennules (fig. 12) are elongated and slender and composed of eight joints, the smallest joints counting from the proximal end are the fifth and seventh, while the second is the largest. The proportional lengths of the various joints are shown approximately by the annexed formula :—

| Lengths of the joints, | - | 15 · 18 · 12 · 11 · 7 · 8 · 5 · 8 |
| Numbers of the joints, | - | 1 · 2 · 3 · 4 · 5 · 6 · 7 · 8 |

The antennæ appear to be two-jointed and moderately stout, but the secondary branch is small and composed of a single joint (fig. 13).

The mandibles are very small and of a cylindrical form, the distal end is obliquely truncated and armed with minute spines ; the palp is also very small, and has the basal part slightly dilated and provided with a minute one-jointed branch (fig. 14).

The second maxillipeds are moderately stout and two-jointed, and armed with an elongated and slender terminal claw (fig. 15).

The first pair of thoracic feet have the inner branches elongated and moderately stout ; the first joint is about three times as long as the entire length of the second and third joints, but these two joints are small and sub-equal. The outer branches are moderately slender, and they are shorter than the first joint of the inner ones (fig. 16).

The next three pairs are all elongated and moderately stout. In the second pair (fig. 17) the inner branches, which are somewhat shorter than the outer, taper towards the distal end, as shown in the drawing, the first joint being more dilated than the second, and the second than the last ; each of the three joints is furnished with a slender seta on its inner margin. The outer branches are slender, and the exterior marginal spines are elongated ; the second joint is also provided with one, and the third with two slender setæ on the inner edge.

In the fourth pair the inner branches are also, as in the second pair, shorter than the second branches, but they are scarcely so stout as those of that pair ; moreover, the first and second joints are each furnished with a seta on the inner margin, while the third bears two setæ. The outer branches do not differ much from the same branches in the second pair (fig. 18).

In the fifth pair the inner produced portion of the basal joint bears four slender setæ on its broadly rounded apex, the second one from the inside being much longer than the others. The secondary joints are sub-cylindrical in general form, but with the ends rounded ; they are each about twice as long as broad and carry five setæ round the outer margin and apex, but the two innermost are considerably longer than the others (fig. 19).

Habitat.—Off Musselburgh, Firth of Forth ; apparently rare.

Remarks.—This is one of those troublesome forms which, while differing in one or other of its structural details from any of the described species I am acquainted with, yet exhibits no single character prominently distinctive, such as we have in *Ameira longicaudata*, T. Scott. The following three characters, however, taken in combination will, I think, enable this species to be distinguished—(1) the structure of the somewhat slender antennules, (2) the comparatively long first joint of the inner branches of the first thoracic legs, along with the short second and third joints, and (3) the form and armature of the fifth pair in the female. The male has not been observed.

Genus *Delavalia*, G. S. Brady.

Delavalia minutissima, T. Scott, sp. nov.   Pl. iv., figs. 3–10.

Description of the Female.—The female of this species resembles the type-form of the genus in its general outline, and also generally in its structural details, but it is the smallest of any species that has yet been described, being scarcely ·4 mm. (about $\frac{1}{67}$ of an inch).

The antennules appear to consist of seven joints; the end joint is about twice as long as the penultimate one, but the others are sub-equal in length, and, as usual, become gradually stouter towards the proximal end (fig. 4).

The antennæ and mouth organs, being so small, were difficult to get hold of, and are not figured, but so far as they could be made out they resembled very closely those of *Delavalia æmula* (T. Scott).

The first pair of thoracic feet resemble in some measure the first pair in *Delavalia robusta*, Brady and Robertson, and of *D. reflexa* of the same authors, but the principal terminal spine of the inner branches is distinctly different, and the spine on the inner distal angle of the second basal joint is remarkably elongated, as shown by the figure (fig. 5).

The next three pairs (figs. 6 to 8) resemble those of *Delavalia æmula*, but are more slender and moderately elongated, and while in the second and third pairs the outer and inner branches are of nearly equal length, the outer branches of the fourth pair are considerably longer than the inner, as shown by figure 8.

The fifth pair, though small, are, in their general character, similar to those of the group to which the species belongs.   The basal joint is furnished with four setæ on the broadly truncate apex; a short and a long seta near its inner aspect and a pair somewhat similar towards its exterior aspect, with a distinct space between the two pairs; the secondary joints are each armed with four setæ on the broadened, truncated end, the two middle setæ being much smaller than the others, as shown by the drawing (fig. 9).

The caudal segments are proportionally more elongated and slender than those of any other of the described species of the genus; these segments, besides being very narrow, are at least equal to the entire length of the last two segments of the abdomen (fig. 10).

Habitat.—Moray Firth; apparently rare.   No males have been observed.

Remarks.—What first attracted my attention towards this species was its small size and the remarkable length of the furcal joints.   It is the smallest species of the group that I have yet observed, and though apparently rare, that may be partly accounted for by its being so easily overlooked.   The specimen from which the figures have been prepared was obtained in a gathering of dredged material collected in the Moray Firth several years ago, but the description of it was delayed in the hope that other specimens might turn up, which would have enabled me to present a more complete series of detail drawings, but this hope has not yet been realised.   The description and figures given here are, however, along with the small size of the copepod, sufficient to distinguish it from those already described, indeed, its extremely long furcal joints would alone mark it out as different, and these taken along with the peculiar armature of the inner branches of the first thoracic feet, and also of the fifth pair, give to the species a character distinct from other *Delavalias*.

Canthocamptus.

Canthocamptus inconspicuus, T. Scott.

> 1900. *Canthocamptus inconspicuus*, T. Scott, 18th Ann. Rept.
> Fishery Board for Scotland, pt. iii., p. 390, pl. xiv., figs. 1–8.

This small Harpactid was obtained in a gathering of Entomostraca collected off Musselburgh in 1894, but only recently examined : this is the first time it has been obtained in the Firth of Forth, and it has only previously been recorded from the Moray Firth. *Canthocamptus inconspicuus* somewhat resembles *C. parvus*, T. and A. Scott, in general appearance, and like that species it has antennules composed of six joints ; but it differs in several particulars, and one of the more obvious differences is the longer furcal joints, and by this character alone it can be distinguished from *E. parvus*.

Canthocamptus parvus, T. and A. Scott.

> 1896. *Canthocamptus parvus*, T. and A. Scott, Ann. Nat. Hist.
> (6), vol. xviii., p. 6, pl. ii., figs. 14–22.

This species has recently been obtained in several gatherings, one of which consisted of small Crustacea collected in the pond at the Sea-fish Hatchery, Bay of Nigg, June 25, 1902. Like the *Canthocamptus* previously mentioned, this one usually occurs very sparingly in any single gathering, but it has apparently a wider distribution, and has been observed not only at different times in the Firth of Forth, where it was first discovered, but also in the Moray Firth and in the Firth of Clyde. In this species the furcal joints are very short, and it thus differs from *C. inconspicuus*. *C. parvus* is usually found near the shore about the roots of algæ, and especially where there is a muddy bottom.

Genus *Neobradya*, T. Scott.

Neobradya pectinifer, T. Scott.

> 1892. *Neobradya pectinifer*, T. Scott, 10th Ann. Rept. Fishery
> Board for Scotland, pt. iii., p. 249, pl. xiii., figs. 19–32.

A single specimen of this rare species occurred in a gathering of small Crustacea collected at the north end of Inchkeith, on November 15th, 1889, but not examined till October, 1902. The species was first observed among some dredged materials collected off St. Monans, and it was afterwards obtained in the Firth of Clyde,* but though its distribution appears to be somewhat extensive, I have only rarely observed it.

Genus *Tetragoniceps*, G. S. Brady.

Tetragoniceps pygmæus, T. Scott, sp. nov. Pl. iv., figs. 11–19.

Some time ago, when re-examining a gathering of small Entomostraca collected near Musselburgh in 1894, I observed odd specimens of a slender copepod very like *Tetragoniceps incertus*, T. Scott—a species described in the Tenth Annual Report of the Fishery Board for Scotland—but rather smaller than that form, and the fact that one or two of them were provided with ovisacs showed that their smaller size could not be ascribed to immaturity, but on account of their likeness to the species named they

---

* Brit. Assoc. Handbook on the Natural History of Glasgow and the West of Scotland (1901), p. 353.

had probably on previous examinations been passed over as being merely a form of it. Recently, however, a more careful scrutiny of these smaller forms has revealed certain structural differences which render their removal from that species necessary, and as this form is distinctly smaller than the other, I propose to call the new species *Tetragoniceps pygmœus*, and the following is a description of it :—

Description of the Female.—The female closely resembles the female of *Tetragoniceps incertus* but is distinctly smaller, measuring only a little over ·5 mm. (about $\frac{1}{47}$ of an inch). (*Tetragoniceps incertus* is about one millimetre in length). The body is very slender, with a small but distinct rostrum (fig. 11).

Antennules almost similar to those of *Tetragoniceps incertus*, differing only to a small degree in the proportional lengths of the different joints (fig. 12). The antennæ and mouth organs are also very similar.

The first four pairs of thoracic feet are also very similar to the same appendages in *Tetragoniceps incertus*, but the first pair are rather more slender, especially the inner branches, while the seta on the inner margin of the first joint appears to be situated nearer the middle of it, the outer branches appear also to be proportionally rather longer (fig. 14).

The fifth pair (fig. 16) are not only smaller, but differ in form and armature ; they are more bluntly rounded at the apex, and instead of terminating in a single stout apical spine as in the species referred to, there is at the base of the larger stone another one, small but distinct ; the arrangement of the supplementary setæ is also different in the two species, as shown by the drawings.

The furcal joints, which are somewhat dilated in the middle, taper towards the distal end ; the principal apical seta of each furcal joint is moderately short, and the outer portion of it terminates somewhat abruptly, but is continued by a slender portion which forms a peculiar loop where the two portions join, as indicated by the figure ; this peculiarity is not found in *Tetragoniceps incertus* (fig. 19).

The male resembles the female, except in the following particulars. The antennules (fig. 13) are modified for grasping, they differ slightly from the male antennules in *Tetragoniceps incertus*.

The inner branches of the third pair of thoracic feet (fig. 15) are small and three-jointed ; the first and second joints are very short, but the inner part of the second is produced into a long bent spiniform process which extends considerably beyond the extremity of the branch ; the last joint is also small and bears a minute terminal spine. In *Tetragoniceps incertus* this branch has a long straight process arising from its inner basal aspect and is furnished with two terminal setæ.

The fifth pair are smaller than those of the female, they also differ in their armature as shown by the drawing (fig. 17). The supplementary foot on the first abdominal segment is furnished with three setæ (fig. 18).

Remarks.—This form, though perhaps it does come somewhat near to *Tetragoniceps incertus*, is not difficult to distinguish from that species, even without dissection, by its smaller size and by the difference in the character of the fifth feet and the furcal joints, and these differences are of course more easily observed when one has both forms under observation.

Genus *Laophonte*, Philippi (1840).

Laophonte gracilis, T. Scott, sp. nov.   Pl. vi., figs. 6–12.

Description of the Female.—The body is slender and sub-cylindrical, and appears to be covered with exceedingly minute hairs. It has a small blunt rostrum, and the furcal joints are scarcely so long as the last abdominal segment (fig. 6). Length about ·67 mm. (about $\frac{1}{37}$ of an inch). The antennules (fig. 7) are short and composed of seven joints; the first three joints are moderately large, but the others are small.

The antennæ and mouth organs are not unlike those of *Laophonte intermedia*, except that the second maxillipeds are long and narrow, and the terminal claw very slender and elongated (fig. 8).

The first pair of thoracic feet are slender, and the first joint of the inner two-jointed branches is elongated and narrow, but the second joint is short and armed with a moderately stout terminal claw ; the inner branches are three-jointed, and just about half as long as the first joint of the inner branches (fig. 9).

The second, third, and fourth pairs, which are somewhat similar to each other in structure, are also slender. The fourth pair is represented by the drawing (fig. 10). In this the outer branches are of moderate length, and composed of three sub-equal joints, but the inner branches are short and two-jointed, the first joint being a small one.

The fifth pair (fig. 11) are comparatively large and foliaceous, and have a general resemblance to those of *Laophonte similis*; the basal joint, which is sub-triangular in outline, is produced interiorly so that its apex reaches to about the middle of the secondary joint ; there are three stout setæ on the inner margin of the basal joint and two smaller apical setæ. The secondary branch is sub-ovate, somewhat longer than broad, and provided with about five setæ on the lower half of the outer margin and apex, the second seta from the inside is of moderate length, but the others are short. One ovisac with several small ova.

Habitat.—In an old quarry at Granton into which the tide ebbs and flows. Rare. This is different from any species known to me.

The following other species of *Laophonte* were also found in the same gathering with the species just described :—

> *Laophonte littorale,* T. Scott.
> *Laophonte intermedia,* T. Scott.
> *Laophonte hispida* (Brady and Robertson).
> *Laophonte thoracica,* Boeck.
> *Laophonte inopinata,* T. Scott.
> *Laophonte curticauda,* Boeck.

Laophonte lamellifera (Claus).

> 1863. *Cleta lamellifera.* Die frei-lebenden Copepoden, p. 123, t. xv., figs. 21–25.

This fine species occurred sparingly in a gathering of copepods washed from some dredgings collected in shallow water off Musselburgh, Firth of Forth. In the same gathering, as well as in another from near the same place, collected by means of a hand-net, between tide marks, another *Laophonte, L. intermedia,* T. Scott,* was much more frequent, and as it somewhat resembles *L. lamellifera* might be mistaken for it, but *L. intermedia* has shorter furcal joints, and the outer branches and second basal joint of the first pair of thoracic feet have a dense covering of minute hairs, which may frequently be seen without dissection.

Laophonte denticornis, T. Scott.

> 1894. 12th Ann. Rept. Fishery Board for Scotland, pt. iii., p. 246, pl. vii., figs. 13–23.

One or two specimens of this species were found in the same gathering

---

* *Thirteenth Annual Report of the Fishery Board for Scotland,* Part III., p. 168 ,
figs 10-20 (1894).

with the last; it is a rare form, and has not hitherto been observed on the south side of the Forth. *Laophonte denticornis* differs from *L. serrata*, Claus, in having the outer branches of the first thoracic feet three-jointed, and in the different form of the fifth feet in both the male and female; the female also wants the posterior dorsal spine which is characteristic of the female of *L. serrata*. *Laophonte hispida* and *thoracica* were also obtained in this Musselburgh gathering.

Genus *Laophontodes*, T. Scott (1894).

Laophontodes typicus, T. Scott.

> 1894.   *Laophontodes typicus*, T. Scott.   12th Ann. Rept. Fishery Board for Scotland, pt. iii., p. 249, pl. viii., figs. 2–8.

This species, which is not difficult to identify, even without dissection, by the peculiar form of the fifth thoracic feet in the female, was moderately frequent in the gathering from the old quarry at Granton in which *Stephos scotti* was obtained.

Genus *Cletodes*, G. S. Brady (1872).

Cletodes neglecta, T. Scott, sp. nov.   Pl. iv., figs. 20–31.

Description of the Female.—Body elongated, narrow, cylindrical; all the segments distinct except the first and second of the abdomen, which are slightly coalescent.   The first three segments of the abdomen have their lateral angles produced into small spiniform processes.   Rostrum short and broadly triangular.   Caudal joints narrow and elongated and equal to nearly one and a half times the length of the last abdominal segment (fig. 20).

The antennules are short and stout and composed of five (or six) joints; the end joint is narrower and rather longer than the others, while the penultimate one is very small; the last four joints are also all setiferous, as shown by the drawing (fig. 21).

Antennæ two-jointed and of moderate length; the end joint is provided with spiniform, marginal and terminal, setæ; a few of the terminal setæ are elongated, but the others are moderately short.   The secondary branches of the antennæ are rudimentary, and are represented by a single short hair as in *Cletodes limicola*, G. S. Brady (fig. 23).

The mandibles are stout, elongated, sub-cylindrical, and armed with a few stout apical teeth.   The palp is composed of a single one-jointed branch, and is provided with several plumose setæ (fig. 24).

The second maxillipeds are composed of two moderately slender joints, and the terminal claw is also slender and elongated (fig. 25).

The first pair of thoracic feet (fig. 26) resemble in their structure and armature the first pair in *Cletodes limicola*; both branches are short, but the three-jointed outer branches are rather longer than the two-jointed inner ones.   In the inner branches the end joint is narrower and about one and a half times longer than the other.   Moreover, the seta on the outer angle as well as the one on the inner angle of the second basal joint are both elongated, the inner one being also plumose.   In the second pair (fig. 27) the outer branches are elongated and slender, but the inner two-jointed branches are short; the second joint of the inner branches is narrow, and fully twice as long as the first joint; two very long hairs spring from its truncate apex, but otherwise it is unarmed; the outer branches are provided with long slender spines on the exterior distal angles of all the three joints; there is also a single slender seta on the lower half of the inner margin of the middle joint. The third pair (fig. 28) are somewhat similar to the second, except that they are rather more slender, and that the second joint of the inner branches bears a small spine on the distal end of its outer margin in addition to the two long terminal setæ. The fourth pair, on the other hand, scarcely differ from the third except that the outer branches are rather more elongated (fig. 29).

The fifth pair resemble very closely the fifth pair of the female in *Cletodes limicola*, so much so that, with the exception of the basal joint being rather more produced, the general configuration and armature of this pair are in the two species almost identical (fig. 30).

The male is very similar to the female. The antennules of the male have, as usual, a modified structure (fig. 22), and the fifth pair of thoracic feet are very small (fig. 31). In the fifth pair of feet the basal joint is nearly rudimentary, and the secondary joint, which is of a narrow cylindrical form, is provided with only two apical setæ, as shown by the drawing.

Habitat.—Moray Firth ; moderately rare.

Remarks.—The form just described has been known to me for a considerable time, but has been left over from year to year, as I was in doubt whether the differences observed were of any real value. As, however, I have not been able to find any described species to which this form could be assigned, I have described it here under a distinct name.

This form belongs to a group of *Cletodes* which are all closely related to each other, and exhibit this relationship by the similarity in the structure of the antennules and of the first pair of thoracic feet, but especially in the structure and armature of the female fifth pair; and perhaps the most typical species of the group is the *Cletodes limicola* of G. S. Brady. In this group the antennules are usually composed of five joints, the penultimate one being very small ; in the first pair of thoracic feet both branches are short, but the inner rather shorter than the outer ; the second joint of the inner branches is also distinctly narrower and considerably longer than the first joint. In the fifth pair the basal joint is small, and only slightly produced interiorly—sometimes not at all—and provided with few, usually two or three, setæ. The secondary joints, on the other hand, are elongated and narrow and usually furnished with five setæ—two at the apex, one on the lower part of the inner margin, and two widely separated on the outer margin. The form just described, while agreeing in some of its structural details with several members of this group, differs in one point or another from them all, so far as they are at present known to me.

Cletodes propinqua, Brady and Robertson.

1875.   *Cletodes propinqua*, B. and R., Brit. Assoc. Report, p. 196.

This curious little Harpactid occurred very sparingly in the same Musselburgh gathering with the *Laophontodes* just referred to. Its bathymetrical distribution appears to extend from the littoral zone down to moderately deep water. The furcal joints are short and pyriform, and seem to be characteristic of the species. One or two other species of *Cletodes*, including *C. limicola*, G. S. Brady,[*] and *C. lata*, T. Scott,[†] were also obtained in the same gathering.

---

[*] *Nat. Hist. Trans. Northumb. and Durham*, vol. iv., p. 438, pl. xxi., figs. 10-17 (1872).
[†] *Tenth Annual Report of the Fishery Board for Scotland*, Part III., p. 257, pl. x., figs. 10-18 (1892).

Genus *Enhydrosoma*, Boeck (1872).

Enhydrosoma gracile, T. Scott, sp. n.   Pl. ii., figs. 16–26 ; pl. iii., fig. 1.

Description of the Female. — Body slender, cylindrical, slightly encurved, but otherwise similar to *E. curvatum*, Brady and Robertson (fig. 16). Length of specimen represented by the drawing about ·45 mm. ($\frac{1}{55}$ of an inch).

Antennules very short, moderately stout, and sparingly setiferous ; they are composed of four joints, the first three being sub-equal in length, but the last is considerably smaller than any of the others (fig. 17). The formula shows approximately the proportional lengths of the various joints :—

$$\frac{1 \quad \cdot \quad 2 \quad \cdot \quad 3 \quad \cdot \quad 4}{22 \quad \cdot \quad 18 \quad \cdot \quad 19 \quad \cdot \quad 11}$$

The antennæ (fig. 18) appear to be somewhat like those of *E. curvatum*. The mandibles (fig. 19) are also similar to those of the same species.

The second maxillipeds (fig. 20) have basal joint short, but the preceding one is elongated, and the terminal claw is slender and moderately short.

The first thoracic feet are somewhat like those of *C. curvatum* ; the inner branches are short and two-jointed, the second joint being only slightly longer than the first, and their extremities, which do not reach to the end of the second joint of the outer branches, bear two elongated slender setæ feathered at the end ; the outer branches are three-jointed ; the first joint is fully as long as the entire length of the next two, but the end joint is shorter than the second one ; each of the three joints is furnished with a moderately long and very slender seta on its outer aspect, while the end one also carries two elongated terminal setæ similar to those on the inner branches (fig. 21).

The second, third, and fourth pairs (figs. 22 and 23), which appear to differ little from each other, have the inner branches very short and composed of two joints, the first being very small, while the end joint is furnished with a few terminal setæ as shown by the figures ; the outer branches are three-jointed and of moderate length and stoutness, the middle joint is slightly shorter than the first or third.

The fifth pair are broadly foliaceous, distinctly two-branched, and both branches are broadly sub-truncate at the end and provided with five moderately stout and elongated setæ, the lengths of which vary to some extent as shown in the drawings ; a single seta springs from a small lobe near the outer distal angle of the outer branch (fig. 24).

The furcal branches are very short (fig. 26).

The ovisac is small and contains very few ova.

The male appears to differ little from the female except in the structure of the antennules (pl. ii., fig. 1), and also to some extent in the form of the fifth feet. The fifth pair in the male consists of a rectangular plate about half as long as broad, and obscurely divided into two portions ; the inner portion is furnished with two moderately elongated setæ on the lower edge, while the outer portion bears three or four setæ as shown by the drawing (pl. i., fig. 25).

Habitat.—Shore at Musselburgh, Firth of Forth ; moderately rare.

Remarks.—This species occurred with a number of other curious forms in a gathering collected in 1894. Several other new species were obtained in this gathering, but some of these have already been described.* This

* See *Thirteenth Annual Report of the Fishery Board for Scotland*, Part III., p. 167, *et seq.* ; *Ann. and Mag. Nat. Hist.* (6), vol. xv., pp. 52-53 ; *Ann. Scot. Nat. Hist.*, January, 1895.

Enhydrosoma is a smaller and more slender form than *E. curvatum* ; the female antennules are apparently only four- instead of five- jointed, the third and fourth joints in *E. curvatum* being in this species completely coalescent. The first pair of thoracic feet are also somewhat different in the two species, but a greater difference is observed in the structure of the female fifth pair ; in *E. curvatum* there is a distinct though small secondary branch, but in the present species the branches, which are sub-equal, do not appear to be distinctly separated.

Enhydrosoma minutum, T. Scott, sp. nov.   Pl. iii., fig. 25 ; pl. vi., figs. 1–5.

Description of the Female.—This is a small but moderately stout species as shown by the drawing (fig. 1, pl. vi.). Its entire length, exclusive of antennules and tail setæ, scarcely reaches to ·4 m.m. (about $\frac{1}{60}$ of an inch).

The antennules (fig. 2, pl. vi.) are composed of five joints, but the fourth is very small ; the armature consists of a number of moderately stout setæ, a few of them being plumose, and the end joint carries a stout terminal spine as shown by the figure.

The mouth organs resemble generally those of *E. gracile*, but the second maxillipeds are comparatively rather stouter (fig. 3, pl. vi.).

The first pair of thoracic feet, which appeared to be somewhat similar in structure to the first pair in *E. gracile*, were accidentally damaged, so that a correct drawing of them could not be prepared.

The second, third, and fourth pairs are all somewhat alike in structure (pl. iii., fig. 25 ; pl. vi., fig. 4), and their outer branches do not differ greatly from the outer branches of the feet similar to them in *E. curvatum*, but the inner branches are very small, they each consist of two joints, the first joint being much shorter than the other, while the end joint tapers towards the distal end, and carries a single elongated terminal seta.

The fifth pair (fig. 5, pl. vi.) are broadly foliaceous and resemble those of *E. gracile* ; but the secondary branches are more distinctly articulated to the basal joints, and the setæ of the two joints appear to be stouter.

The furcal joints are extremely short.   The female appears to carry one ovisac with several moderately large ova.

Habitat.—Aberdeen Bay, Station V., Nov. 12, 1901.   One specimen only.

Remarks. –*Enhydrosoma minutum* differs from the species already described by the difference in the structure of the antennules, and by the form and armature of the inner branches of the second, third, and fourth pairs of feet.   No male specimen has been observed.

Enhydrosoma curvatum (Brady and Robertson).

1875.   *Rhizothrix curvata*, B. and R., Brit. Assoc. Rept., p. 197.
1880.   *Enhydrosoma curvatum*, Brady, Mon. Brit. Copepoda, vol. ii., p. 98, pl. lxxxi., figs. 12–15 ; pl. lxxxii., figs. 11–19.

This also occurred in the gathering from the old quarry near Granton ; it is quite distinct from the two species, *E. gracile* and *E. minutum*, just described ; the difference in the structure of the fifth thoracic feet would alone be sufficient to separate them ; it is, moreover, a somewhat larger species.   There is a previous record of this species from the Forth district, but from a different part of the estuary,* as well as from other places around the Scottish coasts.

* *Eighth Annual Report of the Fishery Board for Scotland*, Part III., p. 319 (1890).

Genus *Nannopus*, Brady (1880).

Nannopus palustris, Brady.

> 1880. *Nannopus palustris*, Brady, Mon. Brit. Copepoda, vol ii.,
> p. 100, pl. lxxvii., figs. 18-20.
> 1902. *Nannopus palustris*, T. Scott, 20th Ann. Rept. Fishery
> Board for Scotland, pt. iii., p. 466, pl. xxiii., figs. 13-25.

This somewhat rare species was observed very sparingly in the gathering from the old quarry near Granton in which several other interesting forms have been obtained, and it also occurred in a hand-net gathering of small Crustacea collected between tide marks at Musselburgh. *Nannopus palustris* appears to be a true littoral or brackish-water species, and is rarely met with in deep water off shore. Another species with a somewhat similar habitat to *Nannopus*, viz. *Palatychelipus littoralis*, G. S. Brady, was also obtained in the vicinity of Musselburgh, where it was observed in 1892,* and was then new to the Scottish coasts. It is now ascertained that the species has a wide distribution, but it does not appear to be anywhere very common.

Genus *Dactylopus*, Claus (1863).

Dactylopus littoralis, T. Scott, sp. nov. Pl. iii., figs. 2–8.

Description of the Female.—Body moderately slender, rostrum short (fig. 2).

Antennules (fig. 3) short and composed of eight joints, the third is shorter than the one which precedes or follows, the fifth and seventh joints are small, while the last is about as long as the combined lengths of the sixth and seventh. The proportional lengths of the various joints are shown approximately by the formula :—

$$\frac{8 \cdot 11 \cdot 9 \cdot 11 \cdot 4 \cdot 6 \cdot 4 \cdot 9}{1 \cdot 2 \cdot 3 \cdot 4 \cdot 5 \cdot 6 \cdot 7 \cdot 8}$$

The second maxillipeds are stout with a moderately stout terminal claw, a somewhat long spine-like seta springs from the inner aspect and near the distal end of the second joint, but, with the exception of another small hair, this joint appears to be devoid of armature of any kind. The first joint is provided with two or three small spine-like hairs at the distal end, while the end joint is almost as narrow as the base of the terminal claw (fig. 4).

The first pair of thoracic feet are moderately short and stout, the first joint of the inner branches is somewhat longer than the outer branches, but the last two joints are small, and the end one is armed with two terminal spines, one being short and stout, the other longer and setiform; the outer are composed of three sub-equal joints similar to those of the species previously described (fig. 5).

The second, third, and fourth pairs are moderately elongated, but the fourth is rather longer than the others ; both branches are three-jointed, and the inner branches are shorter than the outer ones ; the marginal spines of the outer branch are moderately long and slender, and the setæ on both branches elongated and plumose (fig. 6 represents the fourth pair).

The fifth pair are moderately broad and foliaceous, especially the outer

---

* *Tenth Annual Report of the Fishery Board for Scotland*, Part III., p. 205, pl. v., figs. 11-13 (1893).

or secondary joints; these joints, which are of an ovate form and about one and a half times as long as broad, bear five setæ round the lower outer margin and end, the middle one being rather longer than the others; the produced inner portion of the basal joints, which scarcely reach to the end of the outer secondary ones, have the sides slightly rounded and taper to the narrowly-rounded apex; they are each furnished with five setæ— three on the lower inner margin and two close together at the apex, as shown by the following drawing (fig. 7).

The furcal joints are shorter than the last abdominal segment; they each appear to be abruptly truncate, and their principal setæ are elongated and slender (fig. 8).

Habitat.—In a shore gathering collected at Musselburgh, Firth of Forth, in 1894; rather rare.

Remarks.—The copepod described above has as great resemblance to *Stenhelia* as to *Dactylopus* in its general appearance and in some of its structural details, as, for example, in the structure of the first thoracic feet, but in the structure of the antennules, mandibles, and fifth thoracic feet its relationship appears to be closer to *Dactylopus*; it differs from described species in structure of the antennules, in the comparatively stout form of the first thoracic feet, and in the form and armature of the fifth pair. It appears to be a littoral species, as I have only observed it in inshore gatherings.

Dactylopus vararensis, T. Scott, sp. nov.   Pl. iii., figs. 17–24.

Description of the Female.—Body moderately stout, length about ·75 mm. (about $\frac{1}{3}$ of an inch).

Rostrum prominent, slightly incurved (fig. 17).

Antennules short, eight-jointed, the first four and last joints sub-equal in length, the other three short; the first four joints are also considerably stouter than the last four (fig. 18); — the formula shows approximately the proportional lengths of the joints:—

$$\frac{23 \quad 22 \quad 29 \quad 18 \quad 6 \quad 10 \quad 7 \quad 20}{1 \quad 2 \quad 3 \quad 4 \quad 5 \quad 6 \quad 7 \quad 8}$$

The secondary branches of the antennæ are composed of three joints, but the middle one is small.

The mandibles are stout and sub-cylindrical, and the biting part is somewhat oblique and armed with about three strong and several small spiniform teeth; the basal joint of the palp is somewhat dilated, and carries two branches, the proximal one being considerably smaller than the other, as shown in the drawing (fig. 19).

The second maxillipeds have the penultimate joint moderately elongated and narrow, with a fringe of small setæ on the inner aspect of its proximal half; the end joint is scarcely broader than the base of the terminal claw, which is moderately long and slender, and incurved toward the extremity (fig. 20).

In the first pair of thoracic feet, the outer branches, which are composed of three sub-equal joints, are rather longer than the first joint of the inner branches; the first two joints are armed with moderately strong spines on the outer distal angles, while the end joint bears, at the apex, two slender spines and two setæ; the first joint of the outer branches is moderately stout, but scarcely twice as long as the combined lengths of the outer two, which are small, narrow, and sub-equal: the inner branches are armed with a stout terminal claw and two setæ; the first and second joints are also each provided with a seta near the distal end of the inner margin (fig. 21). The second, third, and fourth pair are somewhat similar to those of the species previously described (fig. 22 represents the fourth pair).

The fifth pair are small and foliaceous, the produced inner portion of the basal joint reaches to about the end of the secondary joints and bears five setæ on its broadly rounded end; the two outermost setæ are, like the next two, situated near to each other, but there is a comparatively wide space between each pair, the innermost seta is smaller and spiniform; the outer secondary joints are broadly ovate, being only a little longer than broad, and are each provided with five spines, the middle one of which appears to be more slender than the others, as shown by the drawing (fig. 23).

The furcal joints are very small, being almost rudimentary, while the principal tail setæ are short and stout (fig. 24).

Habitat.—Moray Firth, Station IV., collected 1898; apparently rare.

Remarks.—This species is readily distinguished by the peculiar structure of the first thoracic feet, and the comparatively small fifth feet of the female; the fifth pair is not unlike the fifth pair of *Dactylopus minutus*, Claus, but the first pair is very different. *Dactylopus vararensis* was obtained in a gathering of small Crustacea collected in the Moray Firth in 1878, but as the form could not be recognised at the time it was put aside for further study, and I am still unable to identify it with any described species.

Dactylopus mixtus, T. Scott, sp. nov.   Pl. iii., figs. 9–16.

Description of the Female.—This form has a general resemblance to both *D. tenuiremis*, G. S. Brady, and *D. longirostris*, Claus. The body is moderately robust, and in specimens preserved in spirit the abdomen is considerably reflexed; there is a prominent rostrum, but the furcal joints are very short (fig. 9). Length ·6 mm. (about $\frac{1}{42}$ of an inch).

The antennules (fig. 10) are moderately elongated and composed of eight joints; the second joint is the largest, the fourth and last are also comparatively long, being about half as long again as the one immediately preceding; the third, fifth, sixth, and seventh joints are small. The antennules are thus somewhat like those of *D. longirostris*, Claus, in structure, as shown by the drawing.

The antennæ are provided with three-jointed secondary branches similar to those of *D. strömii*, Baird.

The mandibles are well developed, the biting edge is armed with several spine-like teeth, the two outer ones are stout, but the others are slender; the basal joint of the mandible palp moderately dilated, and bears two small branches towards its distal end; the inner branch is somewhat smaller than the other and is apparently two-jointed; the other branch consists of a single joint (fig. 11).

First maxillipeds somewhat similar to those of *D. strömii*.

The second maxillipeds are of moderate size (fig. 12); the terminal claw springs from a narrow joint about half as long as itself; the second joint is furnished with several small spine-like setæ on the inner margin, and the first joint also carries one or two small hairs at its distal end

The inner branches of the first thoracic feet are elongated, the first joint being longer than the entire outer branches; the second joint is very small, but the second and third together are about equal to half the length of the first joint; the first joint is fringed interiorly with minute slender hairs, while a moderately long seta springs from its inner distal angle, the proximal part of the outer margin is provided with a number of minute spines; the second joint bears a feathered seta interiorly, and a few small hairs on the exterior edge; while the end joint, besides being furnished with a few minute spines on the outer margin, bears also a short but moderately stout terminal claw and two slender hairs—the one very short and the other about twice the length of the claw. The outer branches, which are shorter than the first joint of the inner ones, are composed of three sub-equal joints, the first and second are each armed with a strong dagger-like spine on the outer distal angle, and the second bears also a moderately long seta on its inner distal angle; the end joint is furnished with four spines on the outer margin and apex, but two of the marginal spines are comparatively small, a slender and slightly bent seta also springs from the inner apical angle, as shown in the drawing (fig. 13).

The second, third, and fourth pairs have both branches three-jointed, the inner being shorter than the elongated outer branches. The structure and armature of the second and third pairs are not unlike those of the same appendages in *D. strömii*. In the fourth pair the inner branches scarcely reach beyond the end of the second joints of the outer branches, the first and second joints are each provided with one seta near the distal end of the inner margin, while the third joint bears two marginal and two apical setæ; a small slender spine also springs from near the end of the outer margin. In the second and third pairs of feet the second joint of the inner branches is furnished with *two* setæ on its inner aspect, while the end joint of the second pair carries one marginal and two terminal setæ, and a small and slender terminal spine; but the same joint of the third pair has five marginal and apical setæ in addition to the small apical spine. The outer branches of the fourth pair do not differ much in structure and armature from the outer branches of the second and third pairs; the first and second joints are each provided with a spine on the outer and a plumose seta on the inner distal angles; the third joint bears two small spines on the outer margin and another on the outer angle of the apex, besides setæ on the inner margin and apex, as shown by the drawing (fig. 14).

The fifth pair, which are lamelliform, have the inner produced portion of the basal joint broadly sub-cylindrical, with the apex obliquely truncate and armed with five setæ, the two inner setæ are stout and spiniform, but the other three are more slender; the two outer setæ are close together, but the others are more widely separated; the secondary joint is also broadly sub-cylindrical, scarcely one and a half times longer than broad and obliquely truncate at the end; the three outermost are sub-equal, moderately short and stout; the next two are slender, one being more elongated than the others, while the innermost seta springs from a sub-marginal notch, as shown by the drawing (fig. 15).

The furcal joints are very short, and the principal tail setæ are somewhat dilated at the base. This species carries two ovisacs, as shown in the drawing (fig. 16).

Habitat.—Granton, Firth of Forth (1894). Fishery Board's Hatchery at Bay of Nigg, Aberdeen, November 23rd, 1900.

Remarks.—I was at first inclined to ascribe this form to Claus's *Dactylopus longirostris*, but it differs rather markedly in the structure of the fifth feet of the female. *Dactylopus tenuiremis*, G. S. Brady, also resembles the form just described in its elongated antennules and in one or two other minor details, but it distinctly differs in the proportional lengths of the joints of the outer branches of the first feet and in form of the fifth pair; and I do not know of any other species with which it can be identified.

Dactylopus coronatus, T. Scott.

> 1894. *Dactylopus coronatus,* T. Scott. Twelfth Ann. Report
> Fishery Board for Scotland, pt. iii., p. 255, pl. ix., figs.
> 12–20.

This *Dactylopus* was obtained very sparingly in material dredged in shallow water off Musselburgh ; it has been already taken near the Bass Rock and in Largo Bay, but is nowhere very common.

Dactylopus brevicornis, Claus.

> 1866. *Dactylopus brevicornis,* Claus. Die Copepoden fauna
> von Nizza, p. 29, t. iii., figs. 20–25.
> 1880. *Dactylopus brevicornis,* Brady. Brit. Copep., vol. ii.,
> p. 118, pl. lvii., figs. 10–12 ; lviii., fig. 14.

Several specimens of this small species were obtained in the old quarry near Granton, Firth of Forth. It appears to be a littoral form, but is found also in moderately deep water, and has been recorded from several places round the Scottish coasts, but usually very sparingly. Among other *Dactylopus* from the same gathering was the well-marked *D. flavus,* Claus, and one or two other common forms.

Dactylopus debilis, Giesb.   Pl. v., figs. 20–31.

> 1882. *Dactylopus debilis,* Giesb. Freileb. Copep. d. Kieler
> Fohrde, p. 122, pl. i., figs. 7, 19 *et. seq.*

Description of the Female.—Body slender, and, in spirit specimens, strongly reflexed (fig. 20). The length of the specimen represented by the drawing is only slightly over half a millimetre (about $\frac{1}{48}$ of an inch). The rostrum is moderately prominent, but the furcal joints are very short (fig. 31).

The antennules are of moderate length and composed of eight joints, the first four large but the others considerably smaller (fig. 21). The proportional lengths of the various joints are shown approximately by the annexed formula :—

> Proportional lengths of the joints, $\dfrac{18 \cdot 17 \cdot 13 \cdot 18 \cdot 6 \cdot 8 \cdot 6 \cdot 11}{1 \cdot 2 \cdot 3 \cdot 4 \cdot 5 \cdot 6 \cdot 7 \cdot 8}$
>
> The numbers of the joints,   -

Antennæ short, moderately stout, two-jointed, and furnished with a three-jointed secondary branch (fig. 22).

Mandibles small, the biting edge armed with a number of small teeth ; the basal joint of the palp is dilated and bears two small branches as shown in the drawing (fig. 23).

The second maxillipeds are moderately slender, so also is the elongated terminal claw with which they are armed (fig. 24).

The first pair of thoracic feet are somewhat similar to those of *Dactylopus minutus,* Claus ; the first joint of the inner branches is long and slender, being about three times longer than the combined lengths of the second and third joints, and it is furnished with a small seta near the distal end of the inner margin ; the terminal claw of the inner branches is moderately stout, and there are also two terminal setæ. The outer branches, which are also slender, are rather shorter than the first joint of the inner ones (fig. 25).

The next three pairs are also slender and resemble each other, except in the following particulars :—In the second pair the inner branches are slightly longer than the outer, and while the last joint of the inner branches is provided with a seta on the inner margin, the end joint of the outer branches has no seta similarly situated. In the third pair the inner and outer branches are about equal in length, and in this pair, while the first and second joints of the inner branches are each provided with a single seta on their inner margin, the third joint bears two setæ. On the other hand, the inner margins of only the second and third joints are each provided with one seta (fig. 26). In the fourth pair the inner branches are rather shorter than the outer, and the armature of the inner margins of both branches resembles that of the third pair except that the last joint of the inner branches is furnished with one instead of two setæ on its inner edge (fig. 27).

The fifth pair are moderately large and foliaceous ; the inner produced portion of the basal joint is generally of a sub-cylindrical form, but the distal end tapers to a blunt-pointed apex from which spring two setæ of moderate but unequal length ; the distal half of the inner margin carries also two moderately stout spines, as well as an elongated seta, as shown in the drawing (fig. 29). The secondary joint has a sub-ovate outline, and its extremity extends somewhat beyond the end of the inner produced portion of the basal joint ; it is nearly twice as long as broad and is furnished with five setæ which are arranged round the distal end of the joint as shown by fig. 9 already referred to.

The male somewhat resembles the female, but there are the following important differences in addition to the usual modification in the antennules :—(1) The inner branches of the second pair of thoracic feet are distinctly modified ; these branches in the male appear to be only two-jointed, the first joint is moderately stout but short, the second extends into a prolonged and stout tapering process which reaches considerably beyond the ends of the outer branches ; the first joint also bears one seta on its inner edge, but the elongated second joint is furnished with two (fig. 28).

The fifth feet in the male are small ; the inner portion of the basal joint is broadly cone-shaped and carries two apical setæ; the secondary joint is moderately broad and of a somewhat ovate form, and is provided with five setæ, the two setæ on the outer margin are short and spiniform, the apical seta is elongated and slender, while the two on the inner edge are moderately stout and appear to be plumose (fig. 30). There is also a small trispinous appendage on the first segment of the abdomen.

Habitat.—Off Musselburgh, Firth of Forth ; not common.

Remarks.—This small species seems to agree better with *Dactylopus minutus,* Claus, than with any other member of the genus, but it differs distinctly from that species by the structure of the antennules and of the fifth pair of feet in the female, and by the peculiar character of the second pair of feet in the male.

Genus *Thalestris*, Claus (1863).

Several species belonging to the genus *Thalestris* have been observed in gatherings recently examined, and the following are now recorded for the first time from the Firth of Forth.

Thalestris peltata, Boeck.

> 1864. *Anemophia peltata,* Boeck. Oversigt Norges Copepoder, p. 45.
> 1880. *Thalestris peltata,* Brady, Brit. Copep., vol. ii., p. 138, pl. liii., figs. 11–19.
> 1895. *Thalestris peltata,* T. and A. Scott, Ann. Nat. Hist. (6), vol. xiv., p. 351, pl. xv., figs. 11–15.; pl. xvi., figs. 1–8.

The somewhat aberrant species of *Thalestris* was obtained off Mussel-burgh in shallow water (3–4 fathoms). *Thalestris peltata* appears to be a

I

moderately rare species, but it has so much the appearance of a *Scutellidium* or a *Zaus* that it may have been frequently overlooked. *Thalestris rufocincta*, Norm., and *Thalestris clausii*, Norm., were taken in the same gathering with *T. peltata*.

Genus *Westwoodia*, Dana (1855).

Westwoodia nobilis, Baird.

> 1845. *Arpacticus nobilis*, Baird, Trans. Barw. Nat. Club, vol. ii., p. 155.
> 1880. *Westwoodia nobilis*, Brady, British Copepoda, vol. ii., p. 141, pl. lxiii., figs. 1–13.

This prettily-coloured Harpactid was observed in the same gathering with the *Thalestris* just mentioned. It seems to be a littoral form, and its habitat here agrees with what is stated by Prof. G. S. Brady and Rev. A. M. Norman, but it has also been obtained in water of moderate depth, as off Portincross, Firth of Clyde, where it occurred at depths ranging from ten to thirty fathoms.*

LICHOMOLGIDÆ.

Genus *Paranthessius*, T. Scott, gen. nov.

Antennules short and seven-jointed. Antennæ four-jointed, armed with a stout terminal claw. Mandibles and maxillæ somewhat like those of *Lichomolgus fucicolus*. Anterior maxillipeds small, furnished with strongly curved and elongated terminal claws. The first three pairs of thoracic feet are similar to those of *Lichomolgus*, but in the fourth pair the inner branches appear to be entirely wanting. Fifth pair rudimentary or very small.

Paranthessius dubius, T. Scott, sp. nov.   Pl. vi., figs. 16–24.

A single male specimen of a somewhat curious Lichomolgus-like copepod was obtained in some dredged material sent from the Clyde, and collected on June 13, 1899. It has been left unrecorded hitherto in expectation that other specimens, especially females, might be found, and a more exact knowledge obtained of its structure and affinities. It differs in several particulars from any described genus or species at present known to me, and I therefore submit the following description of it under the name of *Paranthessius dubius*.

The male in its general outline somewhat resembles *Pseudopsyllus elongatus*, a copepod described in my paper in Part III. of the Twentieth Annual Report. The body is elongated and narrow ; the cephalo-thorax is composed of five segments, the first is rather broader than the others and is considerably longer than the combined lengths of the remaining four segments ; these four segments, which are sub-equal in length, become gradually narrower, so that the last is narrower than the first segment of the abdomen. The first abdominal segment is considerably dilated, but the remaining segments are short and narrow ; the furcal plates, which are moderately broad, are about as long as the last two segments of the abdomen (fig. 16). The length of the specimen is fully 2 mm. (about $\frac{1}{12}$ of an inch).

The antennules (fig. 17) are short and moderately stout and composed of seven joints, the second joint is the largest, the third and fourth are

* British Copepoda, vol. ii., p. 142 (1880).

very small, but the remaining three are about equal in length and taper slightly to the distal end. The antennules are also sparingly setiferous, and carry several stout and elongated sensory filaments.

The antennæ are composed of four joints, the first two are large and somewhat dilated, but the third and fourth are narrow—the third being also very short ; the end joint is furnished at the apex with a stout, strongly-hooked claw and several spiniform setæ (fig. 18).

The mandibles and maxillæ resemble very closely the same appendages in *Lichomolgus fucicolus*, G. S. Brady. The mandible is small, with a dilated base, and carries two stout, moderately long, and strongly curved apical appendages and two small basal setæ. The maxillæ are small and digitiform, and at the apex furnished with two slender spiniform setæ (fig. 19).

Second maxillipeds very small, and each armed with a stout, strongly-curved, and moderately elongate terminal claw (fig. 20).

The first feet were damaged, and the inner branches are not figured. The outer branches are three-jointed ; the first joint is short and bears a sabre-like spine on the outer distal angle, but no setæ on the inner margin. The second joint, which is also short, carries a sabre-like spine on the outer distal angle and a moderately long seta on the inner margin. The third joint, which is longer and narrower than those preceding, is furnished with three short sabre-like spines on the outer margin, and a similar but rather longer one at the apex. There are also four moderately long plumose setæ on the inner margin (fig. 21).

The second pair have the outer branches very similar to those of the first pair in structure and armature, except that the third joints have five setæ on the inner edge. The first and second joints of the inner branches have no spines or setæ on the outer margins, but the third joint is provided with a short spine near the distal end of the outer edge, and with two that are longer but of about equal length at the apex. The first joint has one seta on the inner margin, the second two, and the third three. The end ioint is also considerably longer than the first or second (fig. 22).

In the third pair, the first and second joints of the outer branches are similar in structure and armature to the same joints in the second pair ; the third joints are armed with two sabre-like spines on the outer margin, and with two similar terminal spines ; there is also a row of five plumose setæ on the inner margin. The inner branches are provided with one seta on the inner edge of the first joint, and two on the inner edge of the second and third joints. The third joint bears also three moderately long sabre-like spines on its truncate apex, but there are no spines or setæ on the inner margins (fig. 23).

In the fourth pair, the inner branches seem to be entirely obsolete, for on either foot there is no appearance of the endopodites having been broken off.

The outer branches are normal and their armature is very similar to that of the outer branches of the third pair (fig. 24).

The fifth pair are rudimentary, and consist each of a minute digitiform process bearing two small hairs, as shown in fig. 16.

No form that could be regarded as the female of this species has yet been observed.

DESCRIPTION OF THE PLATES.

PLATE II.

Stephos scotti, G. O. Sars.   Diam.

Fig.	1.	Female, side view	× 70.
Fig.	2.	Male, side view	× 70.
Fig.	3.	Fifth pair of feet, female	× 360.
Fig.	4.	Fifth pair of feet, male	× 158.

Parastephos pallidus, G. O. Sars.

Fig.	5.	Female, side view	× 39½.
Fig.	6.	Male, side view	× 39½.
Fig.	7.	Antennule, female	× 70.
Fig.	8.	Foot of second pair	× 79.
Fig.	9.	Fifth pair of feet, female	× 106.
Fig.	10.	Fifth pair of feet, male	× 79.

Pseudophaenna typica, G. O. Sars.

Fig.	11.	Male, side view	× 35.
Fig.	12.	Male, dorsal view	× 35.
Fig.	13.	Antennule, male	× 70.
Fig.	14.	Fifth pair of feet	× 154.
Fig.	15.	Extremity of left foot, greatly magnified.	

Enhydrosoma gracile, T. Scott, sp. nov.

Fig.	16.	Female, side view	× 158.
Fig.	17.	Antennule	× 360.
Fig.	18.	Antenna	× 360.
Fig.	19.	Mandible and palp	× 540.
Fig.	20.	Second maxilliped	× 540.
Fig.	21.	Foot of first pair	× 540.
Fig.	22.	Foot of second pair	× 540.
Fig.	23.	Foot of fourth pair	× 540.
Fig.	24.	Foot of fifth pair, female	× 720.
Fig.	25.	Foot of fifth pair, male.	× 720.
Fig.	26.	Furcal joints and last two segments of abdomen	× 270.

PLATE III.

Enhydrosoma gracile, T. Scott, sp. nov.

Fig.	1.	Antennule, male	× 360.

Dactylopus littoralis, T. Scott, sp. nov.

Fig.	2.	Female, side view	× 70.
Fig.	3.	Antennule	× 360.
Fig.	4.	Second maxilliped	× 720.
Fig.	5.	Foot of first pair	× 360.
Fig.	6.	Foot of fourth pair	× 360.
Fig.	7.	Foot of fifth pair	× 540.
Fig.	8.	Furcal joints and last two segments of abdomen	× 240.

Dactylopus mixtus, T. Scott, sp. nov.   Diam.

Fig. 9. Female, side view   .   × 52.
Fig. 10. Antennule   .   × 270.
Fig. 11. Mandible and palp   .   × 360.
Fig. 12. Second maxilliped   .   × 180.
Fig. 13. Foot of first pair .   .   × 240.
Fig. 14. Foot of fourth pair   .   .   .   .   × 180.
Fig. 15. Foot of fifth pair   .   .   .   .   × 270.
Fig. 16. Furcal joints and last two segments of abdomen   × 133.

Dactylopus vararensis, T. Scott, sp. nov.

Fig. 17. Female, side view   .   .   .   × 79.
Fig. 18. Antennule   .   .   .   .   × 540.
Fig. 19. Mandible and palp   .   .   .   × 720.
Fig. 20. Second maxilliped   .   .   .   × 720.
Fig. 21. Foot of first pair   .   .   .   × 360.
Fig. 22. Foot of fourth pair   .   .   .   × 270.
Fig. 23. Foot of fifth pair   .   .   .   .   × 270.
Fig. 24. Furcal joints and last two segments of abdomen   × 105.

Enhydrosoma minutum, T. Scott, sp. nov.

Fig. 25. Foot of second pair, female   × 540.

———

PLATE IV.

Ectinosoma curticorne, Boeck.

Fig. 1. Furcal joints and last two segments of abdomen   × 270.

Dactylopus mixtus, T. Scott, sp. nov.

Fig. 2. Antenna .   × 270.

Delavalia minutissima, T. Scott, sp. nov.

Fig. 3. Female, lateral view   .   .   .   .   × 158.
Fig. 4. Antennule   .   .   .   .   × 540.
Fig. 5. Foot of first pair   .   .   .   .   × 720.
Fig. 6. Foot of second pair   .   .   .   .   × 540.
Fig. 7. Foot of third pair   .   .   .   .   × 540.
Fig. 8. Foot of fourth pair   .   .   .   .   × 540.
Fig. 9. Foot of fifth pair   .   .   .   .   × 540.
Fig. 10. Furcal joints and last two segments of abdomen   × 270.

Tetragoniceps pygmæus, T. Scott, sp. nov.

Fig. 11. Female, lateral view   .   .   .   .   × 150.
Fig. 12. Antennule   .   .   .   .   .   × 270.
Fig. 13. Antennule, male .   .   .   .   .   × 160.
Fig. 14. Foot of first pair   .   .   .   .   × 540.
Fig. 15. Foot of third pair, male .   .   .   .   × 540.
Fig. 16. Foot of fifth pair, female   .   .   .   × 540.
Fig. 17. Foot of fifth pair, male .   .   .   .   × 540.
Fig. 18. Appendage to first abdominal segment   .   × 540.
Fig. 19. Furcal joints and last two segments of abdomen   × 270.

Cletodes neglecta, T. Scott, sp. nov.   Diam.

Fig. 20.	Female, dorsal view .	× 79.
Fig. 21.	Antennule . .	× 360.
Fig. 22.	Antennule, male . .	× 270.
Fig. 23.	Antenna . .	⋎ 540.
Fig. 24.	Mandible and palp .	× 540.
Fig. 25.	Second maxilliped .	× 720.
Fig. 26.	Foot of first pair .	× 160.
Fig. 27.	Foot of second pair .	× 160.
Fig. 28.	Foot of third pair .	× 270.
Fig. 29.	Foot of fourth pair .	× 270.
Fig. 30.	Foot of fifth pair, female	× 216.
Fig. 31.	Foot of fifth pair, male .	× 540.

PLATE V.

Ameira pusilla, T. Scott, sp. nov.

Fig. 1.	Female, lateral view	× 158.
Fig. 2.	Antennule .	× 540.
Fig. 3.	Antenna . .	× 720.
Fig. 4.	Mandible and palp	× 1080.
Fig. 5.	Second maxilliped	× 1080.
Fig. 6.	Foot of first pair	× 720.
Fig. 7.	Foot of second pair	× 540.
Fig. 8.	Foot of fourth pair	× 360.
Fig. 9.	Foot of fifth pair	× 1080.
Fig. 10.	Furcal joints .	× 270.

Ameira ambigua, T. Scott, sp. nov.

Fig. 11.	Female, lateral view .	× 106.
Fig. 12.	Antennule . .	× 270.
Fig. 13.	Antenna . . .	× 360.
Fig. 14.	Mandible and palp .	× 540.
Fig. 15.	Second maxilliped . .	× 540.
Fig. 16.	Foot of first pair . . .	× 270.
Fig. 17.	Foot of second pair . .	× 270.
Fig. 18.	Foot of fourth pair . .	× 270.
Fig. 19.	Foot of fifth pair . .	× 360.

Dactylopus debilis, Giesbrecht.

Fig. 20.	Female, lateral view .	. × 106.
Fig. 21.	Antennule . .	. × 540.
Fig. 22.	Antenna . .	. × 540.
Fig. 23.	Mandible and palp .	. . × 720.
Fig. 24.	Second maxilliped .	. . × 720.
Fig. 25.	Foot of first pair . .	. . × 540.
Fig. 26.	Foot of third pair .	. . × 540.
Fig. 27.	Foot of fourth pair .	. . × 540.
Fig. 28.	Foot of second pair, male	. . × 540.
Fig. 29.	Foot of fifth pair, female	. . × 720.
Fig. 30.	Foot of fifth pair, male .	. . × 1080.
Fig. 31.	Furcal joints . .	. greatly enlarged.

PLATE VI.

Enhydrosoma minutum, T. Scott, sp nov.

Fig. 1.	Female, lateral view	. . × 159.
Fig. 2.	Antennule	. . × 540.
Fig. 3.	Second maxilliped	. . × 720.
Fig. 4.	Foot of fourth pair	greatly magnified.
Fig. 5.	Foot of fifth pair	. . × 760.

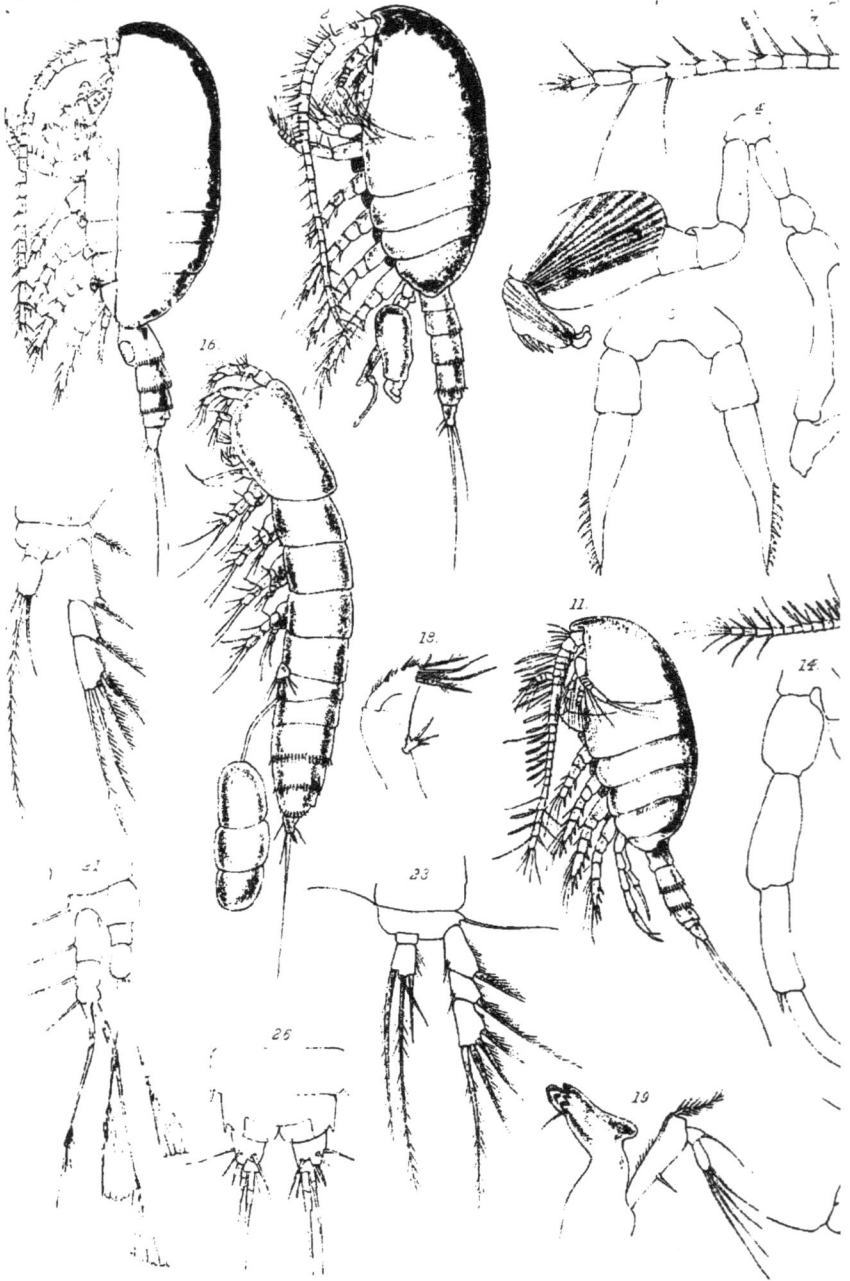

FIGS. 1-4.—*Stephos scotti* (G. O. Sars). FIGS. 5-10.—*Parastephos pallidus* (G. O. Sars).

PLATE II.

dophænna typica (G. O. Sars). Figs. 16-26.—*Enhydrosoma gracile* (T. Scott).

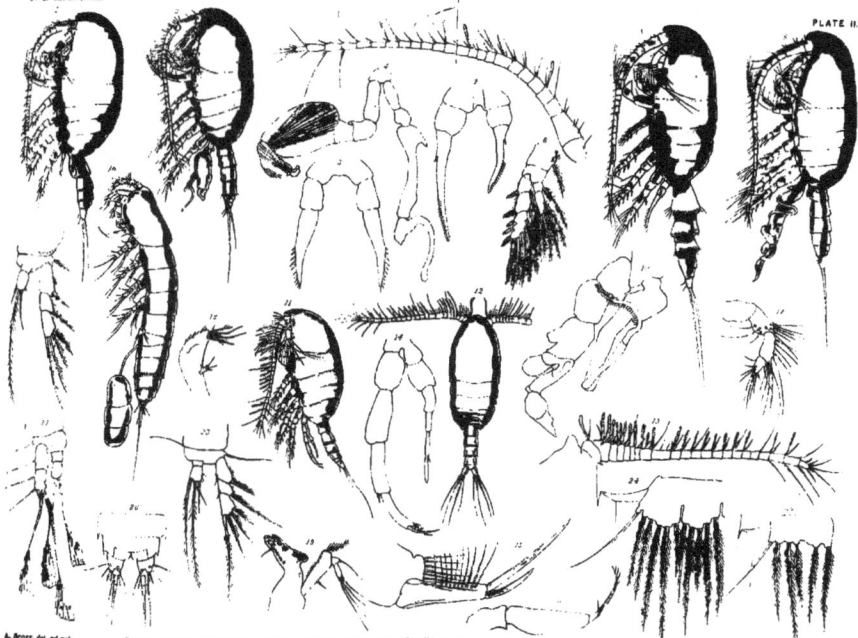

PLATE II.

A. Scott, del. ad nat.   FIGS. 1–4.—*Scophus scotti* (G. O. Sars).   FIGS. 5–10.—*Parastephos pallidus* (G. O. Sars).   FIGS. 11–15.—*Pseudophaenna typica* (G. O. Sars).   FIGS. 16–26.—*Eukyphasma gracilis* (T. Scott).

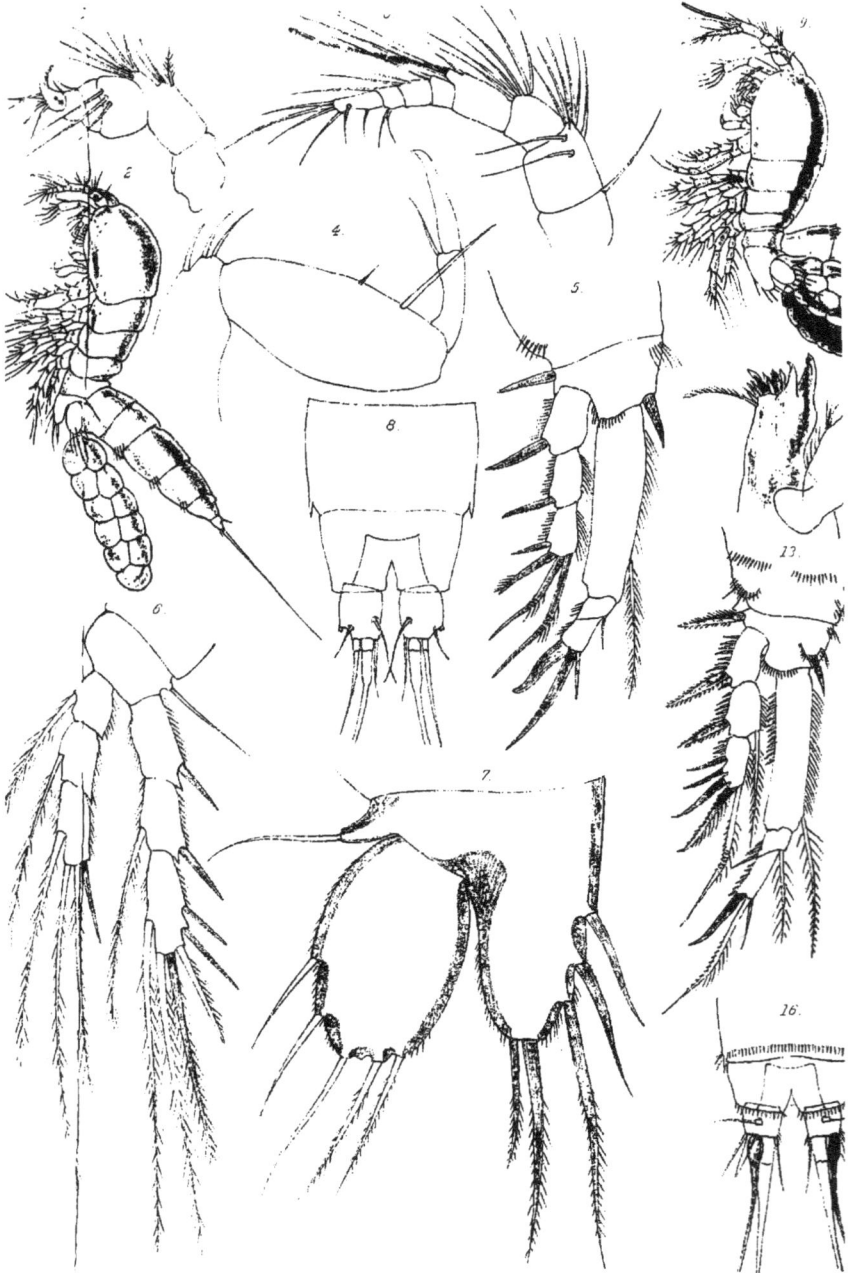

Fig. 1.—*Enhydrosoma gracile* (T. Scott). Figs. 2-8.—*Dactylopus littoralis* (T. Scott).
Fig. 25.—*Enhydro*

PLATE III.

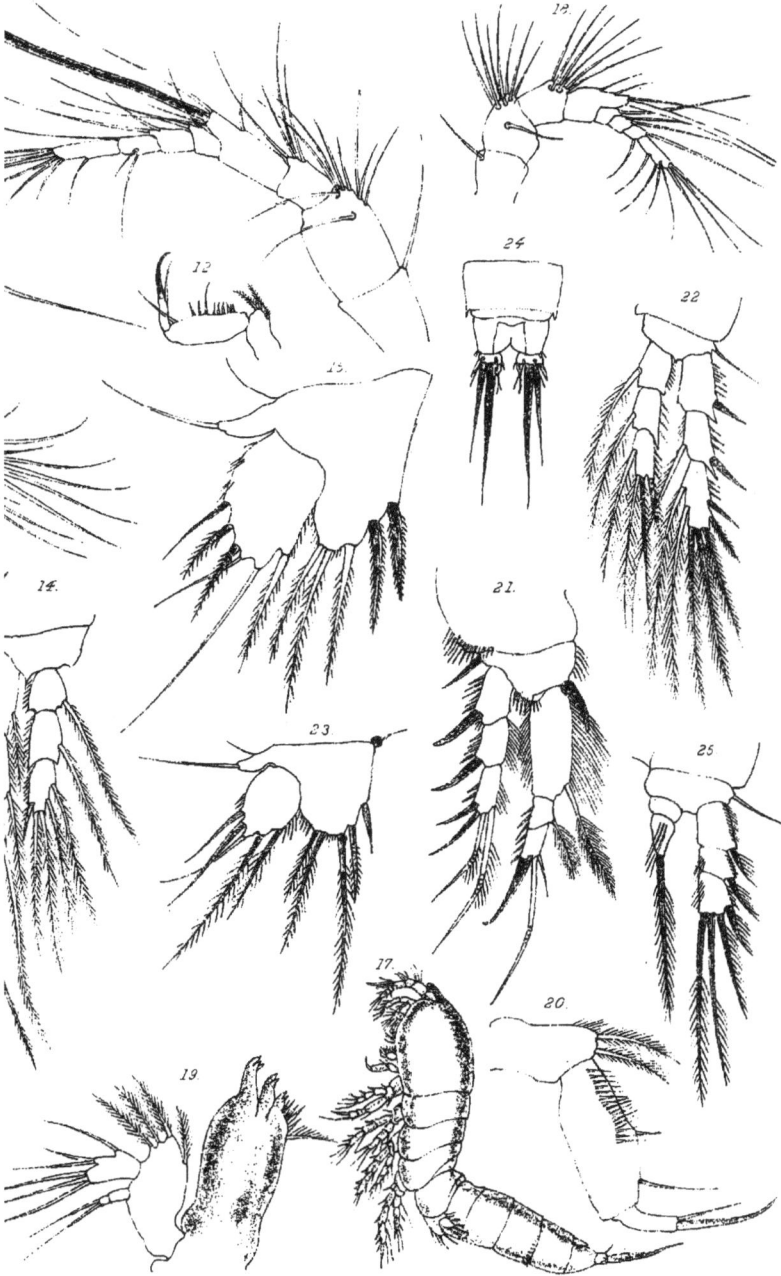

Dactylopus mixtus (T. Scott). FIGS. 17-21.—*Dactylopus vararensis* (T. Scott).
(T. Scott).

PLATE I

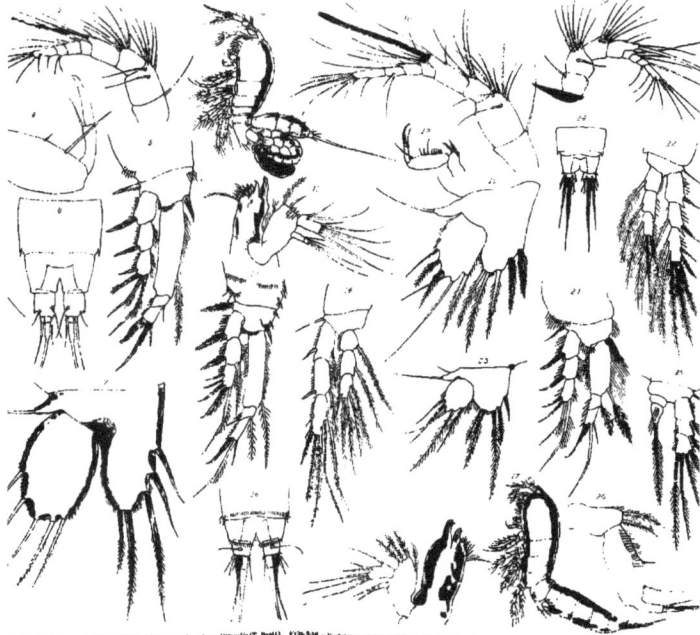

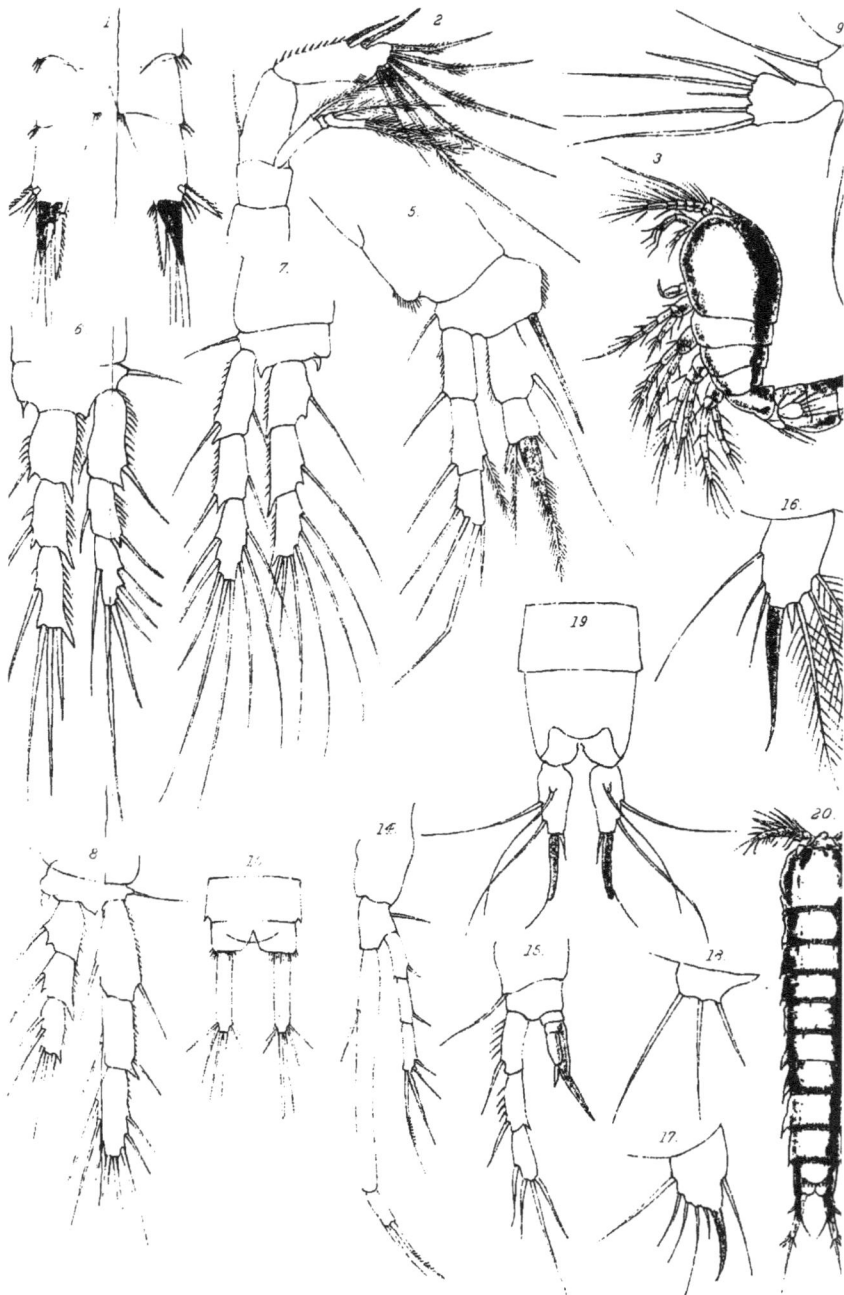

Fig. 1.—*Ectinosoma curticorne* (Bœck). Fig. 2.—*Dactylopus rararensis* (T. Scott). F

PLATE IV.

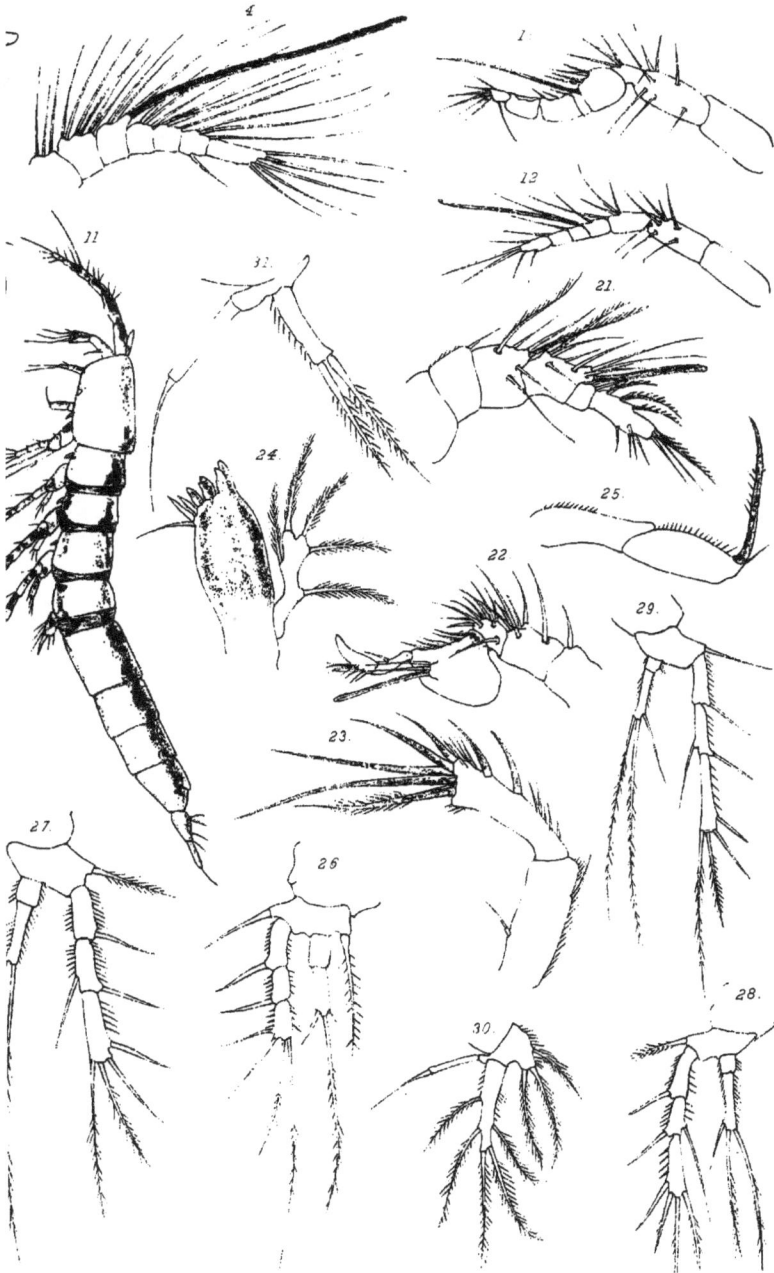

PLATE IV.

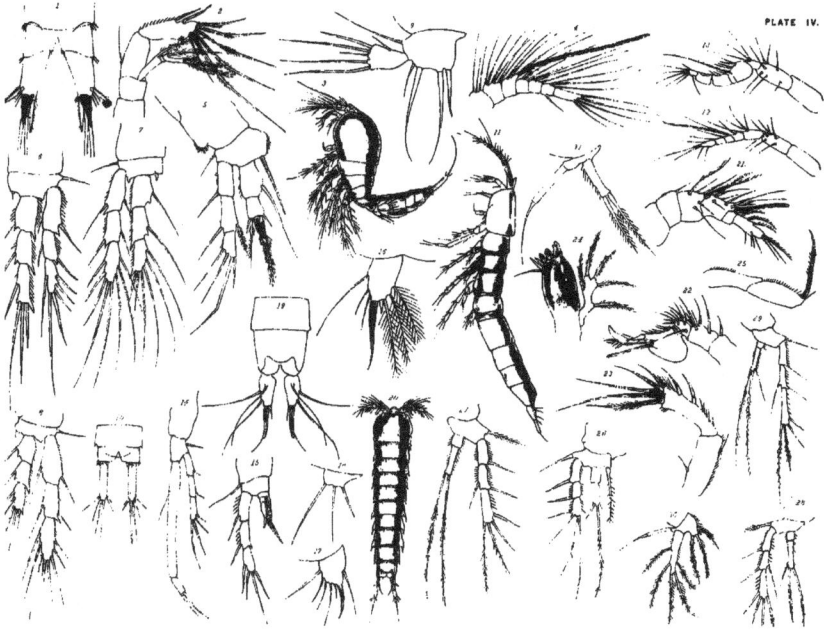

FIGS 1-10.—*Ameira pusilla* (T. Scott). FIGS. 11-19.—*Ameira ami*

PLATE V.

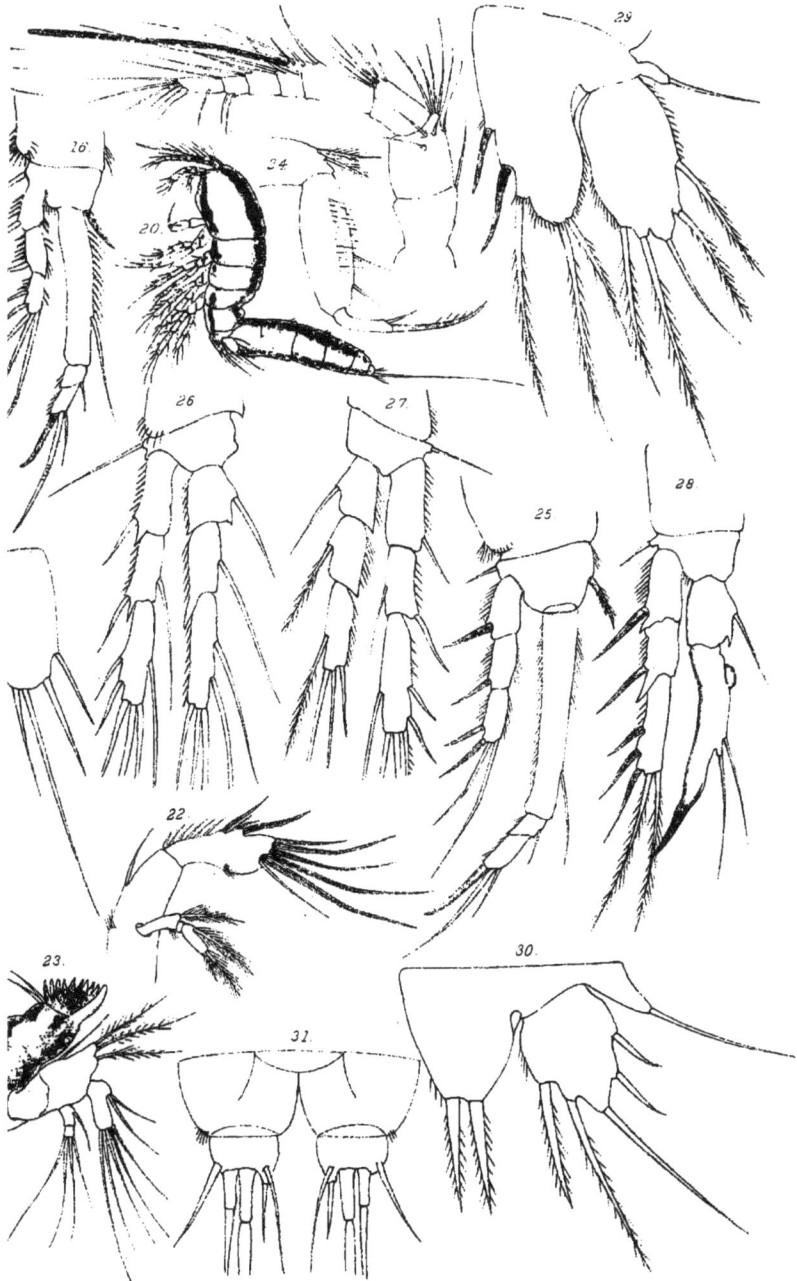

Figs. 20-31.—*Dactylopus parvus* (T. Scott).

PLATE V.

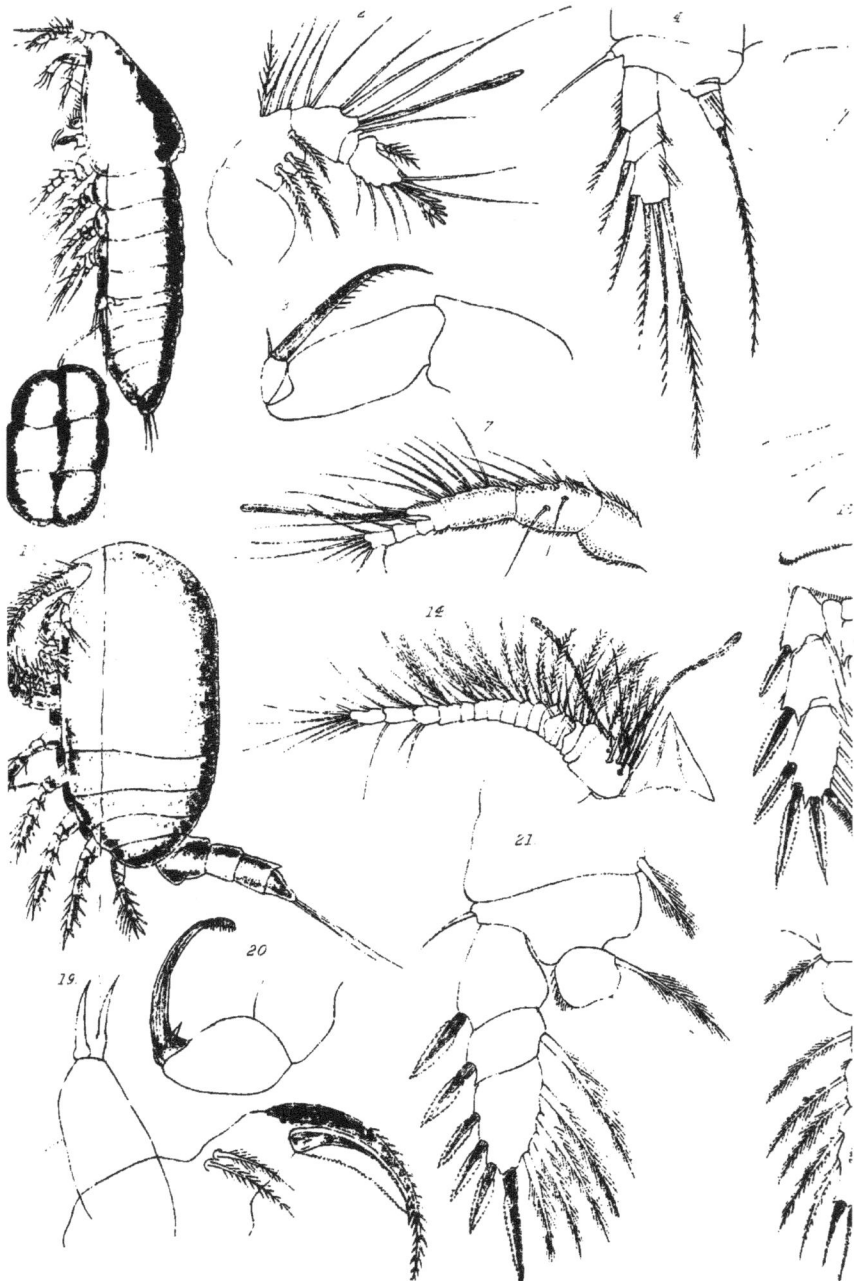

A. Scott, del. ad nat.   Figs. 1-5.—*Euhydrosoma minutum* (T. Scott).   Figs. 6-12.—*Laophonte gracilis* (T. Scott).   Figs.

PLATE VI.

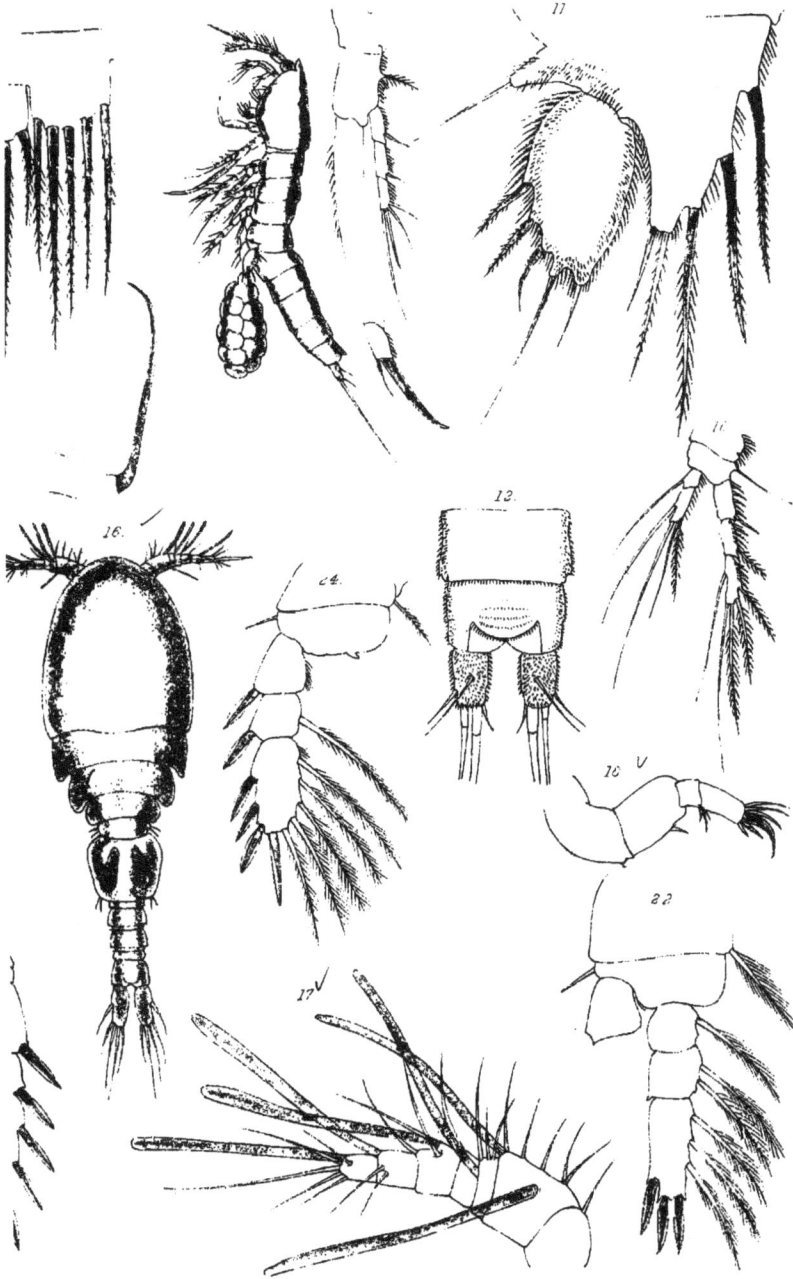

cyclops obtusatus (Brady and Robertson). Figs. 16-24.—*Paranthessius dubius* (T. Scott)

PLATE VI.

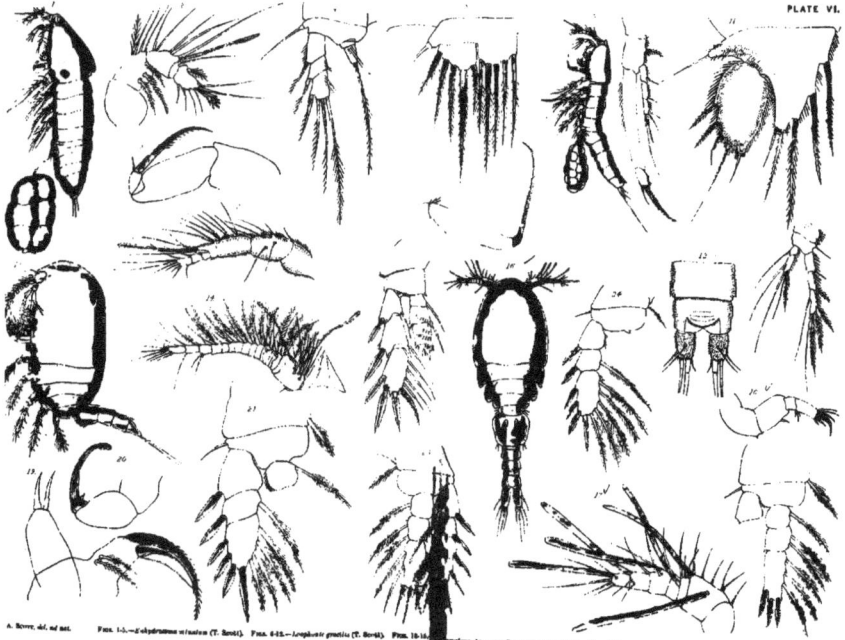

A. Scott, del. ad nat.   Figs. 1-5.—*Eckydrosoma minutum* (T. Scott).   Figs. 6-12.—*Acrophanic gracilis* (T. Scott).   Figs. 13-15.—*...pelops obtusatus* (Brady and Robertson).   Figs. 16-24.—*Paranthessius dubius* (T. Scott)

Laophonte gracilis, T. Scott, sp. nov.   Diam.

6. Female, lateral view	. .	× 79.
7. Antennule .	. .	× 360.
8. Second maxilliped	. .	× 540.
9. Foot of first pair .	. . .	× 360.
10. Foot of fourth pair	. . .	× 270.
11. Foot of fifth pair	. . . .	× 760.
12. Caudal joints with last two segments of abdomen		× 240.

Pseudocyclops obtusatus, Brady and Robertson.

13. Female, lateral view		× 79.
14. Antennule and rostrum .		× 180.
15. Foot of fifth pair .		× 270.

Paranthessius dubius, T. Scott, sp. nov.

16. Male, dorsal view	. . .	× 35.
17. Antennule	. . . .	× 53.
18. Antenna .	. . .	× 79.
19. Mandible and maxilla . (*m*—maxilla)		× 540.
20. Second maxilliped	. .	× 158.
21. Foot of first pair, outer branch .	.	× 158.
22. Foot of second pair, inner branch	.	× 106.
23. Foot of third pair	. .	× 106.
24. Foot of fourth pair	. .	× 106.

E FURTHER OBSERVATIONS ON THE
) OF FISHES, WITH A NOTE ON THE
) OBSERVED IN THE STOMACH OF A
1ON PORPOISE.

——— ——

*n Twenty-first Annual Report of the Fishery Board for
:otland—Part III.—Published July 20, 1903.*]

[*Reprint from Twenty-first Annual Report of the Fishery Board for Scotland—Part III.—Published July 20, 1903.*]

VI.—SOME FURTHER OBSERVATIONS ON THE FOOD OF FISHES, WITH A NOTE ON THE FOOD OBSERVED IN THE STOMACH OF A COMMON PORPOISE. By Thomas Scott, LL.D., F.L.S.

In my paper on the food of fishes published in Part III. of last year's Report,[*] I gave the results of the examination of fishes belonging to fifty-six different species. In the present paper twenty-two species are represented, sixteen of which are teleosteans and the others Rays and Dog-fishes. Their names are as follow :—

Sebastes norvegicus (Ascan.).	The Norway Haddock.
Trigla gurnardus, Lin.	The Grey Gurnard.
Lampris luna (Gmelin).	The King Fish.
Anarrichichas lupus, Lin.	The Cat or Wolf-fish.
Lumpenus lampetræformis (Walbourn).	The Sharp-tailed Lumpenus.
Mugil chelo, Cuvier.	The Grey Mullet.
Labrus mixtus, Lin.	The Striped Wrasse.
Gadus luscus, Lin.	The Whiting Pout or Brassie.
„ *esmarkii*, Nilsson.	The Norway Pout.
Molua molva, Lin.	The Ling.
Onos cimbrius, Lin.	The Four-bearded Rockling.
Ammodytes tobianus, Lin.	The Lesser Sand-eel.
Drepanopsetta plattessoides (Fabr.).	The Long Rough Dab.
Pleuronectes cynoglossus, Lin.	The Witch-sole.
Argentina sphyræna, Lin.	The Hebridean Smelt.
„ *silas* (Ascanius).	The Greater Argentine.
Raia batis, Lin.	The Grey Skate.
„ *fullonica*, Lin.	The Shagreen or Fuller's Ray.
„ *radiata*, Donovan.	The Starry Ray.
„ *circularis*, Couch.	The Cuckoo or Sandy Ray.
Squalus acanthias, Lin.	The Picked Dog-fish.
Scylliorhinus canicula (Lin.).	The Lesser Spotted Dog-fish or Rough-hound.

These fishes are referred to in the sequel in the order in which they are given here.

At the end of the notes on the food of these fishes I describe the results obtained by the examination of the food found in the stomach of a common Porpoise cast ashore last year in the vicinity of the Laboratory.

Sebastes norvegicus (Ascanius).

Three *Sebastes*, measuring $11\frac{1}{4}$, $11\frac{1}{2}$, and $12\frac{1}{2}$ inches respectively, and captured in the North Sea in December 1901, had apparently been all feeding on soft animal substances (probably Annelids), for though each of their stomachs contained a quantity of food, there was nothing to show satisfactorily what it consisted of. Had it consisted of Crustaceans, shell-fish, or fish, even considerably digested, the remains of the harder parts, or some of them, would have afforded an indication of the nature of the food.

---

[*] *Twentieth Annual Report of the Fishery Board for Scotland*, Part III., p. 486.

Trigla gurnardus, Lin.

The stomachs of four Grey Gurnards were examined in March 1902. The fishes measured $10\frac{1}{5}$, $10\frac{3}{4}$, $11\frac{5}{8}$, and $11\frac{3}{4}$ inches respectively; one of the stomachs contained six specimens of *Crangon allmanni* and the remains of a young Clupeoid; another contained fragments of *Crangon* and the remains of small Clupeoids; in the stomach of the third were the remains of Crustaceans and small fishes, but too imperfect for identification; while the fourth contained nothing that could be identified.

Lampris luna (Gmelin).

A King-fish, *Lampris luna,* was captured at Shetland on October 20th, 1900, and was forwarded to the Fishery Board's Laboratory at Bay of Nigg. I had the privilege of examining the stomach of this fish, and found that it had been living exclusively and largely on Cephalopods; unfortunately none of the Cephalopods were perfect enough for identification, the soft parts being scarcely recognisable. The horny

jaws of the creatures had, however, been able to resist to a large extent the solvent action of the digestive fluids, otherwise the determination of the food would have been almost impossible. The number of Cephalopod jaws found in this stomach was 108, and, as each Cephalopod has one pair of jaws, the number of these molluscs which had been recently captured by the King-fish would therefore amount to fifty-four. A few of the jaws were of a moderately large size, but the majority were apparently those of small specimens. Amongst the digested matter contained in this stomach were also a few things that looked like the partially-dissolved cartilaginous shells of Cuttlefishes, but they were so imperfect that no use could be made of them for the purpose of identification.* The jaws, after being mounted on a slide, were photographed by Dr. Williamson, and the accompanying figure is reproduced from the photograph.

Anarrhichas lupus, Lin.

The stomachs of eight Cat- or Wolf- fishes were examined ; the sh were captured in the Moray Firth on May 16th, 1902, and were all of moderately large size. The following is a note of the contents of each of the eight stomachs:—

(1.) Fragments of a large Crab, *Cancer pagurus*, and of *Ophiura* sp.

(2.) Part of a large *Buccinum undatum*, and several large specimens of *Ophiura ciliaris*.

(3.) Fragments of several large *Solen siliqua*, the shell of a *Natica* containing a small hermit Crab, and a specimen of *Hyas coarctatus*.

(4.) Fragments of a moderately large *Cancer pagurus*, of *Solen siliqua*, and of *Cardium echinatum*, and a specimen of *Aphrodite aculeata*.

(5.) Remains of *Solen siliqua, Natica* sp., *Eupagurus bernhardus*, and *Hyas coarctatus*.

(6.) Fragments of *Eupagurus bernhardus*, and of several *Ophiuræ*.

(7.) Fragments of *Natica* sp., and *Venus lincta*, and of a large *Eupagurus bernhardus*. Eighty-two specimens of Star-fishes. *Ophiura ciliaris* (Linn.). Some fragments of *Echinocardium* sp. (probably *E. cordatum*). A specimen of *Aphrodite aculeata* and a fragment of a Zoophyte.

(8.) Remains of five *Natica* sp., and of *Littorina littorea*. Two *Eupagurus bernhardus*, and forty-four specimens of *Ophiura ciliaris*, all more or less complete.

Lumpenus lampetræformis (Walbaum).

The food observed in the stomach of a Sharp-tailed Lumpenus captured on the Fisher Bank consisted almost entirely of small Crustacea, the following species of which were identified :—*Leucon nasica, Diastylis resima, Bythocythere simplex, Macrocypris minna, Cytheropteron* sp., and *Robertsonia tenuis.* Two specimens of *Cyclichna nitidula*, and one or two *Operculina ammoides*, a species of Foraminifera, were also noticed. The *Lumpenus* is a fish that appears to live on or near the bottom, and it is to be expected that demersal organisms rather than pelagic will constitute the chief part of its food.

* Dr. T. Wemyss Fulton, in his "Ichthyological Notes" in Part III. of the *Nineteenth Report* of the Fishery Board for Scotland, also incidentally refers to the large number of Cephalopod mandibles observed in the stomach of the King-fish.

Mugil chelo, Cuvier.

A Grey Mullet about 14½ inches in length, captured by the salmon fishers at the Bay of Nigg on March 14th, 1902, and which was kindly handed over to the Laboratory for examination. It belonged to the same species as those obtained in the Bay last year, viz. *Mugil chelo.* There was some food in the stomach of this specimen, but it was too much digested for satisfactory identification.

Another specimen, captured on the 11th of June, had also in its stomach very little food that could be identified, the only organisms satisfactorily distinguished were one or two *Temora longicornis* and a number of specimens of the Cypris stage of *Balanus* sp.

Labrus mixtus, Lin.

A specimen 12½ inches in length was captured 15 miles north-east from Tiumpan Head, Lewis, in 70 fathoms, in May 1902, and sent to the Laboratory at the Bay of Nigg for examination. The stomach of this specimen contained nothing perfect enough for identification, but in the intestines were found the vertebræ of fishes, fragments of molluscan shells, and some small rounded stones.

Several specimens of *Clavella labracis,* v. Beneden, were obtained on the gills of this *Labrus.*

Gadus luscus, Lin.

Several specimens of the Whiting Pout, varying in length from 7¾ to 11½ inches and captured off Aberdeen in January last year, appear to have been feeding chiefly on Crustaceans. The food found in the stomach of the smallest specimen (7¾ in.) consisted of the remains of Annelids, belonging apparently to the Chætopodæ, and of fragments of Schizopoda and Amphidoda, but the only organism that could be satisfactorily identified was a male specimen of *Erichthonius hunteri* (Spence Bate). A Whiting Pout 9½ inches in length had in its stomach four small Cephalopods (*Rossia ? macrosoma,* Delle Chiage), and a specimen of *Pandalus montagui,* Leach. In the stomach of another 10 inches long were the fragments of what appeared to be *Spirontocaris securifrons,* Norman. Fragments of what looked like *Schistomysis inermis* were observed in the stomach of another 10½ inches in length; while in the stomach of the largest of these Whiting Pouts were found *Crangon allmanni,* Kinahan, *Pandalus montagui,* Leach, and *Pandalina brevirostris* (Rathke)—the length of the fish was 11½ inches.

Gadus esmarkii, Nilsson.

A considerable number of Norway Pouts captured in the North Sea have been examined, but as there was a good deal of similarity in the contents of their stomachs, only a few are particularised here. Small Crustaceans were largely represented amongst the contents of their stomachs, but Schizopods, *Parathemisto* and pelagic Copepods were more frequently observed than other members of that group, as shown by the following sample of the fishes, which ranged in length from about 5½ to 6¾ inches.

LENGTH OF FISH.	CONTENTS OF STOMACH.
inches	Numerous small Schizopods, genus and species doubtful, *Temora longicornis*, few.
5¾ ,,	Numerous small Crustaceans, which look like *Temora longicornis*, but too imperfect to be satisfactorily determined.
5¾ ,,	Several *Parathemisto oblivia*.
6 ,,	This contained nothing that could be identified.
6½ ,,	Remains of small Schizopoda and a number of *Temora longicornis*.
6¾ ,,	Numerous examples of *Parathemisto* and a minute Isopod—the male of a species belonging to the Chelifera.

Molua molva, Lin.

A number of Ling were examined, the food of which consisted chiefly of small fishes. It has been observed that the Ling, more than any other gadoid, is in the habit, when captured, of ejecting not only its food but also its stomach, turning it inside out just as one turns the finger of a glove, so that when visiting the market it is not uncommon to see Ling with their stomachs protruding from their mouths.

Onus cimbrius, Lin.

Twenty specimens of the Four-Bearded Rockling captured on the Bressay Shoal at a depth of 75 fathoms, on December 11th, 1901, were examined, and the contents of their stomachs recorded. As this species was not included amongst those in my previous paper on fish food, I give a more detailed account of the food observed in this sample from Bressay Shoal. Their sizes ranged from 6¾ to 11¾ inches, and their food, as shown in the appended tabular account, consisted chiefly of small Crustacea :—

SIZE OF THE FISH.	CONTENTS OF THE STOMACH.
6¾ inches.	*Pseudocuma cercaria*, *Metopa nasuta*, and some other Crustacean remains.
7¼ ,,	*Metopa nasuta* and remains of some other Crustacea.
7¼ ,,	Remains of Amphipods, but the species doubtful.
7⅜ ,,	*Metopa nasuta* and remains of other Crustacea.
7⅝ ,,	Crustacean remains, but too imperfect for identification.
8 ,,	Contents of stomach similar to the last.
8½ ,,	*Erythrops* sp., *Eudorella* sp., *Halimedon parvimanus* ; *Aceros phyllonyx* ; *Cylichna* sp.
9 ,,	*Leucon nasica* (male) and some mucus.
9 ,,	Fragments of *Calocaris macandreæ* ; *Erythrops* sp., *Leucon nasica* ; *Aceros phyllonyx* ; *Phystisica marina* ; a young Dragonet, 15 mm. long.
9⅛ ,,	*Metopa nasuta* and a small lamellibranch shell.
9½ ,,	This stomach contained only a little mucus.
9⅝ ,,	Remains of two small flat fishes, and fragments of small Crustaceans.
10¼ ,,	*Metopa nasuta*, and remains of some other Crustaceans.
10¼ ,,	The food of this stomach consisted of fragments of Crustacea, but too imperfect for identification.
10¾ ,,	Fragments of *Aceros phyllonyx*, *Eudorella* sp., and Annelids.
11¼ ,,	The only food observed in this stomach consisted of the remains of Chætopod Annelids.
11½ ,,	*Campylaspis* sp. (male) ; *Halimedon parvimanus* ; *Metopa rubrovittata* and *Aceros phyllonyx*.
11½ ,,	Fragments of *Aceros phyllonyx* and the remains of some other Crustaceans.
11½ ,,	This stomach contained nothing that could be identified.
11¾ ,,	Remains of Chætopod Annelids only.

Ammodytes tobianus, Lin.

Several immature specimens of the lesser Sand-launce, captured in the North Sea and measuring from 5 to 7 inches in length, were examined, but the only organisms observed in their stomachs were one or two small fragments of Zoophytes.

Drepanopsetta plattessoides (Fabr.).

The examination of twenty-two Long Rough Dabs, chiefly of small size, yielded the following results:—Four contained nothing that could be identified; *Boreophausia* sp. was found in one; *Leptomysis gracilis* (two specimens) occurred in one; and the remains of a Schizopod, the genus and species of which were doubtful, were observed in one. The remains of small Echinoderms, including a minute Echinus, the plates, pedicellariæ, and fragments of the arms of Brittle Star-fishes were obtained in the stomach of twelve of the fishes examined, while the remains of small Annelid tubes were observed in nine. A few specimens of Foraminifera, such as *Globigerina*, *Discorbina*, etc., probably derived from the worm-tubes, were also observed, but these only occurred in three stomachs. One of the fishes measured about seven inches, but the others ranged from three-and-a-half to about four-and-a-quarter inches in length.

Pleuronectes cynoglossus, Linn.

The stomachs of two Pole-dabs or Witches captured on the Fisher Bank were examined, and found to contain a considerable quantity of food; the contents of both were much alike and consisted almost entirely of small Crustaceans, and the following are the species identified:—*Diastylis resima*, *Lamprops rosea*, *Maera loveni* (fragments), *Ampelisca* sp., and the remains of one or two other Cumaceans and Amphipods. Fragments of one or two small Annelids were the only other organisms observed.

Argentina sphyræna, Lin.

A number of Argentines were captured to the eastward of the Shetland Islands in December 1901, and the subsequent examination of their stomachs showed that they had been living chiefly on small Crustacea, Star-fish, and Annelids, but the contents of a considerable proportion of the stomachs were indistinguishable. The following tabulated results will show the nature and amount of the food observed. The lengths of the fishes are in centimetres:—

[TABLE.

LENGTH OF THE FISH.	CONTENTS OF THE STOMACH.
20·0 centimetres.	The contents of the stomach consisted entirely of the remains of brittle starfishes (*? Amphiura*) and a small quantity of mucus.
20·5 ,,	This stomach contained nothing that could be identified.
20·5 ,,	The only organisms observed in this stomach were a single *Parathemisto oblivia* and and a *Metopa*, the species of which is doubtful.
20·5 ,,	Five *Parathemisto oblivia*, a *Metopa* (sp. ?) and some Annelid remains.
20·5 ,,	Nothing that could be identified was observed in this stomach.
21·0 ,,	Five *Parathemisto oblivia* and some mucus.
21·0 ,,	The remains of brittle starfishes (probably *Amphiura*) were the only objects that could be determined.
21·0 ,,	This stomach contained nothing that could be distinguished.
21·0 ,,	The objects observed in this stomach were a *Philine*, probably *P. nitida*, but the shell had become too much digested for identification, and the remains of a few Chætopod Annelids.
21·0 ,,	Two *Parapleustes latipes* (M. Sars) and fragments of another species of Amphipod.
23·0 ,,	One *Parapleustes latipes*, a minute (young) *Astropecten irregularis*, and the remains of small Chætopod Annelids.
24·0 ,,	Four specimens of *Parathemisto oblivia* were the only organisms that could be determined in this stomach.
24·0 ,,	In this stomach there was nothing that could be identified.

Eight smaller Argentines captured on the Great Fisher Bank in June 1902 were also examined. They measured from 17 to 20 centimetres in length ; the food in the stomachs of three specimens was too much decomposed for identification, two others contained fragments of Annelids, and three the remains of small Crustacea—the only form identified being a young *Pandalus*.

Argentina silus (Ascanius).

Two specimens of the Greater Argentine—one from the Fisher Bank, the other from about 57 miles north-west of the Outer Skerries, and captured in April and June 1902, were examined ; they each measured about thirteen inches from the base of the tail to the anterior extremity. The only organisms in the stomach of the one from Fisher Bank, perfect enough to be identified, were a number of *Calanus*, while the food observed in the stomach of the other consisted chiefly of the remains of *Nyctiphanes*.

Raia circularis, Couch.

The stomach of a Cuckoo Ray captured at Station VI., Firth of Clyde, on October 25th, 1901, and sent to the Laboratory from the s.s. "Garland," was examined on January 16th, 1902, and the following Crustaceans, etc., were observed in it :—Remains of one or two *Hyas coarctatus*, fragments of *Stenorhynchus* ; a whole specimen of *Corystes cassivelaunus* ; twenty-two specimens of *Spirontocaris pusiolus* ; seven specimens of *Pandalina brevirostris*, one *Virbius varians* ; nine specimens of *Ampelisca spinipes* ; fragments of *Amphidotus* sp., a small *Solen* sp., and a small Butterfish, *Pholis Gunnellus* ; there were also a few specimens of the Annelid species, *Ammotrypane aulogoster*, and fragments of one or two other forms that could not be identified. The size of this specimen of *Raia circularis* was not stated. Another specimen of the same kind of Ray captured in the North Sea and examined on March 14th had some remains of round-fishes in its stomach, but they were too much digested for identification ; this specimen measured seventeen-and-a-half inches across the pectoral fins.

Raia batis, Lin.

A specimen of Grey Skate, measuring sixteen-and-a-quarter inches across the pectoral fins, had in its stomach the remains of *Crangon allmanni*, but apparently nothing else.

Raia fullonica, Lin.

A specimen of Fuller's Ray, measuring fifteen-and-a-half inches across the pectoral fins, had also been feeding on Crustacea, but the species could not be determined.

Raia radiata, Donovan.

The stomach of a small Starry Ray was found to contain only the remains of fish too imperfect for identification.

The three fishes referred to above, which had been obtained from a trawling steamer working in the North Sea, were examined in March 1902.

Squalis acanthias, Lin.

A considerable number of Picked Dog-fishes have been examined for ecto- and ento- parasites, and as their stomachs were also examined, I append some observations on the food observed in the stomachs of these specimens. The number of specimens referred to here is twenty-two. They were captured in the North Sea, and forwarded to the Laboratory during the month of March 1902. Eight were examined on the 7th, four on the 14th, and ten on the 26th of the month.

SIZE OF THE FISH.	CONTENTS OF THE STOMACH.
3 feet 6½ inches.	Nothing that could be identified.
2 ,, 8½ ,,	Only the remains of round fishes.
2 ,, 7½ ,,	A moderately large Herring.
2 ,, 9½ ,,	Only two small Dabs.
2 ,, 10¼ ,,	The food was very much digested, but consisted apparently of fishes, as the two ear-stones of a small Whiting were obtained.
2 ,, 10½ ,,	Part of a full-grown Herring was observed.
2 ,, 9 ,,	Part of a young Coal-fish.
3 ,, 1½ ,,	The remains of fishes, but the species was not determined.
2 ,, 6 ,,	The remains of one or two Long Rough Dabs.
2 ,, 8½ ,,	The remains of fishes, but the species doubtful.
3 ,, 0 ,,	The contents of this stomach was similar to the last.
2 ,, 5 ,,	Contents similar to the last.
2 ,, 8½ ,,	Fragments of two *Gadus esmarkii*.
2 ,, 11 ,,	Two nearly whole *Gadus esmarkii* and remains of others, also a portion of the testes of a moderately large Gadoid.
2 ,, 6 ,,	The only objects in this stomach that could be distinguished were a few fish eggs.
2 ,, 8 ,,	One small *Gadus esmarkii* and the remains of another.
2 ,, 11 ,,	A Herring about 9½ inches long.
2 ,, 7½ ,,	Fragments of a Lemon Sole about 8 or 9 inches long, a Gadoid about 6½ inches, and the remains of a fish, not determined, about 9 or 10 inches long.
2 ,, 6 ,,	Fragments of a moderately large Herring and a small Gadoid.
2 ,, 6¾ ,,	Remains of a Gadoid and a Sand-eel.
2 ,, 7½ ,,	Fragments of a moderately large Herring, and some other fish remains.
2 ,, 8¼ ,,	Fragments of a large Herring, and other fish remains.

All these Dog-fishes were females. On the gills of several of them *Eudactylines* were moderately frequent, while *Tetrarhynchi* were observed in the stomachs and intestines of all but a few of those examined.

Scylliorhinus canicula, Lin.

The following three specimens of Lesser Spotted Dog-fishes were obtained among the Picked Dog-fishes just referred to. The food observed in their stomachs consisted entirely of fishes as under :—

SIZE OF FISH.	CONTENTS OF STOMACH.
2 feet 4½ inches.	A Herring 8¾ inches in length.
2 ,, 4¼ ,,	Fragments of a Herring apparently of moderate size.
2 ,, 5¾ ,,	Remains of fishes too imperfect to be determined.

A number of other fishes have been examined, including the Greater Fork-Beard *Phycis blennoides* (Brun.), the Twaite Shad, *Clupea finta,* Cuvier, and the Conger Eel, *Conger niger* (Risso), but as their stomachs did not contain any matter that could be identified they are not specially referred to in this paper. It may be remarked also that several fresh-water Perch, *Perca fluviatilis,* Rondeletius, kindly sent to me by Dr. Williamson from Marlee Loch, Forfarshire, and which were examined to ascertain the nature of their food, were found to have been living almost exclusively on insect larvæ. No parasites were observed on the gills of these fishes, but roundish sacs were frequent on the wall of the body cavity and appeared to contain encysted Cestoids.

NOTE ON THE FOOD OBSERVED IN THE STOMACH OF A COMMON PORPOISE.

The following description of the contents of the stomach of a Common Porpoise captured in the Bay of Nigg in the vicinity of the Laboratory may be of interest, as serving to show how destructive these Cetaceans may be when they get among a shoal of fishes.

The specimen referred to had become entangled in the nets of the salmon fishers at the Bay of Nigg, and having in this way been prevented from coming to the surface for respiration had been suffocated. It was captured on the 18th of June 1902, and measured about 3 feet 9 inches in length, and it appeared to be healthy and in good condition, except that some of the passages of the liver were crowded with brownish-coloured thread-worms ; what appeared to be the same kind of worms were also found encysted in various parts of the liver, while many of them, in a "free" condition, were found in the stomach.

The only food found in the stomach consisted of the partly digested remains of fishes which, for the most part, appeared to be Whitings. Besides the remains of the soft parts of the fishes, no fewer than two hundred and eighty earstones (or otoliths) were obtained ; fully two hundred and forty of them were almost certainly those of Whitings, the majority of which represented fishes of moderate size—probably about eight inches or so in length. Twelve other otoliths were small and of an oblong form, they were not so attenuated at the ends as the typical Whiting earstone, and appeared to belong to the young of some other Gadoid ; the remainder—about twenty-two in number—were extremely small, and somewhat resembled the earstones of Sand-eels.

One or two of the largest of the Whiting earstones measured ten millimetres in length and a number of them nine millimetres, but the average
length would be about eight millimetres. A considerable number of the
earstones were found scattered over the surface of the stomach mixed up
with the soft partly digested matter, but by far the larger number were
found neatly packed together in the narrow distal end of the stomach;
these earstones were remarkably clean and perfect. A few of the
smaller of the Whiting earstones were slightly eroded by the solvent
action of the digestive fluids. It may be mentioned that the intestines,
which were of great length, contained very little matter, and no parasites
were in them or in the other viscera except those already referred to.

Usually a Whiting has only two earstones, so that the two hundred and
forty found in the stomach and which almost certainly belonged to
Whitings represented one hundred and twenty fishes, and if each pair of
the remainder represented a fish, the earstones found in this stomach
would represent one hundred and forty fishes. But, while making every
allowance for the voracity of these cetaceans, it is hardly likely that this

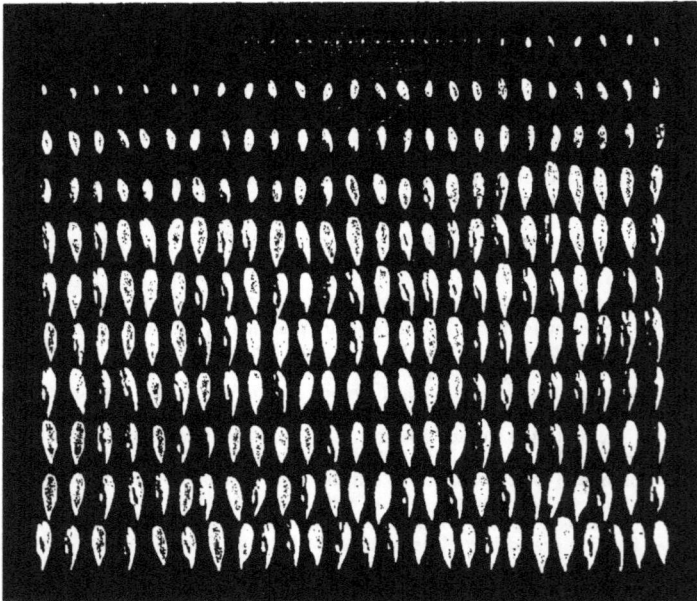

Porpoise had taken all these fishes at a single meal; but, judging from
the perfect condition of the majority of the earstones, they could not have
been long in the stomach. The annexed woodcut is reproduced from a
photograph of the earstones as arranged and mounted on a slide.